SPACE TELESCOPE SCIENCE INSTITUTE

SYMPOSIUM SERIES: 17

Series Editor S. Michael Fall, Space Telescope Science Institute

THE LOCAL GROUP AS AN ASTROPHYSICAL LABORATORY

The Local Group of galaxies consists of the Milky Way and all of its neighbors. The proximity of these galaxies allows for detailed studies of the processes that have led to the formation of these galaxies, their structures, and their evolution. In particular, studies of the Local Group can test predictions of structure formation that are based on dark energy and cold dark matter. This book presents a collection of review papers, written by world experts, on some of the most important aspects of Local Group Astrophysics. It is an invaluable resource for both professional researchers and graduate students in this cutting-edge area of research.

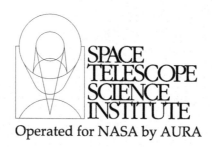

SPACE
TELESCOPE
SCIENCE
INSTITUTE

Operated for NASA by AURA

The Local Group as an astrophysical laboratory

Proceedings of the
Space Telescope Science Institute Symposium,
held in Baltimore, Maryland
May 5–8, 2003

Edited by
MARIO LIVIO
Space Telescope Science Institute, Baltimore, MD 21218, USA

THOMAS M. BROWN
Space Telescope Science Institute, Baltimore, MD 21218, USA

Published for the Space Telescope Science Institute

Operated for NASA by AURA

CAMBRIDGE UNIVERSITY PRESS
Cambridge, New York, Melbourne, Madrid, Cape Town,
Singapore, São Paulo, Delhi, Tokyo, Mexico City

Cambridge University Press
The Edinburgh Building, Cambridge CB2 8RU, UK

Published in the United States of America by Cambridge University Press, New York

www.cambridge.org
Information on this title: www.cambridge.org/9780521175333

First published 2006
First paperback edition 2011

A catalogue record for this publication is available from the British Library

ISBN 978-0-521-84759-9 Hardback
ISBN 978-0-521-17533-3 Paperback

Contents

Participants

Aloisi, Alessandra	The Johns Hopkins University
Avera, Randy	Randolph Publishing/NASA
Babiuc, Maria Cristina	Technical University "Gh. Asachi" Iasi
Beckwith, Steve	Space Telescope Science Institute
Bennett, David	University of Notre Dame
Bergeron, Pierre	Université de Montréal
Bianchi, Luciana	The Johns Hopkins University
Blair, William	The Johns Hopkins University
Blitz, Leo	Radio Astronomy Laboratory, University of California
Bresolin, Fabio	Institute for Astronomy
Brown, Thomas	Space Telescope Science Institute
Bullock, James	Harvard-Smithsonian Center for Astrophysics
Cacciari, Carla	INAF – Osservatorio Astronomico di Bologna
Chaname, Julio	Chaname Ohio State University
Chandar, Rupali	Space Telescope Science Institute
Christian, Carol	Space Telescope Science Institute
Crowther, Paul	University College London
de Grijs, Richard	University of Cambridge
De Marchi, Guido	Space Telescope Science Institute
Drozdovsky, Igor	University of Pittsburgh
Durrell, Patrick	Penn State Univeristy
Evans, Christopher	Isaac Newton Group of Telescopes
Flint, Kathleen	Carnegie Institution of Washington – DTM
Forestell, Amy	University of Texas at Austin
Freeman, Ken	Mount Stromlo Observatory
Fruchter, Andrew	Space Telescope Science Institute
Fullerton, Alex	The Johns Hopkins University
Gallagher, Jay	University of Wisconsin – Madison
Garnett, Donald	Steward Observatory, University of Arizona
Gebhardt, Karl	University of Texas at Austin
Ghex, Andrea	University of California, Los Angeles
Gieles, Mark	Sterrekundig Institut
Godon, Patrick	Space Telescope Science Institute
Goudfrooij, Paul	Space Telescope Science Institute
Grebel, Eva	Max Planck Institute for Astronomy
Greyber, Howard	
Grocholski, Aaron	University of Florida
Guhathakurta, Puragra	University of California at Santa Cruz, Lick Observatory
Hanes, Dave	Queen's University
Harris, Jason	Space Telescope Science Institute
Harris, William	McMaster University
Hartnett, Kevin	NASA/Goddard Space Flight Center
Hauser, Mike	Space Telescope Science Institute
Heap, Sara	NASA/Goddard Space Flight Center
Heckman, Timothy	The Johns Hopkins University
Holtzman, Jon	New Mexico State University
Hopp, Ulrich	Universitäts-Sternwarte Munich

Hornschemeier, Ann	The Johns Hopkins University
Jeletic, James	NASA/Goddard Space Flight Center
Jogee, Shardha	Space Telescope Science Institute
Kalirai, Jason	University of British Columbia
Kepley, Amanda	University of Wisconsin – Madison
Kratsov, Andrey	University of Chicago
Lamers, Henny	Astronomical Institute, Utrecht University
Landsman, Wayne	NASA/Goddard Space Flight Center/SSAI
Leckrone, David	NASA/Goddard Space Flight Center
Lee, Myung Gyoon	DTM/CIW and Seoul National University
Leisy, Pierre	IAC/ING Daniel Lennon Isaac Newton Group of Telescopes
Livio, Mario	Space Telescope Science Institute
Lockman, Felix	NRAO
Macri, Lucas	NOAO
Maíz, Jesús	Space Telescope Science Institute
Majewski, Steven	University of Virginia
Margon, Bruce	Space Telescope Science Institute
Massa, Derck	NASA/Goddard Space Flight Center
Massey, Philip	Lowell Observatory
Mateo, Mario	University of Michigan
McLean, Brian	Space Telescope Science Institute
Nanduri, Vidyardhi	Cosmology Research Center
Neill, Don	Columbia University
Niedner, Mal	NASA/Goddard Space Flight Center
Nota, Antonella	Space Telescope Science Institute
Oey, Sally	Lowell Observatory
Onken, Christopher	Ohio State University
Palma, Christopher	Penn State University
Panagia, Nino	Space Telescope Science Institute
Peng, Eric	Rutgers University
Points, Sean	Northwestern University
Profitt, Charles	CSC/CUA/STScI
Putman, Mary	University of Colorado – CASA Center
Rasmussen, Andrew	Columbia University
Ree, Chang Hee	Yonsei University of Korea/Caltech
Regan, Michael	Space Telescope Science Institute
Reid, Neill	Space Telescope Science Institute
Reitzel, David	University of California at Los Angeles
Rejkuba, Marina	European Southern Observatory
Rey, Soo-Chang	Yonsei University of Korea/Caltech
Rhee, George	NMSU
Rich, Michael	University of California at Los Angeles
Richer, Harvey	University of British Columbia
Romaniello, Martino	European Southern Observatory
Sahu, Kailash	Space Telescope Science Institute
Salim, Samir	University of California – Los Angeles
Sarajedini, Ata	University of Florida
Sellwood, Jerry	Rutgers University
Sembach, Kenneth	Space Telescope Science Institute

Shaya, Ed	NASA/Goddard Space Flight Center
Shen, Juntai	Rutgers University
Siegel, Michael	Space Telescope Science Institute
Sjouwerman, Lorant	NRAO
Soderblom, David	Space Telescope Science Institute
Stanghellini, Letizia	Space Telescope Science Institute
Steigman, Gary	Ohio State University
Steinmetz, Matthias	Astrophysikalisches Institut Potsdam
Susa, Hajime	Rikkyo University
Swaters, Rob	JHU/Space Telescope Science Institute
Sweigart, Allen	NASA/Goddard Space Flight Center
Thilker, David	The Johns Hopkins University
Tosi, Monica	INAF – Osservatorio Astronomico di Bologna
van den Bergh, Sidney	Dominican Astrophysical Observatory
van der Marel, Roeland	Space Telescope Science Institute
Verner, Ekaterina	Goddard Space Flight Center/CUA
Villaver, Eva	Space Telescope Science Institute
Vladilo, Giovanni	Osservatorio Astronomico di Trieste – INAF
Walborn, Nolan	Space Telescope Science Institute
Wesemael, Francois	Université de Montréal
Whitmore, Brad	Space Telescope Science Institute
Wyse, Rosemary	The Johns Hopkins University
Young, Teresa	
Zezas, Andreas	Harvard-Smithsonian Center for Astrophysics

Preface

The Space Telescope Science Institute Symposium on "The Local Group as an Astrophysical Laboratory" took place during 5–8 May 2003.

The Local Group is in some sense the universe in a nutshell. The processes of galaxy mergers and interactions are the bread and butter of hierarchical structure formation. These processes can be studied in unsurpassed detail in the Local Group. Starburst regions in the LMC provide spectacular local versions of their high-redshift counterparts. While black holes are believed to reside at the centers of most galaxies, the best determination of the mass of a central black hole has been achieved in our own Galaxy (through the orbits of individual stars). In addition, the Local Group provides a rich census of star formation histories and of stellar populations. In short, before we attempt to understand the Universe, understanding our own backyard is a good start.

These proceedings represent only a part of the invited talks that were presented at the symposium. We thank the contributing authors for preparing their manuscripts.

We thank Sharon Toolan of ST ScI for her help in preparing this volume for publication.

Mario Livio
Thomas M. Brown
Space Telescope Science Institute
Baltimore, Maryland

History of the Local Group

By SIDNEY VAN DEN BERGH

Dominion Astrophysical Observatory, Herzberg Institute of Astrophysics, National Research Council of Canada, 5071 West Saanich Road, Victoria, BC, Canada V9E 2E7; email: sidney.vandenbergh@nrc-cnrc.gc.ca

It is suggested that M31 was created by the early merger and subsequent violent relaxation of two or more massive metal-rich ancestral galaxies within the core of the Andromeda subgroup of the Local Group. On the other hand, the evolution of the main body of the Galaxy appears to have been dominated by the collapse of a single ancestral object that subsequently evolved by capturing a halo of small metal-poor companions. It remains a mystery why the globular cluster systems surrounding galaxies like M33 and the LMC exhibit such striking differences in evolutionary history. It is argued that the first generation of globular clusters might have been formed nearly simultaneously in all environments by the strong pressure increase that accompanied cosmic reionization. Subsequent generations of globulars may have formed during starbursts that were triggered by collisions and mergers of gas-rich galaxies.

The fact that the [G]alactic system is a member of a group is a very fortunate accident. Hubble (1936, p. 125)

1. Introduction

According to Greek mythology, the goddess of wisdom, Pallas Athena, clad in full armor, emerged from Zeus's head after Hephaestus split it open. In much the same way the Local Group sprang forth suddenly, and almost complete, in Chapter VI of *The Realm of the Nebulae* (Hubble 1936, pp. 124–151). Hubble describes the Local Group as "a typical small group of nebulae which is isolated in the general field." He assigned (in order of decreasing luminosity) M31, the Galaxy, M33, the Large Magellanic Cloud, the Small Magellanic Cloud, M32, NGC 205, NGC 6822, NGC 185, IC 1613 and NGC 147 to the Local Group, and regarded IC 10 as a possible member. In the two-thirds of a century since Hubble's work, the number of known Local Group members has increased from 12 to 36 (see Table 1) by the addition of almost two dozen low-luminosity galaxies. Recent detailed discussions of individual Local Group galaxies are given in Mateo (1998), Grebel (2000) and van den Bergh (2000a). Hubble (1936, p. 128) pointed out that investigations of the Local Group were important for two reasons: (i) "[T]he members have been studied individually, as the nearest and most accessible examples of their particular types, in order to determine the[ir] internal structures and stellar contents," and (ii) "the [G]roup may be examined as a sample collection of nebulae, from which criteria can be derived for further exploration."

Small galaxy groups, like the Local Group, are quite common. From inspection of the prints of the *Palomar Sky Survey*, van den Bergh (2002a) has estimated that 16% of nearby galaxies are located in such small groups. Hubble (1936, p. 128) emphasized that: "The groups [such as the Local Group] are aggregations drawn from the general field, and are not additional colonies superposed on the field." From its observed radial velocity dispersion of 61 ± 8 km s^{-1}, the Local Group is found to have a virial mass of $(2.3 \pm 0.6) \times 10^{12}$ $M_\odot$ (Courteau & van den Bergh 1999). The zero-velocity surface of the Local Group has a radius of 1.18 ± 0.15 Mpc, but $\sim$80% of the Local Group members are actually situated within 0.4 Mpc of the barycenter of the Local Group, which is located between M31 and the Galaxy. From its virial mass and the integrated

Name	Alias	DDO Type†	D(kpc)	M_V
M31	NGC 224	Sb I–II	760	−21.2
Milky Way	Galaxy	Sbc I–II:	8	−20.9
M33	NGC 598	Sc II–III	795	−18.9
LMC	...	Ir III–IV	50	−18.5
SMC	...	Ir IV/IV–V	59	−17.1
M32	NGC 221	E2	760	−16.5
NGC 205	...	Sph	760	−16.4
IC 10	...	Ir IV:	660	−16.3
NGC 6822	...	Ir IV–V	500	−16.0
NGC 185	...	Sph	660	−15.6
IC 1613	...	Ir V	725	−15.3
NGC 147	...	Sph	660	−15.1
WLM	DDO 221	Ir IV–V	925	−14.4
Sagittarius	...	dSph(t)	24	−13.3
Fornax	...	dSph	138	−13.1‡
Pegasus	DDO 216	Ir V	760	−12.3
Leo A	DDO69	Ir V	800	−12.2¶
SagDIG	...	Ir V	1180	−12.0‖
Leo I	Regulus	dSph	250	−11.9
And I	...	dSph	810	−11.8
And II	...	dSph	700	−11.8
Aquarius	DDO 210	V	1025	−11.3
Pegasus II	And VI	dSph	815	−10.5††
And V	...	dSph	810	−10.2
And III	...	dSph	760	−10.2
Cetus	...	dSph	775	−10.2‡‡
Leo II	...	dSph	210	−10.1
Pisces	LGS 3	dIr/dSph	620	−9.8¶¶
Phoenix	...	dIr/dSph	395	−9.8
Sculptor	...	dSph	87	−9.8
Tucana	...	dSph	895	−9.6‖‖
Cassiopeia	And VII	dSph	690	−9.5
Sextans	...	dSph	86	−9.5
Carina	...	dSph	100	−9.4
Draco	...	dSph	79	−8.6
Ursa Minor	...	dSph	63	−8.5

† Uncertain values are marked by a colon (:)
‡ from Majewski et al. (2003)
¶ from Dolphin et al. (2002)
‖ from Lee & Kim (2000)
†† from Pritzel et al. (2002)
‡‡ from Whiting et al. (1999)
¶¶ from Miller et al. (2001)
‖‖ from Walker (2003)

TABLE 1. Members of the Local Group

luminosity of Group members of $4.2 \times 10^{10}\ L_\odot$, the dynamical mass-to-light ratio of the Local Group is found to be $M/L_V = 44 \pm 12$ in solar units. Such a high M/L value is an order of magnitude larger than the mass-to-light ratio in the solar neighborhood of the Galaxy. This supports the conclusion by Kahn & Woltjer (1959) that the mass of the Local Group is dominated by invisible matter.

2. Neighborhood of the Local Group

The Local Group is situated in the outer reaches of the Virgo supercluster. The nearest neighbor of the Local Group is the small Antlia group (van den Bergh 1999a). This tiny cluster is located at a distance of only 1.7 Mpc from the barycenter of the Local Group, i.e., well beyond the zero-velocity surface of the Group. The Antlia group has a mean radial velocity of $+114 \pm 12$ km s^{-1}. The number of galaxies brighter than $M_V = -11.0$ in the Local Group is 22, compared to only four such objects in the Antlia group. However, because the Antlia group contains no supergiant galaxies like M31 and the Milky Way system, its integrated luminosity is $\sim$150 times smaller than that of the Local Group. Since the Local Group is a relatively small cluster, this result suggests that clusters of galaxies have a range in luminosities (and masses?) that extend over at least four orders of magnitude.

At a distance of $\sim$3.9 Mpc, the Centaurus group—which contains M83 (NGC 5236) and Centaurus A (NGC 5128)—is the nearest massive cluster (van den Bergh 2000b). If one assumes NGC 5128 and NGC 5236 to be members of a single cluster, one obtains a total virial mass of 1.4×10^{13} $M_\odot$ and a zero-velocity radius of 2.3 Mpc for the Centaurus group. In other words, the zero-velocity surfaces of the Centaurus group and the Local Group would almost touch each other. However, Karachentsev et al. (2002) conclude that the galaxies surrounding NGC 5128 and M83, respectively, actually form dynamically distinct clusterings. If that is indeed the case, the total mass of these clusters is reduced to only 3×10^{12} $M_\odot$, and the radius of the zero-velocity surface around Cen A is less than 1.3 Mpc.

A second massive nearby cluster contains IC 342 and the highly obscured elliptical Maffei 1. McCall (1989) concluded that "it is likely that IC 342 and Maffei 1 had a significant impact on the past dynamical evolution of the major members of the Local Group." More recently, Fingerhut et al. (2003a,b) have, however, found that the Galactic absorption in front of Maffei 1 is lower than was previously believed. As a result, the distances to Maffei 1 and its companions are too large for these objects to have had significant dynamical interactions with individual Local Group members since the Big Bang.

If the peculiar velocities of galaxies are induced by gravitational interactions, then one might expect massive field galaxies to have a lower velocity dispersion than dwarfs. Data on nearby field galaxies do not appear to support this expectation (Whiting 2003). Whiting (2003) finds the mean radial velocity dispersion among field galaxies within 10 Mpc to be 113 km s^{-1}. A much lower dispersion of $\sim$30 km s^{-1} for the local Hubble flow has, however, been found by Karachentsev et al. (2002). It is noted in passing that the former value appears to be significantly larger than the radial velocity dispersion of 61 ± 8 km s^{-1} found by Courteau & van den Bergh (1999) within the Local Group itself.

3. Subclustering in the Local Group

To first approximation, the Local Group is a binary system with massive clumps of galaxies centered on M31 and on the Galaxy. van den Bergh (2000, p. 290) estimates a mass $M(A) = (1.15–1.5) \times 10^{12}$ $M_\odot$ for the Andromeda subgroup of the Local Group compared to $M(G) = (0.46–1.25) \times 10^{12}$ $M_\odot$ for the Galactic subgroup. Sakamoto, Chiba & Beers (2003) have given somewhat higher mass estimates. If Leo I is included, they find $M(G) = (1.5–3.0) \times 10^{12}$ $M_\odot$, compared to $M(G) = (1.1–2.2) \times 10^{12}$ $M_\odot$ if it is assumed that Leo I is not a member of the Galactic subgroup. Recent proper motion observations by Piatek et al. (2002) suggest that the Fornax dwarf spheroidal galaxy may

Environment	Sph+dSph	dSph/dIr	Ir
Isolated	1	2	6
?	1	1	1
In subcluster	15	0	0

TABLE 2. Environments of dwarf members of the Local Group

not, as previously thought, be a distant satellite of the Galaxy. Instead, the data appear to indicate that Fornax is a free-floating member of the Local Group that is presently near periGalacticon. Within each of the two main Local Group subclusters there are additional subclumps such as the M31 + M32 + NGC 205 triplet, the NGC 147 + NGC 185 binary, and the LMC + SMC binary. For the entire Local Group one finds, at the 99% confidence level, that low-luminosity early-type dSph galaxies are more concentrated in subclumps than late-type dIr galaxies. In other words, most dIr galaxies appear to be free-floating members of the Local Group, whereas the majority (but not all) of dSph galaxies seem to be directly associated with either M31 or the Galaxy. It is not yet clear if the mean ages of the stellar populations in dSph galaxies are themselves a function of location. van den Bergh (1995) has tentatively suggested that star formation in dSph galaxies in the dense regions close to M31 and the Galaxy might typically have started earlier than star formation in remote dSph galaxies. The data in Table 2 show a strong correlation between the morphological types of faint galaxies with $M_V > -16.5$ (i.e., objects fainter than M32) and their environment. This dependence was first noticed by Einasto et al. (1974). Almost all Sph + dSph galaxies are seen to be associated with the two dense subclusters within the Local Group, whereas most Ir galaxies appear to be more or less isolated Group members. It seems quite possible (cf. Skillman et al. 2003) that the faint dIr and dSph galaxies have similar progenitors, and the observed differences between them are due to environmental factors favoring gas loss (from those dwarfs) that occurred in dense environments, i.e., near giant galaxies.

Within the Local Group, the M31, M32, M33 subgroup has a total luminosity $L_V = 3. \times 10^{10} L_\odot$. This is significantly greater than that of the subgroup centered on the Milky Way System, which has $L_V = 1.1 \times 10^{10} L_\odot$. The luminosities of the M31 and Galactic subgroups account for 71% and 24% of the total Local Group luminosity, respectively. It should, however, be noted that some uncertainty in the luminosity ratio of M31 to that of the Galaxy is introduced by the fact that both of these systems are viewed edge-on. As a result, the internal absorption corrections (which may be quite large) are uncertain. Nevertheless, the notion that M31 is more massive than the Galaxy receives some support from the observation that M31 appears to have 450 ± 100 globular clusters, compared to only 180 ± 20 such clusters associated with the Galaxy (Barmby et al. 2000). Furthermore, the bulge mass of M31 is $3.6 \times 10^{10} M_\odot$ (Freeman 1999), which is almost twice as large as the $2 \times 10^{10} M_\odot$ mass of the Galactic bulge. On the basis of these results one might expect the total mass of the M31 subgroup of the Local Group to be two or three times larger than that of the Galaxy subgroup. Surprisingly, this does not appear to be the case. Using radial velocity observations, Evans et al. (2000) conclude that: "There is no dynamical evidence for the widely held belief that M31 is more massive—it may even be less massive." Gottesman et al. (2002) also concluded from dynamical arguments that the mass of M31 "is unlikely to be as great as that of our own Milky Way." These authors even make the heretical suggestion that M31 might not have a massive halo at all! Either the well-known perfidity of small-number statistics has mislead us about the relative masses of M31 and the Galaxy, or the mass-to-light ratio in the Milky Way system is much higher than that of the Andromeda galaxy. If the latter conclusion is correct, then one would have to accept the existence of significant galaxy-to-galaxy variations in the ratio of visible to dark matter among giant spirals.

4. The halos of M31 and the galaxy

It has been known for many years that the halo of M31 contains some globular clusters that are much more metal-rich than those in the Galactic halo (van den Bergh 1969). Perhaps the best known example is the luminous globular cluster Mayall II. Furthermore, the color-magnitude diagrams for individual M31 halo stars (Mould & Kristian 1986, Pritchet & van den Bergh 1988, Durrell, Harris & Pritchet 1994) all show that (i) halo stars have a wide range in metallicity, and (ii) the mean metallicity of stars in the halo of M31 is surprisingly high. The mean values of [Fe/H] for M31 halo stars obtained by Mould & Kristian, by Pritchet & van den Bergh and by Durrell, Harris & Pritchet are $<$[Fe/H]$>= -0.6, -1.0$ and -0.6, respectively. (It is noted in passing that the halo of M31 does contain a metal-poor component which includes clusters such as Mayall IV, some RR Lyrae variables, and non-variable horizontal-brach stars; Sarajedini & Van Duyne 2001.) The observation that the stars in the halo of M31 appear, on average, to be much more metal-rich than those in the Galactic halo, suggests that these two giant spiral galaxies had quite different evolutionary histories (Durrell, Harris & Pritchet 1994). The higher metallicity of M31 halo stars indicates that the building blocks of the Andromeda halo had much higher masses than those of the Galactic halo. Simulations of Murali et al. (2002) show that a significant fraction of the mass that was originally in such merging ancestral galaxies will end up as intergalactic debris and, presumably, also in an extended halo of the final merged object.

An independent check on the metallicities of M31 giants is provided by the recent spectroscopic observations undertaken by Reitzel & Guhathakurta (2002). Their spectra of stars in the halo of M31 have a mean metallicity of $<$[Fe/H]$>= -1.3$. This value is significantly lower than that derived from photometric observations of stars in the halo of the Andromeda galaxy. A possible explanation for this difference is that insufficient correction was made for the fact that old metal-rich red giants are fainter (and therefore more difficult to observe) than are old metal-poor giants. Non homogeneity of the Andromeda halo might also have contributed to the observed difference in the mean metallicities derived from photometry and spectroscopy. For example, Reitzel & Guhathakurta (2002) find four stars of solar metallicity in the halo of M31. They suggest that these objects might represent metal-rich debris from an accretion event. Other evidence for such accretion events is provided by Ibata et al. (2001) and Ferguson et al. (2002). Recently, Yanny et al. (2003) also found possible evidence for such a tidal stream in the Galaxy, located beyond the plane of the Milky Way at a distance of $\sim$20 kpc from the Galactic center.

An interesting clue regarding the origin of the difference between the Andromeda and the Galactic halos is provided by the observations of Pritchet & van den Bergh (1994; see Figure 1), which show that M31 has an $R^{1/4}$ profile out to $\sim$20 kpc from its nucleus. Such a structure is likely to have resulted from violent relaxation. This suggests that the overall morphology of the Andromeda galaxy was determined by violent relaxation resulting from the early merger of two (or more) massive metal-rich ancestral objects. On the other hand, the main body of the Milky Way system may have been assembled à la Eggen, Lynden-Bell & Sandage (1964), with its metal-poor halo forming via late infall and capture of "small bits and pieces" (Searle & Zinn 1978). However, it appears that these

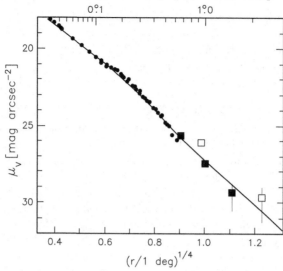

FIGURE 1. Profile of M31 derived from star counts. The figure shows that an $R^{1/4}$ law provides a reasonable fit to the observations. This favors the interpretation that M31 was formed from violent relaxation following mergers.

fragments differed in a rather fundamental way from those producing dwarf spheroidal galaxies. Shetrone, Côté, & Sargent (2001) find that dSph galaxies are iron-rich and have $0.02 \leq [\alpha/\text{Fe}] \leq 0.13$, compared to typical Galactic values of $[\alpha/\text{Fe}] \sim 0.28$ dex over the same range of metallicities. This shows that the bulk of the Galactic halo stars cannot have formed in dwarf spheroidal galaxies that subsequently disintegrated. In particular, Fulbright (2002) finds that less than 10% of local metal-poor stars with $[\text{Fe/H}] < -1.2$ have α-to-iron abundance ratios similar to those found in dSph galaxies. More generally, Tolstoy et al. (2003) conclude that the observed element abundance patterns make it difficult to form a significant proportion of the stars observed in our Galaxy in the small galaxies that subsequently merged to form the disk, bulge, and inner halo of the Milky Way.

Bekki, Harris & Harris (2003) studied the distribution of stars of various metallicities after the merger of two spirals. However, their model is not likely to be applicable to the early merger of the ancestral objects of M31, because extended disks would be destroyed by frequent tidal interactions at large look-back times. *Hubble Space Telescope* images show that galaxies with obvious disks mostly have $z < 1.5$, whereas the vast majority of the objects seen at $z > 1.5$ appear to have either compact or chaotic morphologies (van den Bergh 2002b). So the Bekki et al. model, in which extended disks merge, is probably inappropriate for galaxies at $z > 1.5$, i.e., for mergers that took place more than nine Gyr ago.

In their pioneering study of the metallicities of individual stars in galactic halos, Mould & Kristian (1986) also observed stars in the halo of the late-type spiral M33. Their color-magnitude diagram suggested that the halo of M33 was very metal-poor and had $<[\text{Fe/H}]> = -2.2$. This value is more than an order of magnitude lower than that for stars in the halo of the Andromeda galaxy.

In summary, M31 may have formed from the early merger of the two or three most massive galaxies in the core of the Andromeda subgroup of the Local Group. The less massive Andromeda companions, such as M32 and NGC 205, may represent objects in the core of the Andromeda subgroup which had such low masses they managed to survive individually.

5. History of globular cluster systems

Due to differences in evolutionary history, the halo of M31 contains relatively metal-rich globular clusters, whereas the Galactic halo does not. Another difference between the M31 and Galactic globular cluster systems has been noted by Rich et al. (2002), who find that M31 does not appear to contain globular clusters with extremely blue horizontal branches, such as M92 in the Galactic halo. Another major difference between globular cluster systems is provided by M33 and the LMC. These two late-type disk galaxies have comparable luminosities ($M_V = -18.9$ and -18.5, respectively), but radically different globular cluster systems. Surprisingly, the LMC globulars, which are both very old and quite metal-poor, appear to have disk kinematics (Schommer et al. 1993). On the other hand, the metal-poor globular clusters associated with M33 seem to have halo-like kinematics (Schommer et al. 1991). Sarajedini et al. (1998) found that eight out of ten globular clusters in their M33 halo sample had significantly redder horizontal branches than Galactic globulars of similar metallicity. This difference might be interpreted as a second parameter effect. Alternatively, and perhaps more plausibly, the M33 globulars may be a few Gyr younger than their Galactic counterparts. In the latter interpretation, the M33 halo globular clusters exhibit an unexpectedly large age dispersion of ∼3–5 Gyr. It is presently a mystery why the M33 halo globular clusters would have formed a few Gyr later than typical Galactic halo and LMC disk globulars. Some light might eventually be shed on these questions by observations of the radial velocities of RR Lyrae stars in the LMC, and perhaps also in the near future, of RR Lyrae stars in M33. The main differences between the globular cluster systems of M33 (Sc) and the Galaxy (Sbc) could perhaps be understood (van den Bergh 2002b) by assuming that late-type galaxies take significantly longer to arrive at their final morphology than do spirals of earlier morphological types. However, the great age of the LMC cluster system appears to conflict with this simple explanation. The observations of Mould & Kristian (1986) appear to show that the field stars in the halo of M33 are extremely metal-poor and have $<$[Fe/H]$>= -2.2$. It would be important to confirm this result by new photometry and to compare this value with the mean metallicity of stars in the outer halo of the Large Magellanic Cloud.

Not unexpectedly, the majority of Local Group dwarf galaxies are surrounded by small families of metal-poor globular clusters. However, it is puzzling that the Sagittarius dwarf has one globular cluster companion (Terzan 7, [Fe/H] $= -0.36$) that is quite metal-rich. How could such a relatively high metallicity have been built up within a dwarf galaxy? The only other Galactic halo ($R_{gc} > 10$ kpc) globular clusters known to have metallicities higher than [Fe/H] $= -1.0$ are Pal 1 and Pal 12. The latter object is, itself, suspected of also being associated with the disintegrating Sagittarius dwarf (Irwin 1999). This speculation is supported by the observations of Martínez-Delgado et al. (2002), who have found that Pal 12 is possibly embedded in tidal debris of the Sagittarius dwarf.

Rosenberg et al. (1998a,b) have also shown that the cluster Pal 1 is significantly younger than most other Galactic globular clusters. It is presently not clear which kind of evolutionary scenarios would allow halo clusters like Pal 1 and Pal 12 to attain such relatively high metallicities. From its present luminosity and morphological type, the dwarf elliptical galaxy M32 would have been expected to be embedded in a swarm of 10–20 globular clusters. It is therefore puzzling that not a single globular cluster appears to be associated with this galaxy. Perhaps some of the innermost M32 clusters were dragged into its compact luminous nucleus by dynamical friction. Also, loosely bound outer globulars that were originally associated with M32 might have been detached by tidal interactions with the main body of M31. Such detached M32 clusters would remain in the halo of M31 and might be recognized by being unusually compact. It would be very worthwhile to undertake a systematic search for such M32 clusters with small R_h values in the halo of M31.

For the vast majority of galaxies, the specific globular cluster frequency S is less than 10 (Harris & van den Bergh 1981). However, the Fornax dwarf, which has five globulars associated with it, has $13 < S < 26$. Recent work by Kleyna et al. (2003) appears to indicate that the Ursa Minor dwarf may have an even higher S value. If a dynamically cold clustering of stars that these authors find in UMi is a globular cluster (or a disintegrated cluster), then $S \sim 400$ for the UMi system. Taken at face value, this result suggests that the fraction of the light of dwarf spheroidals in the form of globular clusters may be much higher in dwarf spheroidal galaxies than it is in more luminous (massive) systems.

If the Milky Way system had collapsed in the fashion advocated by Eggen, Lynden-Bell & Sandage (1962), but with some gaseous dissipation, then one would have expected the stars and globular clusters in the Galactic halo to exhibit a radial metallicity gradient. On the other hand, the halo of the Milky Way system would not be expected to have such a metallicity gradient if, as envisioned by Searle & Zinn (1978), it had formed by the accretion of many "bits and pieces." Using the 1999 version of the globular cluster catalog of Harris (1996), van den Bergh (2003) found a possible hint for the existence of such a radial metallicity gradient among Galactic halo ($R_{gc} > 10$ kpc) globular clusters. However, the reality of this a gradient is not supported by the more recent data contained in the 2003 version of Harris's catalog. On the other hand, the clusters in the main body of the Galaxy, i.e., those with $R_{gc} < 10$ kpc, do appear to show a radial abundance gradient. Globulars at $R_{gc} < 4.0$ kpc are, on average, found to be more metal-rich than those having $4.0 \leq R_{gc} < 10$ kpc. A Kolmogorov-Smirnov test shows that there is only a 4% probability that the metallicities of these inner and outer globular clusters samples were drawn from the same parent population. Taken at face value, the existence of a Galactic metallicity gradient between 4 kpc and 10 kpc favors the suggestion that the ELS model provides an adequate description of the formation of the main body of the Galactic halo, whereas the SZ model predictions agree with the observed lack of a metallicity gradient in the region with $R_{gc} > 10$ kpc.

It is a curious (and unexplained) fact that the distribution of flattening values of globular clusters differs significantly from galaxy to galaxy. Both open and globular clusters in the LMC are, for example, generally more flattened than their Galactic counterparts. Furthermore, the luminous clusters in the Large Cloud are typically more flattened than the less luminous ones (van den Bergh 1983). Finally, it is of interest to note that the most luminous globular clusters in many Local Group galaxies also seem to be among the most flattened. Examples are the globular Mayall II ($\epsilon = 0.22$) in M31, ω Centauri ($\epsilon = 0.19$) in the Galaxy, and NGC 1835 ($\epsilon = 0.21$) in the LMC.

6. The luminosity function of the Local Group

The presently known members of the Local Group exhibit a flat luminosity function with slope $\alpha = -1.1 \pm 0.1$ (Pritchet & van den Bergh 1999). This value is significantly lower than the slope $\alpha = -1.8$ that is predicted by the cold dark matter theory (Klypin et al. 1999). The low frequency of faint Local Group dwarfs has been confirmed in a recent hunt for new faint Local Group members by Whiting, Hau & Irwin (2002). This observed deficiency of faint LG members suggests that many of the progenitors of low mass galaxies were destroyed before they had a chance to form significant numbers of

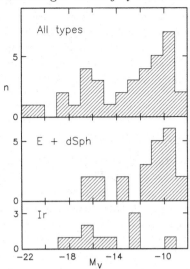

FIGURE 2. Luminosity function of the Local Group. The data suggest, but do not prove, that dIr luminosity distribution is less steep than that for dSph galaxies.

stars. Alternatively, it might be assumed that the missing faint galaxies can be identified with compact high-velocity clouds. However, Maloney & Putman (2003) have recently shown that such objects, if they were located at distances of ∼1 Mpc, would be largely ionized. Therefore, these authors conclude that the compact high-velocity clouds are not at cosmologically significant distances, but that they are instead associated with the Galactic halo. All attempts to search for evidence of star formation in compact high-velocity clouds (e.g., Simon & Blitz 2002) have so far remained unsuccessful. It is therefore assumed that the deficiency of faint Local Group members is real. Figure 2 appears to show (van den Bergh 2000a, p. 281) that the luminosity function of dSph galaxies in the Local Group is steeper than for dIr galaxies. This conclusion should, however, be regarded as provisional, because a Kolmogorov-Smirnov test shows the difference between the luminosity distributions of Local Group dSph and dIr galaxies is only significant at the 75% confidence level. If the luminosity function of dSph galaxies is, indeed, steeper than for dIr galaxies, then future discoveries are most likely to turn up very faint dSph (rather than dIr) members of the Local Group. It would clearly be very important to undertake sky surveys in two (or more) colors to search for the signatures of the color-magnitude diagrams of extremely faint (and so far undiscovered) resolved dwarf members of the Local Group.

Three distinct explanations might be invoked to account for the apparent excess of faint dSph galaxies among presently known Local Group members: (i) Perhaps gas was more likely to escape from faint (low-mass) galaxies than from more massive objects. As a result, low-mass galaxies would most often end up as gas-free dSph galaxies. (ii) Alternatively, the mass spectrum with which galaxies form might depend on environmental density in such a way that high density regions (i.e., the neighborhood of M31 and the Galaxy) form a larger fraction of low-mass objects (the majority of which end up as dSph galaxies). The latter assumption would be consistent with the work of Trentham et al. (2001) and Trentham & Tully (2003), who found that the galaxian luminosity function of the dense Virgo cluster is much steeper than for the less dense Ursa Major cluster and the Local Group. On the other hand, the view that dense regions produce galaxies with steep galaxian luminosity functions appears to conflict with the result of Sabatini et al. (2003), who find that the Virgo cluster luminosity function seems to be steeper in the low density outer regions than it is in the high density core of this cluster. This observation might, however, be accounted for by assuming that tides preferentially destroy fragile dwarfs in the cores of dense clusters. (iii) Finally, and perhaps most plausibly, the gas in the progenitors of the missing dwarfs might have been photoevaporated during reionization.

7. Morphological evolution of Local Group members

It is difficult to tease out information on the morphological evolution of galaxies from the distribution of stars of various ages. The central bulges of giant spirals, such as M31 and the Galaxy, are dominated by old stars. This supports the notion that these objects were built up inside out, with their bulges forming first and the disk possibly being accreted at a later time. It would be very important to establish how old the first (presumably quite metal-poor) generation of Galactic disk stars is. This problem is made more intractable because such thin disk stars have to be disentangled from stars that are physically associated with—and embedded within—an older thick disk population. In fact, tidal interactions might pump energy into (and hence thicken) an initially thin disk of very metal-poor stars. *Hubble Space Telescope* observations of galaxies at large look-back times suggest that most disk star formation occurs at $z \lesssim 1.5$, i.e., during the last 9 Gyr (van den Bergh 2002b). One reason for the paucity of disks at larger redshifts is, presumably, that such extended structures would be destroyed by tidal forces during the frequent encounters between galaxies at high redshifts. From the disk kinematics of two slightly metal-poor stars, Liu & Chaboyer (2000) find a thin disk age of 9.7 ± 0.6 Gyr. Such an age is consistent with the ages of spiral disks, inferred from the fact that the *HST* images of distant galaxies start to show obvious disks at $z \sim 1.5$.

Since bars are generally assumed to have formed from global instabilities in disks, one would not expect to see barred galaxies at $z > 1.5$. If bars cannot form from initially chaotic protodisks, then bar formation might be delayed to even smaller look-back times. This suspicion appears to be confirmed by observations which seem to suggest that the frequency of barred galaxies declines precipitously beyond redshifts of $z \sim 0.7$ (van den Bergh et al. 1996, 2002). If this conclusion is correct, then one would expect the Bar of the LMC to be younger than 6 Gyr. This conclusion is consistent with (but not proved by) the observation that bursts of star formation occurred in the Bar of the Large Cloud 4–6 Gyr and 1–2 Gyr ago (Smecker-Hane et al. 2002).

8. The history of star and cluster formation

Almost all Local Group galaxies are found to contain some very old stars like RR Lyrae variables. This shows that these galaxies started to form stars quite early in the history of the Universe, i.e., more than $\sim$10 Gyr ago. The best candidate for a "young" Local Group member is the dwarf irregular Leo A. However, Dolphin et al. (2002) discovered a few RR Lyrae variables in this galaxy which have ages > 9 Gyr. Furthermore, recent observations by Schulte-Ladbeck et al. (2002) show that this object also contains some metal-poor red horizontal branch stars. This clearly demonstrates that Leo A is not a young galaxy.

M32 and many of the Local Group dSph galaxies have not experienced recent star formation. On the other hand, the gas-rich spiral and irregular Group members are still forming stars at a significant rate. Qualitative data on the past rate of star formation in such galaxies can be obtained from their integrated colors and the intensity ratios of various spectral lines. However, the hope that the age distribution of star clusters in galaxies could provide more detailed information on the past rate of star formation has been shattered by the work of Larsen & Richtler (1999, 2000), which appears to show that the rate of cluster formation varies as a rather high power of the rate of star formation. In other words, there is not a one-to-one correspondence between the rate of cluster formation and the general rate of star formation. In the Local Group, this phenomenon is beautifully illustrated by the difference between the quiescent dwarf irregular IC 1613, which contains few star clusters of any kind (i.e., Baade 1963, p. 231; van den Bergh 1979), and the Large Magellanic Cloud, which is presently quite actively forming both stars and clusters. It seems likely that the present specific frequency of globular clusters in galaxies was mainly determined by their peak rates of star formation, with elevated peak rates resulting in high present specific cluster frequencies. So far, only fragmentary information is available on the luminosity evolution of individual Local Group members. Few star clusters in the LMC have ages between the 3.2 Gyr age of the oldest open clusters and the ~13 Gyr age of the LMC globular clusters (Rich, Shara & Zurek 2001). This probably means that the Large Cloud experienced a quiescent period that extended for ~10 Gyr. During this "dark age," no violent bursts of star formation (which could have triggered the formation of star clusters) occurred. However, it is quite likely that a trickle of star formation (such as that which presently occurs in IC 1613) continued during the dark ages between 3.2 Gyr and 13 Gyr ago. This speculation is supported by the data of Da Costa (2002) which seem to show that the metallicity in the LMC increased between the beginning and the end of the dark age, i.e., between the termination of globular cluster formation ~13 Gyr ago, and the beginning of open cluster formation 3.2 Gyr ago. A possibly more complicated scenario is hinted at by the work of Smecker-Hane et al. (2002), who conclude that star formation in the Bar of the LMC was episodic, while the rate of star formation remained more or less constant within the disk of the Large Cloud. However, an important caveat is that the rate of star formation in the LMC disk was so low that the data do not provide strong constraints on its star formation history.

The Carina dSph galaxy seems to have experienced a major burst of star formation 7 Gyr ago (Hurley-Keller, Mateo & Nemeic 1998). However, at maximum this object probably only reached $M_V \sim -16$, making it too faint to have become what Babul & Ferguson (1996) have called a "boojum."

There has been a long controversy among astronomers regarding the nature (or even the existence of) a fundamental difference between open clusters and globular clusters. The present consensus is that all cluster populations initially formed with a power law mass spectrum, and globular clusters are simply the oldest and most massive population component that was best able to withstand the erosion caused by the destruction of lower mass clusters via evaporation, encounters with massive interstellar clouds, and disk/bulge shocks. However, a different scenario has been proposed by van den Bergh (2001). He suggested that there have, in fact, been two (perhaps quite distinct) epochs of cluster formation. During the first of these, globular clusters might have formed as halo gas was being compressed by shocks driven inward by ionization fronts generated during cosmic reionization at $5 \lesssim z \lesssim 15$. Such effects would presumably be greatest in the halos of small protogalaxies that are relatively easy to ionize. A second generation of massive clusters might have formed by the compression (and subsequent collapse) of giant molecular clouds, triggered by the heating of the interstellar medium induced by collisions between gas rich protogalaxies. A similar view has recently been expressed by Schweizer (2003), who also argues that the first generation of globular clusters formed nearly simultaneously with pristine molecular clouds heated and shocked by the strong pressure increase that accompanied cosmological reionization. Schweizer argues that this hypothesis might also account for the similarity of metal-poor globular clusters in all types of galaxies and environments. Both van den Bergh (2001) and Schweizer (2003) argue that second generation globular clusters were mainly formed during subsequent collisions and mergers between galaxies.

9. Intergalactic matter

From dynamical arguments Kahn & Woltjer (1959) first showed that the Local Group can only be stable if it contains a significant amount of invisible matter. Using radial velocity observations of Local Group members, Courteau & van den Bergh (1999) have estimated that the Local Group has a total mass of $(2.3 \pm 0.6) \times 10^{12} \ M_\odot$, from which the mass-to-light ratio (in solar units) is found to be $M/L_V = 44 \pm 14$. This high value shows that the total mass of the Local Group exceeds that of the visible parts of its constituent galaxies by an order of magnitude. In their 1959 paper, Kahn & Woltjer suggested that this "missing mass" in the Local Group might be in the form of hot (5×10^5 degrees) low density (1×10^{-4} protons cm^{-3}) gas, which would be difficult to detect observationally. Hui & Haiman (2003) have recently shown that the thermal history of such gas has probably been quite complex. In recent years, the notion that hot gas is responsible for the missing mass in the Local Group has been overshadowed by the idea that this missing mass is actually in the form of cold dark matter (Blumenthal et al. 1984). However, recent *Far Ultraviolet Explorer* satellite observations of the absorption lines of O VI (which have radial velocities of only a few hundred km s^{-1}), suggest that hot gas may provide a non-negligible contribution to the missing mass in the Local Group (Nicastro et al. 2003). Alternatively, Sternburg (2003) feels the hot gas clouds observed by Nicastro et al. might just be reprocessed metal-enriched gas that was ejected from the neighborhood of the Galactic plane in supernova-driven fountains. If the latter suggestion is correct, one might expect that the hot clouds would be relatively metal-rich. Alternatively, they would be expected to be metal-poor if they are composed of hot primordial gas.

10. The end

The final fate of the Local Group has been discussed by Forbes et al. (2000), who conclude that dynamic friction will eventually result in the merger of M31 and the Galaxy. This merged object will resemble an elliptical galaxy with $M_V \sim -21$ that contains $\sim$700 globular clusters. In a Universe that continues to expand forever this object will, in the distant future, be the only remaining visible object in the Universe (Bennett et al. 2003, Spergel et al. 2003).

11. Summary

• Both the high metallicity of the M31 halo and the $R^{1/4}$ luminosity profile of the Andromeda galaxy suggest that this object might have formed from the early merger and subsequent violent relaxation of two (or more) relatively massive metal-rich ancestral objects.

• The main body of the Galaxy may have formed in the manner suggested by Eggen, Lynden-Bell & Sandage (1962), whereas its halo is more likely to have been assembled by accretion of "bits and pieces" in the manner first suggested by Searle & Zinn (1978).

• It is profoundly puzzling that the old metal-poor globular clusters in the LMC appear to have been formed in an early disk, whereas the globulars associated with M33 seem to have originated in a slightly younger halo.

• It is speculated that the oldest generation of globular clusters in the Universe might have formed as halo gas was compressed and heated in shocks that were driven inwards by ionization fronts generated during cosmic reionization. On the other hand, second generation globular clusters formed as a result of the heating of molecular clouds during collisions between gas-rich galaxies. It is emphasized that the history of star formation in galaxies is often very different from the history of cluster formation.

• It is presently not understood how globular clusters like Terzan 7 (which is associated with the Sagittarius dwarf) were able to attain a relatively high metallicity. Neither do we know why the globular clusters associated with some galaxies are much more flattened than are those in others.

• It is suggested that the specific globular frequency of galaxies was mainly determined by the peak rate of star formation during evolution.

• All Local Group galaxies appear to contain a very old population component, i.e., all nearby galaxies started to form stars just after the galaxies were formed. In other words, there are no truly young galaxies in the Local Group.

• The Local Group has a mass of $(2.3\pm0.6)\times10^{12}\ M_\odot$, a luminosity $L_V = 4.2\times10^{10}\ L_\odot$, and a zero-velocity radius of 1.18 ± 0.16 Mpc. Most of the mass and luminosity of the Local Group appears to be concentrated in two subgroups that are centered on M31 and the Galaxy, respectively. There is presently a lively controversy about which of these two subgroups is the most massive. If the Galactic subgroup turns out to be more massive than the M31 group, then the ratio of dark to visible matter must differ significantly from group to group.

• It is not yet clear if hot low density gas provides a significant contribution to the total mass of Local Group galaxies.

It is a pleasure to thank Oleg Gnedin, and Eva Grebel for helpful comments on an early draft of the present paper. I also thank David Duncan for preparing the figures.

REFERENCES

BAADE, W. 1963 *Evolution of Stars and Galaxies*. Harvard University Press.

BABUL, A. & FERGUSON, H. C. 1996 *ApJ* **458**, 100.

BARMBY, P., HUCHRA, J. P., BRODIE, J. P., FORBES, D. A., SCHRODER, L. L., & GRILLMAIR, C. J. 2000 *AJ* **119**, 727.

BEKKI, K., HARRIS, W. E., & HARRIS, G. L. H. 2003 *MNRAS* **338**, 587.

BENNETT, C. L., ET AL. 2003 *ApJS* **148**, 1.

BLUMENTHAL, G. R., FABER, S. M., PRIMACK, J. R., & REES, M. J. 1984 *Nature* **311**, 517.

COURTEAU, S. & VAN DEN BERGH, S. 1999 *AJ* **118**, 337.

DA COSTA, G. S. 2002. In *Extragalactic Star Clusters* (eds. D. Geisler, E. K. Grebel & D. Minniti). IAU Symposium No. 207, p. 83. ASP.

DOLPHIN, A. E., ET AL. 2002 *AJ* **123**, 3154.

DURRELL, P. R., HARRIS, W. E., & PRITCHET, C. J. 1994 *AJ* **108**, 2114.

EGGEN, O. J., LYNDEN-BELL, D., & SANDAGE, A. R. 1962 *ApJ* **136**, 748.

EINASTO, J., SAAR, E., KAASIK, A., & CHERNIN, A. D. 1974 *Nature* **251**, 111.

EVANS, N. W. & WILKINSON, M. I. 2000 *MNRAS* **316**, 929.

EVANS, N. W., WILKINSON, M. I., GUHATHAKURTA, P., GREBEL, E. K., & VOIGT, S. S. 2000 *ApJ* **540**, 9.

FERGUSON, A. M. N., IRWIN, M. J., IBATA, R. A., LEWIS, G. F. & TANVIR, N. R. 2002 *AJ* **124**, 1452.

FINGERHUT, R. L., ET AL. 2003a *ApJ*, **587**, 672.

FINGERHUT, R. L., ET AL. 2003b, in preparation.

FORBES, D. A., MASTERS, K. L., MINNITI, D., & BARMBY, P. 2000 *A&A* **358**, 471.

14 S. van den Bergh: *History of the Local Group*

FREEMAN, K. C. 1999. In *The Stellar Content of Local Group Galaxies* (eds. P. Whitelock & R. Cannon). IAU Symposium No. 192, p. 383. ASP.

FULBRIGHT, J. P. 2002 *AJ* **123**, 404.

GOTTESMAN, S. T., HUNTER, J. H., & BOONYASAIT, V. 2002 *MNRAS* **337**, 34.

HARRIS, W. E. & VAN DEN BERGH, S. 1996 *AJ* **86**, 1627.

HUBBLE, E. 1936 *The Realm of the Nebulae*. Oxford University Press.

HUI, L. & HAIMAN, Z. 2003 *ApJ* **596**, 9.

HURLEY-KELLER, D., MATEO, M., & NEMEC, J. 1998 *AJ* **115**, 1840.

IBATA, R., IRWIN, M., LEWIS, G., FERGUSON, A. N. M., & TANVIR, N. 2001 *Nature* **412**, 49.

IRWIN, M. J. 1999. In *Stellar Content of the Local Group* (eds. P. Whitelock & R. D. Cannon). IAU Symposium No. 192, p. 409. ASP.

KAHN, F. D. & WOLTJER, L. 1959 *ApJ* **130**, 705.

KARACHENSEV, I. D., ET AL. 2002 *A&A* **385**, 21.

KLEYNA, J. T., WILKINSON, M. I., GILMORE, G., & EVANS, N. W. 2003 *ApJ*, **588**, L21; Erratum-ibid. **589**, L59.

KLYPIN, A., KRAVTSOV, A., VALENZUELA, O., & PRADA, F. 1999 *ApJ* **522**, 82.

LARSEN, S. S. & RICHTLER, T. 1999 *A&A* **345**, 59.

LARSEN, S. S. & RICHTLER, T. 2000 *A&A* **354**, 836.

LEE, M. G. & KIM, S. C. 2000 *AJ* **119**, 777.

LIU, W. M. & CHABOYER, B. 2000 *ApJ* **544**, 818.

MAJEWSKI, S. R., SKRUTSKIE, M. F., WEINBERG, M. D., & OSTHEIMER, J. C. 2003 *ApJ* **599**, 1082.

MALONEY, P. R. & PUTMAN, M. E. 2003; *ApJ* **589**, 270.

MARTÍNEZ-DELGADO, D., ZINN, R., CARRERA, R. & GALLART, C. 2002 *ApJ* **573**, L19.

MATEO, M. 1998 *ARAA* **36**, 435.

MCCALL, M. L. 1989 *AJ* **97**, 1341.

MILLER, B. W., DOLPHIN, A. E., LEE, M. G., KIM, S. C., & HODGE, P. 2001 *ApJ* **562**, 713.

MOULD, J. R. & KRISTIAN, J. 1986 *ApJ* **305**, 591.

MURALI, C., KATZ, N., HERNQUIST, L., WEINBERG, D. H., & DAVE, R. 2002 *ApJ* **571**, 1.

NICASTRO, F., ET AL. 2003 *Nature* **421**, 719.

PIATEK, S., ET AL. 2002 *AJ* **124**, 3198.

PRITCHET, C. J. & VAN DEN BERGH, S. 1994 *AJ* **107**, 1730.

PRITCHET, C. J. & VAN DEN BERGH, S. 1999 *AJ* **118**, 883.

PRITZEL, B. J., ARMANDROFF, T. E., JACOBY, G. H., & DA COSTA, G. S. 2002 *AJ* **124**, 1464.

REITZEL, D. B. & GUHATHAKURTA, P. 2002 *AJ* **124**, 234.

RICH, R. M., CORSI, C. E., BALLAZZINI, M., FREDERICI, L., CACCIARI, C., & FUSI PECCI, F. 2002. In *Extragalactic Star Clusters* (eds. D. Geisler, E. K. Grebel, & D. Minniti). IAU Symposium No. 207, p. 140. ASP.

RICH, R. M., SHARA, M. M., & ZUREK, D. 2001 *AJ* **122**, 842.

ROSENBERG, A., PIOTTO, G., SAVIANE, I., APARICIO, A., & GRATON, R. 1998a *AJ* **115**, 658.

ROSENBERG, A., SUIANE, I., PIOTTO, G., APARICIA, A., & ZAGGIA, S. R. 1998b *AJ* **115**, 648.

SABATINI, S., DAVIES, J., SCARAMELLA, R., SMITH, R., BAES, M., LINDER, S. M., ROBERTS, S., & TESTA, V. 2003 *MNRAS* **341**, 981.

SAKAMOTA, T., CHIBA, M., & BEERS, T. C. 2003 *A&A*, **397**, 899.

SARAJEDINI, A., GEISLER, D., HARDING, P., & SCHOMMER, R. 1998 *ApJ* **508**, L37.

SARAJEDINI, A. & VAN DUYNE, J. 2001 *AJ* **122**, 2444.

SCHOMMER, R. A., CHRISTIAN, C. A., CALDWELL, N., BOTHUN, G. D., & HUCHRA, J. 1991 *AJ* **101**, 873.

SCHOMMER, R. A., OLSZEWSKI, E. W., SUNTZEFF, N. B., & HARRIS, H. C. 1992 *AJ* **103**, 447.

SCHULTE-LADBECK, R. E., HOPP, U., DROZDOVSKY, I. O., GREGGIO, L., & CRONE, M. M. 2002 *AJ* **124** 896.

SCHWEIZER, F. 2003. In *New Horizons in Globular Cluster Astronomy* (eds. G. Piotto, G. Meylan, G. Djorgovski, & M. Riello). ASP Conf. Proc. 296, p. 467. ASP.

SEARLE, L. & ZINN, R. 1978 *ApJ* **104**, 1472.

SHETRONE, M. D., CÔTÉ , P., & SARGENT, W. L. W. 2001 *ApJ* **548**, 592.

SIMON, J. D. & BLITZ, L. 2002 *apJ* **574**, 726.

SKILLMAN, E. D., TOLSTOY, E., COLE, A. A., DOLPHIN, A. E., SAHA, A., GALLAGHER, J. S., DOHM-PALMER, R. C., & MATEO, M. 2003 *ApJ,* **596**, 253.

SMECKER-HANE, T. A., COLE, A. A., GALLAGHER, J. S., & STETSON, P. B. 2002 *ApJ* **566**, 239.

STERNBERG, A. 2003 *Nature*, **421**, 708.

TOLSTOY, E., VENN, K. A., SHETRONE, M., PRIMAS, F., HILL, V., KAUFFER, A., & SZEIFERT, T. 2003 *AJ* **125**, 707.

TRENTHAM, N., TULLY, B., & VEREIJEN, M. 2001 *MNRAS* **325**, 385.

TRENTHAM, N. & TULLY, R. B. 2002, *MNRAS*, **335**, 712.

VAN DEN BERGH, S. 1969 *ApJS* **19**, 145.

VAN DEN BERGH, S. 1971 *Nature* **231**, 35.

VAN DEN BERGH, S. 1979 *ApJ* **230**, 95.

VAN DEN BERGH, S. 1983 *PASP* **95**, 839.

VAN DEN BERGH S. 1995. In *The Local Group: Comparative and Global Properties* (eds. A. Layden, R. C. Smith, & J. Storm). ESO Workshop No. 51, p. 3. ESO.

VAN DEN BERGH, S. 1999a *ApJ* **517**, L97.

VAN DEN BERGH, S. 1999b *AJ* **117**, 221.

VAN DEN BERGH, S. 2000a *The Galaxies of the Local Group*. Cambridge University Press.

VAN DEN BERGH, S. 2000b *AJ* **119**, 609.

VAN DEN BERGH, S. 2001 *ApJ* **559**, L113.

VAN DEN BERGH, S. 2002a *AJ* **124**, 782.

VAN DEN BERGH, S. 2002b *PASP* **114**, 797.

VAN DEN BERGH, S. 2003 *ApJ* **590**, 590, 797.

VAN DEN BERGH, S., ABRAHAM, R. G., ELLIS, R. S., TANVIR, N. R., SANTIAGO, B. X., & GLAZEBROOK, K. S. 1996 *AJ* **112**, 359.

VAN DEN BERGH, S., ABRAHAM, R. G., WHYTE, L. F., MERRIFIELD, M. R., ESKRIDGE, P. B., FROGEL, J. A., & POGGE, R. 2002 *AJ* **123**, 2924.

WALKER, A. R. 2003. In *Stellar Candles for the Extragalactic Distance Scale* (eds. D. Alloin & W. Gieren). Lecture Notes in Physics 635, p. 265. Springer.

WHITING, A. B. 2003 *ApJ* **587**, 186.

WHITING, A. B., HAU, G. K. T., & IRWIN, M. 1999 *AJ* **118**, 2767.

WHITING, A. B., HAU, G. K. T., & IRWIN, M. 2002 *ApJS* **141**, 123.

YANNY, B., ET AL. 2003 *ApJ* **588**, 824; Erratum-ibid. 2004 **605**, 575.

Primordial nucleosynthesis

By GARY STEIGMAN

Department of Physics, The Ohio State University, Columbus, OH 43210, USA

The primordial abundances of deuterium, helium-3, helium-4, and lithium-7 probe the baryon density of the Universe only a few minutes after the Big Bang. Of these relics from the early Universe, deuterium is the baryometer of choice. After reviewing the current observational status of the relic abundances (a moving target!), the baryon density determined by big bang nucleosynthesis (BBN) is derived. The temperature fluctuation spectrum of the cosmic background radiation (CBR), established several hundred thousand years later, probes the baryon density at a completely different epoch in the evolution of the Universe. The excellent agreement between the BBN- and CBR-determined baryon densities provides impressive confirmation of the standard model of cosmology, permitting the study of extensions of the standard model. In combination with the BBN- and/or CBR-determined baryon density, the relic abundance of ^{4}He provides an excellent chronometer, constraining those extensions of the standard model which lead to a nonstandard early-Universe expansion rate.

1. Introduction

As the hot, dense, early Universe rushed to expand and cool, it briefly passed through the epoch of big bang nucleosynthesis (BBN), leaving behind as relics the first complex nuclei: deuterium, helium-3, helium-4, and lithium-7. The abundances of these relic nuclides were determined by the competition between the relative densities of nucleons (baryons) and photons and, by the universal expansion rate. In particular, while deuterium is an excellent baryometer, ^{4}He provides an accurate chronometer. Nearly 400 thousand years later, when the cosmic background radiation (CBR) had cooled sufficiently to allow neutral atoms to form, releasing the CBR from the embrace of the ionized plasma of protons and electrons, the spectrum of temperature fluctuations imprinted on the CBR encoded the baryon and radiation densities, along with the universal expansion rate at that epoch. As a result, the relic abundances of the light nuclides and the CBR temperature fluctuation spectrum provide invaluable windows on the early evolution of the Universe along with sensitive probes of its particle content.

The fruitful interplay between theory and data has been key to the enormous progress in cosmology in recent times. As new, more precise data became available, models have had to be refined or rejected. It is anticipated this process will—indeed, should—continue. Therefore, this review of the baryon content of the Universe as revealed by BBN and the CBR is but a signpost on the road to a more complete understanding of the history and evolution of the Universe. By highlighting the current successes of the present "standard" model along with some of the challenges to it, I hope to identify those areas of theoretical and observational work which will contribute to continuing progress in our endeavor to understand the Universe, its past, present, and future.

2. A BBN primer

Discussion of BBN can begin when the Universe is already a few tenths of a second old and the temperature is a few MeV. At such early epochs, the Universe is too hot and dense to permit the presence of complex nuclei in any significant abundances and the baryons (nucleons) are either neutrons or protons whose relative abundances are determined by the weak interactions

$$p + e^- \longleftrightarrow n + \nu_e, \; n + e^+ \longleftrightarrow p + \bar{\nu}_e, \; n \longleftrightarrow p + e^- + \bar{\nu}_e \; . \qquad (2.1)$$

The higher neutron mass favors protons relative to neutrons, ensuring proton dominance. When the weak interaction rates (Eq. 2.1) are fast compared to the universal expansion rate (and in the absence of a significant chemical potential for the electron neutrinos), $n/p \approx \exp(-\Delta m/T)$, where Δm is the neutron-proton mass difference and T is the temperature ($T_\gamma = T_e = T_\nu = T_\mathrm{N}$ prior to $e^\pm$ annihilation). If there were an *asymmetry* between the number densities of ν_e and $\bar{\nu}_e$ ("neutrino degeneracy"), described by a chemical potential μ_e (or, equivalently, by the dimensionless degeneracy parameter $\xi_e \equiv \mu_e/T$) then, early on, $n/p \approx \exp(-\Delta m/T - \xi_e)$. For a *significant* positive chemical potential ($\xi_e \gtrsim 0.01$; more ν_e than $\bar{\nu}_e$) there are fewer neutrons than for the "standard" case (SBBN) which, as described below, leads to the formation of less ^{4}He.

The first step in building complex nuclei is the formation of deuterons via $n+p \longleftrightarrow D+\gamma$. Sufficiently early on, when the Universe is very hot ($T \gtrsim 80$ keV), the newly-formed deuterons find themselves bathed in a background of gamma rays (the photons whose relics have cooled today to form the CBR at a temperature of 2.7 K) and are quickly photo-dissociated, removing the platform necessary for building heavier nuclides. Only below $\sim$80 keV has the Universe cooled sufficiently to permit BBN to begin, leading to the synthesis of the lightest nuclides D, ^{3}He, ^{4}He, and ^{7}Li. Once BBN begins, D, ^{3}H, and ^{3}He are rapidly burned (for the baryon densities of interest) to ^{4}He, the light nuclide with the largest binding energy. The absence of a stable mass-5 nuclide, in combination with Coulomb barriers, suppresses the BBN production of heavier nuclides; only ^{7}Li is synthesized in an astrophysically interesting abundance. All the while the Universe is expanding and cooling. When the temperature has dropped below $\sim$30 keV, at a time comparable to the neutron lifetime, the thermal energies of the colliding nuclides is insufficient to overcome the Coulomb barriers, the remaining free neutrons decay, and BBN ends.

From this brief overview of BBN it is clear that the relic abundances of the nuclides produced during BBN depend on the competition between the nuclear and weak interaction rates (which depend on the baryon density), and the universal expansion rate (quantified by the Hubble parameter H), so that the relic abundances provide early-Universe baryometers and chronometers.

2.1. Early-Universe expansion rate

The Friedman equation relates the expansion rate (measured by the Hubble parameter H) to the energy density (ρ): $H^2 = \frac{8\pi G}{3}\rho$ where, during the early, "radiation-dominated" (RD) evolution the energy density is dominated by the relativistic particles present ($\rho = \rho_\mathrm{R}$). For SBBN, prior to $e^\pm$ annihilation, these are: photons, $e^\pm$ pairs and, three flavors of left-handed (i.e. one helicity state) neutrinos (and their right-handed, antineutrinos).

$$\rho_\mathrm{R} = \rho_\gamma + \rho_e + 3\rho_\nu = \frac{43}{8}\rho_\gamma \; , \qquad (2.2)$$

where ρ_γ is the energy density in CBR photons. At this early epoch, when $T \lesssim$ few MeV, the neutrinos are beginning to decouple from the γ-$e^\pm$ plasma and the neutron to proton ratio, crucial for the production of primordial ^{4}He, is decreasing. The time-temperature relation follows from the Friedman equation and the temperature dependence of ρ_γ

$$\mathrm{Pre} - \mathrm{e}^\pm \text{ annihilation} : \; t \, T_\gamma^2 = 0.738 \text{ MeV}^2 \text{ s} \; . \qquad (2.3)$$

To a very good (but not exact) approximation the neutrinos (ν_e, ν_μ, ν_τ) are decoupled when the $e^\pm$ pairs annihilate as the Universe cools below $m_e c^2$. In this approximation the neutrinos don't share in the energy transferred from the annihilating $e^\pm$ pairs to the CBR photons so that in the post-$e^\pm$ annihilation universe the photons are hotter than the neutrinos by a factor $T_\gamma / T_\nu = (11/4)^{1/3}$, and the relativistic energy density is

$$\rho_{\rm R} = \rho_\gamma + 3\rho_\nu = 1.68\rho_\gamma \; . \tag{2.4}$$

The post-$e^\pm$ annihilation time-temperature relation is

$$\text{Post} - \text{e}^\pm \text{ annihilation}: \quad t\,T_\gamma^2 = 1.32 \text{ MeV}^2 \text{ s} \; . \tag{2.5}$$

2.1.1. *Additional relativistic energy density*

One of the most straightforward variations of the standard model of cosmology is to allow for an early (RD) nonstandard expansion rate $H' \equiv SH$, where $S \equiv H'/H = t/t'$ is the *expansion rate factor*. One possibility for $S \neq 1$ is from the modification of the RD energy density (see Eqs. 2.2 & 2.4) due to "extra" relativistic particles X: $\rho_{\rm R} \rightarrow \rho_{\rm R} + \rho_X$. If the extra energy density is normalized to that which would be contributed by one additional flavor of (decoupled) neutrinos (Steigman, Schramm, & Gunn 1977), $\rho_X \equiv \Delta N_\nu \rho_\nu$ ($N_\nu \equiv 3 + \Delta N_\nu$), then

$$S_{\rm pre} \equiv (H'/H)_{\rm pre} = (1 + 0.163\Delta N_\nu)^{1/2}; \quad S_{\rm post} \equiv (H'/H)_{\rm post} = (1 + 0.135\Delta N_\nu)^{1/2} \; . \tag{2.6}$$

Notice that S and ΔN_ν are related *nonlinearly*. It must be emphasized that it is S and not ΔN_ν that is the fundamental parameter in the sense that *any* term in the Friedman equation which scales as radiation, decreasing with the fourth power of the scale factor, will change the standard-model expansion rate ($S \neq 1$). For example, higher-dimensional effects such as in the Randall-Sundrum model (Randall & Sundrum 1999a) may lead to either a speed-up in the expansion rate ($S > 1$; $\Delta N_\nu > 0$) or, to a slow-down ($S < 1$; $\Delta N_\nu < 0$); see, also, Randall & Sundrum (1999b), Binetruy et al. (2000), Cline et al. (2000).

2.2. *The baryon density*

In the expanding Universe, the number densities of all particles decrease with time, so that the magnitude of the baryon density (or that of any other particle) has no meaning without also specifying *when it is measured*. To quantify the universal abundance of baryons, it is best to compare $n_{\rm B}$ to the CBR photon density n_γ. The ratio, $\eta \equiv n_{\rm B}/n_\gamma$ is very small, so that it is convenient to define a quantity of order unity,

$$\eta_{10} \equiv 10^{10}(n_{\rm B}/n_\gamma) = 274\,\Omega_{\rm B}h^2 \equiv 274\,\omega_{\rm B} \; , \tag{2.7}$$

where $\Omega_{\rm B}$ is the ratio (at present) of the baryon density to the critical density and h is the present value of the Hubble parameter in units of 100 km s^{-1}Mpc^{-1} ($\omega_{\rm B} \equiv \Omega_{\rm B}h^2$).

3. BBN abundances

The relic abundances of D, ^{3}He, and ^{7}Li are *rate limited*, determined by the competition between the early Universe expansion rate and the nucleon density. Any of these three nuclides is, therefore, a potential baryometer; see Figure 1.

In contrast to the synthesis of the other light nuclides, once BBN begins ($T \lesssim 80$ keV) the reactions building ^{4}He are so rapid that its relic abundance is not rate limited. The primordial abundance of ^{4}He is limited by the availability of neutrons. To a very good approximation, its relic abundance is set by the neutron abundance at the beginning

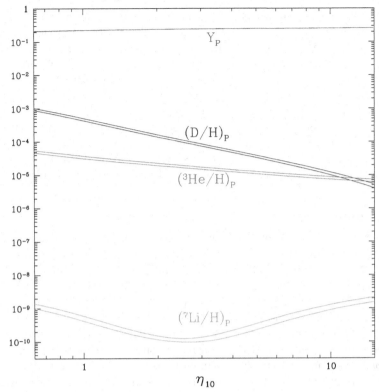

FIGURE 1. The SBBN-predicted abundances of D, ^{3}He, and ^{7}Li by number with respect to hydrogen, and the ^{4}He mass fraction Y_P, as a function of the nucleon (baryon) abundance parameter η_{10}. The bands reflect the theoretical uncertainties ($\pm 1\sigma$) in the BBN predictions.

of BBN. As a result, the primordial mass fraction of ^{4}He, Y_P, while being a relatively insensitive baryometer (see Figure 1), is an excellent, early-Universe chronometer.

The qualitative effects of a nonstandard expansion rate on the relic abundances of the light nuclides may be understood with reference to Figure 1. For the baryon abundance range of interest the relic abundances of D and ^{3}He are *decreasing* functions of η; in this range, D and ^{3}He are being destroyed to build ^{4}He. A faster than standard expansion ($S > 1$) provides less time for this destruction so that more D and ^{3}He will survive. The same behavior occurs for ^{7}Li at low values of η, where its abundance is a decreasing function of η. However, at higher values of η, the BBN-predicted ^{7}Li abundance *increases* with η, so that less time available results in less production and a *smaller* ^{7}Li relic abundance. Except for dramatic changes to the early-Universe expansion rate, these effects on the relic abundances of D, ^{3}He, and ^{7}Li are subdominant to their variations with the baryon density. Not so for ^{4}He, whose relic abundance is very weakly (logarithmically) dependent on the baryon density, but very strongly dependent on the early-Universe expansion rate. A faster expansion leaves more neutrons available to build ^{4}He; to a good approximation $\Delta Y \approx 0.16\,(S-1)$. It is clear then that if ^{4}He is paired with any of the other light nuclides, together they can constrain the baryon density (η or $\Omega_B h^2 \equiv \omega_B$) and the early-Universe expansion rate (S or ΔN_ν).

As noted above in §2, the neutron-proton ratio at BBN can also be modified from its standard value in the presence of a significant electron-neutrino asymmetry ($\xi_e \gtrsim 0.01$). As a result, Y_P is also sensitive to any neutrino asymmetry. More ν_e than $\bar\nu_e$ drives the neutron-to-proton ratio down (see Eq. 2.1), leaving fewer neutrons available to build ^{4}He; to a good approximation $\Delta Y \approx -0.23\,\xi_e$ (Kneller & Steigman 2003). In contrast, the relic abundances of D, ^{3}He, and ^{7}Li are very insensitive to $\xi_e \neq 0$, so that when paired with ^{4}He, they can simultaneously constrain the baryon density and the electron-neutrino asymmetry. Notice that if *both* S and ξ_e are allowed to be free parameters, another observational constraint is needed to simultaneously constrain η, S, and ξ_e. While neither ^{3}He nor ^{7}Li can provide the needed constraint, the CBR temperature anisotropy spectrum, which is sensitive to η and S, but not to ξ_e, can (see Barger et al. 2003b). This review will concentrate on combining constraints from the CBR and SBBN ($S = 1$, $\xi_e = 0$) and also for $S \neq 1$ ($\xi_e = 0$). For the influence of and constraints on electron neutrino asymmetry, see Barger et al. (2003b) and further references therein.

4. Relic abundances

BBN constraints on the universal density of baryons and on the early-Universe expansion rate require reasonably accurate determinations of the relic abundances of the light nuclides. As already noted, D, ^{3}He, and ^{7}Li are all potential baryometers, while ^{4}He is an excellent chronometer. The combination of the availablility of large telescopes and advances in detector technology has made it possible to obtain abundance estimates at various sites in the Galaxy and elsewhere in the Universe with unprecedented precision (statistically). However, the path to accurate primordial abundances is littered with systematic uncertainties which have the potential to contaminate otherwise exquisite data. It is, therefore, fortunate that the relic nuclides follow very different post-BBN evolutionary paths and are observed in diverse environments using a wide variety of astronomical techniques. Neutral deuterium is observed in absorption in the UV (or, in the optical when redshifted) against background, bright sources (O or B stars in the Galaxy, QSOs extragalactically). Singly-ionized helium-3 is observed in emission in Galactic H II regions via its spin-flip transition (the analog of the 21 cm line in neutral hydrogen). The helium-4 abundance is largely determined by observations of recombination lines of ionized (singly and doubly) ^{4}He compared to those of ionized hydrogen in Galactic and, especially, extragalactic H II regions. Observations of ^{7}Li, at least those at low metallicity (nearly primordial) are restricted to absorption in the atmospheres of the oldest, most metal-poor stars in the halo of the Galaxy. The different evolutionary histories (described below) combined with the differrent observational strategies provide a measure of insurance that systematic errors in the determination of one of the light element abundances are unlikely to propagate into other abundance determinations.

4.1. *Deuterium: The baryometer of choice*

The deuteron is the most weakly bound of the light nuclides. As a result, any deuterium cycled through stars is burned to ^{3}He and beyond. Thus, its post-BBN evolution is straightforward: deuterium observed anywhere, anytime, provides a *lower* bound to the primordial D abundance. For "young" systems, in the sense of little stellar evolution (e.g. sites at high redshift and/or with very low metallicity), the observed D abundance should reach a plateau at the primordial value. Although there are observations of deuterium in the solar system and the interstellar medium (ISM) of the Galaxy which provide interesting lower bounds to its primordial abundance, the observations of relic D in a few, high redshift, low metallicity, QSO absorption line systems (QSOALS) are of most value in estimating its primordial abundance.

While its simple post-BBN evolution is the greatest asset for relic D, the identical absorption spectra of D I and H I (except for the velocity/wavelength shift resulting from the

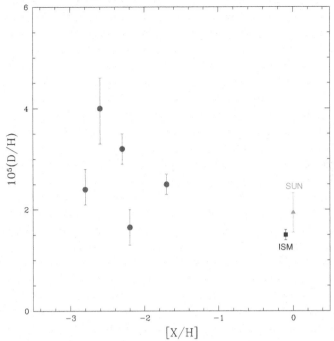

FIGURE 2. Deuterium abundances, by number with respect to hydrogen D/H, versus metallicity (relative to solar on a log scale) from observations (as of early 2003) of QSOALS (filled circles). "X" is usually silicon or oxygen. Shown for comparison are the D abundances inferred for the local ISM (filled square) and the solar system (presolar nebula: "Sun"; filled triangle).

heavier reduced mass of the deuterium atom) is a severe liability, limiting significantly the number of useful targets in the vast Lyman-alpha forest of the QSO absorption spectra (see Kirkman et al. 2003 for a discussion). It is essential in choosing a target QSOALS that its velocity structure be "simple" since a low column density H I absorber, shifted by $\sim$81 km s^{-1} with respect to the main H I absorber (an "interloper") would masquerade as D I absorption. If this is not recognized, a too high D/H ratio would be inferred. Since there are many more low-column density absorbers than those with high H I column densities, absorption systems with somewhat lower H I column density (e.g. Lyman-limit systems: LLS) are more susceptible to this contamination than the higher H I column density absorbers (e.g. damped Lyα absorbers: DLA). However, while the DLA have many advantages over the LLS, a precise determination of the H I column density requires an accurate placement of the continuum, which could be compromised by interlopers. This might lead to an overestimate of the H I column density and a concomitant underestimate of D/H (J. Linsky, private communication). As a result of these complications, the path to primordial D using QSOALS has not been straightforward, and some abundance claims have had to be withdrawn or revised. Presently there are only five QSOALS with reasonably firm deuterium detections (Kirkman et al. 2003 and references therein); these are shown in Figure 2 along with the corresponding solar system and ISM D abundances. It is clear from Figure 2, that there is significant dispersion among the derived D abundances at low metallicity which, so far, mask the anticipated deuterium plateau. This suggests that systematic errors of the sort described here may have contaminated some of the determinations of the D I and/or H I column densities.

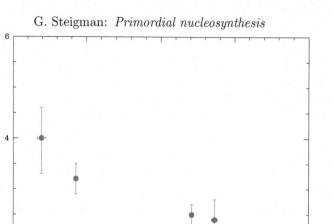

FIGURE 3. Deuterium abundances versus the H I column densities for the corresponding QSOALS shown in Figure 2.

To explore the possibility that such systematic effects, which would be correlated with the H I column density, may be responsible for at least some of the dispersion revealed in Figure 2, it is useful to plot the same QSOALS data versus the H I column density; this is shown in Figure 3. Indeed, there is the suggestion from this very limited data set that the low column density absorbers (LLS) have high D/H, while the high column density systems (DLA) have low abundances. However, on the basis of extant data it is impossible to decide which, if any, of these systems has been contaminated; there is no justification for excluding any of the present data. Indeed, perhaps the data is telling us that our ideas about post-BBN deuterium evolution need to be revised.

To proceed further using the current data, I follow the lead of O'Meara et al. (2001) and Kirkman et al. (2003) and adopt for the primordial D abundance the weighted mean of the D abundances for the five lines of sight (Kirkman et al. 2003); the dispersion in the data is used to set the error in y_D: $y_D = 2.6 \pm 0.4$. It should be noted that using the same data Kirkman et al. (2003) derive a slightly higher mean D abundance: $y_D = 2.74$. The difference is traced to their first finding the mean of $\log(y_D)$ and then using it to compute the mean D abundance ($y_D \equiv 10^{\langle \log(y_D) \rangle}$).

The BBN-predicted relic abundance of deuterium depends sensitively on the baryon density, $y_D \propto \eta^{-1.6}$, so that a ~10% determination of y_D can be used to estimate the baryon density to ~6%. For SBBN ($S = 1$ ($N_\nu = 3$), $\xi_e = 0$), the adopted primordial D abundance corresponds to $\eta_{10}(\text{SBBN}) = 6.10^{+0.67}_{-0.52}$ ($\Omega_B h^2 = 0.0223^{+0.0024}_{-0.0019}$), in spectacular agreement with the Spergel et al. (2003) estimate of $\eta_{10} = 6.14 \pm 0.25$ ($\Omega_B h^2 = 0.0224 \pm 0.0009$) based on WMAP and other CBR data (ACBAR and CBI) combined with large scale structure (2dFGRS) and Lyman-alpha forest constraints. Indeed, if the Spergel et al. (2003) estimate is used for the BBN baryon density, the BBN-predicted deuterium abundance is $y_D = 2.57 \pm 0.27$ (where a generous allowance of ~8% has been

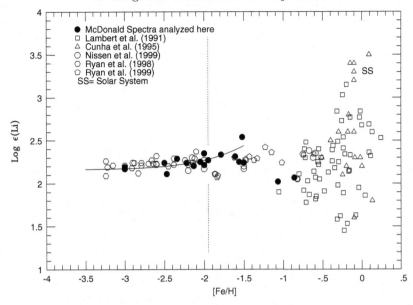

FIGURE 4. Lithium abundances, $\log \epsilon(\mathrm{Li}) \equiv [\mathrm{Li}] \equiv 12+\log(\mathrm{Li/H})$ versus metallicity (on a log scale relative to solar) from a compilation of stellar observations by V. V. Smith.

made for the uncertainty in the BBN prediction at fixed η; for the Burles, Nollett, & Turner (2001) nuclear cross sections and uncertainties the result is $y_D = 2.60^{+0.20}_{-0.18}$).

4.2. *Helium-3*

Unlike D, the post-BBN evolution of ^{3}He and ^{7}Li are quite complex. ^{3}He is destroyed in the hotter interiors of all but the least massive (coolest) stars, but it is preserved in the cooler, outer layers of most stars. In addition, hydrogen burning in low mass stars results in the production of significant amounts of *new* ^{3}He (Iben 1967; Rood 1972; Dearborn, Steigman, & Schramm 1986; Vassiliadis & Wood 1993; Dearborn, Steigman, & Tosi 1996). To follow the post-BBN evolution of ^{3}He, it is necessary to account for all these effects—quantitatively—in the material returned by stars to the interstellar medium (ISM). As indicated by the existing Galactic data (Geiss & Gloeckler 1998; Bania, Rood, & Balser 2002), a very delicate balance exists between net production and net destruction of ^{3}He in the course of the evolution of the Galaxy. As a consequence, aside from noting an excellent qualitative agreement between the SBBN predicted and observed ^{3}He abundances, ^{3}He has—at present—little role to play as a quantitatively useful baryometer. In this spirit, it is noted that an uncertain estimate of the primordial abundance of ^{3}He may be inferred from the observation of an outer-Galaxy (less evolved) H II region (Bania et al. 2002): $y_3 \equiv 10^5(^3\mathrm{He/H}) = 1.1 \pm 0.2$.

4.3. *Lithium-7*

A similar scenario may be sketched for ^{7}Li. As a weakly bound nuclide, it is easily destroyed when cycled through stars except if it can be kept in the cooler, outer layers. The high lithium abundances observed in the few "super-lithium-rich red giants" provide direct evidence that at least some stars can synthesize post-BBN lithium and bring it to the surface. But, an unsolved issue is how much of this newly-synthesized lithium is actually returned to the ISM rather than mixed back into the interior and destroyed.

With these caveats in mind, in Figure 4 lithium abundances are shown as a function of metallicity from a compilation by V. V. Smith (private communication). Since the

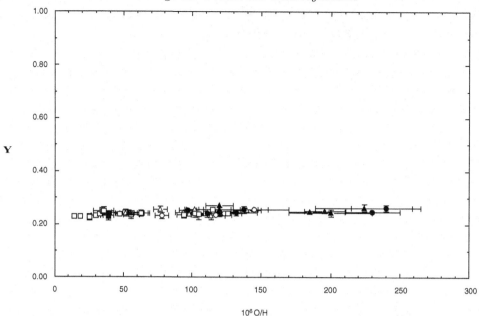

FIGURE 5. The ^{4}He mass fraction Y derived from observations of extragalactic H II regions of low metallicity versus the corresponding H II region oxygen abundances (from K. A. Olive).

quest for nearly primordial lithium is restricted to the oldest, most metal-poor stars in the Galaxy, stars that have had the most time to redistribute—and destroy or dilute—their surface lithium abundances, it is unclear whether the "plateau" at low metallicities is representative of the primordial abundance of lithium. Although it seems clear that the lithium abundance in the Galaxy has increased since BBN, a quantitatively reliable estimate of its primordial abundance eludes us at present. Given this state of affairs, the most fruitful approach is to learn about stellar structure and evolution by comparing the BBN-predicted lithium abundance to those abundances inferred from observations of the oldest stars, rather than to attempt to use the stellar observations to constrain the BBN-inferred baryon density. Concentrating on the low-metallicity, nearly primordial data, it seems that [Li] $\equiv$ 12+log(Li/H) $\approx$ 2.2 $\pm$ 0.1. This estimate will be compared to the BBN-predicted lithium abundance using D as a baryometer and, to the BBN-predicted lithium abundance using the CBR-inferred baryon density. Any tension between these BBN-predicted abundances and that inferred from the Galactic data may provide hints of nonstandard stellar astrophysics.

4.4. *Helium-4: The BBN chronometer*

The good news about ^{4}He is that, as the second most abundant nuclide, it may be observed throughout the Universe. The bad news is that its abundance has evolved since the end of BBN. In order to infer its primordial value it is therefore necessary to track the ^{4}He abundance determinations (mass fraction Y_P) as a function of metallicity or, to limit observations to very low metallicity objects. Although, as for D, there are observations of ^{4}He in the ISM and the solar system, the key data for determining its primordial abundance comes from observations of metal-poor, extragalactic H II regions. A compilation of current data (courtesy of K. A. Olive) is shown in Figure 5 where the ^{4}He mass fraction is plotted as a function of the oxygen abundance; note that the solar oxygen abundance, O/H $\approx 5 \times 10^{-4}$ (Allende-Prieto, Lambert, & Asplund 2001) is off-scale in this figure. These are truly low metallicity H II regions.

It is clear from Figure 5 that the data exist to permit the derivation of a reasonably accurate estimate (statistically) of the primordial ^{4}He mass fraction Y_P, with or without any extrapolation to zero-metallicity. What is not easily seen in Figure 5 given the Y_P scale, is that Y_P derived from the data assembled from the literature by Olive & Steigman (1995) and Olive, Skillman, & Steigman (1997) ($Y_P = 0.234 \pm 0.003$) is marginally inconsistent (at $\sim 2\sigma$) with the value derived by Izotov, Thuan, & Lipovetsky (1997) and Izotov & Thuan (1998) from their nearly independent data set ($Y_P = 0.244 \pm 0.002$). In addition, there are a variety of systematic corrections which might modify *both* data sets (Steigman, Viegas, & Gruenwald 1997; Viegas, Gruenwald, & Steigman 2000; Olive & Skillman 2001; Sauer & Jedamzik 2002; Gruenwald, Steigman, & Viegas 2002; Peimbert, Peimbert, & Luridiana 2002).

Unless/until the differences in Y_P derived by different authors from somewhat different data sets is resolved and the known systematic errors are corrected for (the unknown ones will always hang over us like the sword of Damocles), the following compromise, adopted by Olive, Steigman, & Walker (2000), may not be unreasonable. From Olive & Steigman (1995) and Olive, Skillman, & Steigman (1997), the 2σ range for Y_P is 0.228–0.240, while from the Izotov, Thuan, & Lipovetsky (1997) and Izotov & Thuan (1998) data the 2σ range is $Y_P = 0.240$–0.248. Thus, although the current estimates are likely dominated by systematic errors, they span a $\sim 2\sigma$ range from $Y_P = 0.228$ to $Y_P = 0.248$. Therefore, as proposed by Olive, Steigman, & Walker (2000), we adopt here a central value for $Y_P = 0.238$ and a $\sim 1\sigma$ uncertainty of 0.005: $Y_P = 0.238 \pm 0.005$. Given the approximation (see §3) $\Delta Y \approx 0.16\,(S - 1)$, for $\sigma_{Y_P} \approx 0.005$ the uncertainty in S is ≈ 0.03 (corresponding to an uncertainty in ΔN_ν of ≈ 0.4).

5. The baryon density from SBBN

For SBBN, where $S = 1$ ($N_\nu = 3$) and $\xi_e = 0$, the primordial abundances of D, ^{3}He, ^{4}He, and ^{7}Li are predicted as a function of only one free parameter, the baryon density parameter (η or $\Omega_B h^2 \equiv \omega_B$). As described above (see §4.1), D is the baryometer of choice. From SBBN and the adopted relic abundance of deuterium, $y_D = 2.6 \pm 0.4$, $\eta_{10} = 6.1^{+0.7}_{-0.5}$ ($\Omega_B h^2 = 0.022 \pm 0.002$).

Having determined the baryon density to $\sim 10\%$ using D as the SBBN baryometer, it is incumbent upon us to compare the SBBN-predicted abundances of the other light nuclides with their relic abundances inferred from the observational data. For this baryon density, the predicted primordial abundance of ^{3}He is $y_3 = 1.04 \pm 0.10$, in excellent agreement with the primordial value of $y_3 = 1.1 \pm 0.2$ inferred from observations of an outer-Galaxy H II region (Bania et al. 2002). Within the context of SBBN, D and ^{3}He are completely consistent.

The first challenge to SBBN comes from ^{4}He. For the SBBN-determined baryon density the predicted ^{4}He primordial mass fraction is $Y_P = 0.248 \pm 0.001$, to be compared with our adopted value from extragalactic H II regions (Olive, Steigman & Walker 2000) of $Y_P^{OSW} = 0.238 \pm 0.005$. Agreement is only at the $\sim 2\sigma$ level. Given the unresolved systematic uncertainties in determining Y_P from the H II region data, it is not clear at present whether this is a challenge to SBBN or to our understanding of H II region recombination spectra. As will be seen below, this tension between SBBN D and ^{4}He can be relieved for nonstandard BBN if the assumption that $S = 1$ ($N_\nu = 3$) is relaxed.

The conflict with the inferred primordial abundance of lithium is even more challenging to SBBN. For $y_D = 2.6 \pm 0.4$, [Li] $= 2.65^{+0.10}_{-0.12}$. This is to be compared to the estimate

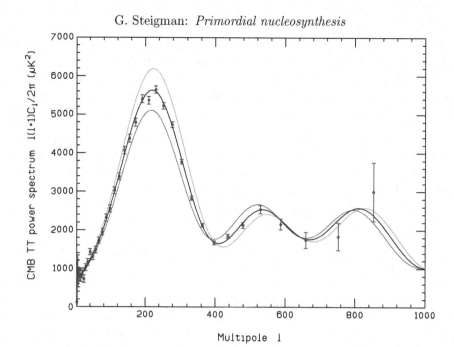

FIGURE 6. The CBR temperature fluctuation anisotropy spectra for three choices of the baryon density parameter $\omega_B = 0.018$, 0.023, 0.028, in order of increasing height of the first peak. Also shown are the WMAP data points.

(see Figure 4) of $[\text{Li}] = 2.2 \pm 0.1$ based on a sample of metal-poor, halo stars. The conflict is even greater with the Ryan et al. (2000) estimate of $[\text{Li}] = 2.09^{+0.19}_{-0.13}$ derived from an especially selected data set. Unlike the tension between SBBN and the D and ^4He abundances, the conflict between D and ^7Li cannot be resolved by a nonstandard expansion rate (nor, by an electron neutrino asymmetry). Most likely, the resolution of this conflict is astrophysical since the metal-poor halo stars from which the relic abundance of lithium is inferred have had the longest time to mix their surface material with that in their hotter interiors, diluting or destroying their prestellar quota of lithium (see, e.g. Pinsonneault et al. 2002 and references to related work therein).

At present SBBN in combination with the limited data set of QSOALS deuterium abundances yields a $\sim$10% determination of the baryon density parameter. Consistency between the inferred primordial abundances of D and ^3He lends support to the internal consistency of SBBN, but the derived primordial abundances of ^4He and ^7Li pose some challenges. For ^4He the disagreement is only at the $\sim2\sigma$ level and the errors in the observationally inferred value of Y_P are dominated by poorly quantified systematics. However, if the current discrepancy is real, it might be providing a hint at new physics beyond the standard model (e.g. nonstandard expansion rate and/or nonstandard neutrino physics). Before considering the effects on BBN of a nonstandard expansion rate ($S \neq 1$; $N_\nu \neq 3$), we will compare the SBBN estimate of the baryon density parameter with that from the CBR.

6. The baryon density from the CBR

Some 400 kyr after BBN has ended, when the Universe has expanded and cooled sufficiently so that the ionized plasma of protons, alphas, and electrons combines to form neutral hydrogen and helium, the CBR photons are set free to propagate throughout the

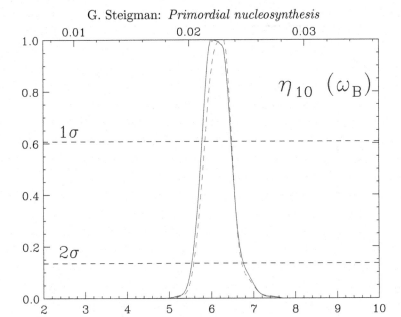

FIGURE 7. The normalized likelihood distributions for the baryon density parameter η_{10} derived from SBBN and the primordial abundance of deuterium (solid curve; see §4.1) and from the CBR using WMAP data alone (dashed curve). The bottom horizontal axis is the baryon-to-photon ratio parameter η_{10}; the top axis is the baryon density parameter $\omega_B = \Omega_B h^2$.

Universe. Observations of the CBR today reveal the anisotropy spectrum of temperature fluctuations imprinted at that early epoch. The so-called acoustic peaks in the temperature anisotropy spectrum arise from the competition between the gravitational potential and the pressure gradients. An increase in the baryon density increases the inertia of the baryon-photon fluid shifting the locations and the relative heights of the acoustic peaks. In Figure 6 are shown three sets of temperature anisotropy spectra for cosmological models which differ only in the choice of the baryon density parameter ω_B. Also shown in Figure 6 are the WMAP data from Bennett et al. (2003). It is clear from Figure 6 that the CBR provides a very good baryometer—independent of that from SBBN and primordial deuterium. Based on the WMAP data alone, Barger et al. (2003a) find that the best fit value for the density parameter is $\eta_{10} = 6.3$ ($\omega_B = 0.023$) and that the 2σ range extends from $\eta_{10} = 5.6$ to 7.3 ($0.020 \leqslant \omega_B \leqslant 0.026$). This is in excellent (essentially perfect!) agreement (as it should be) with the CBR-only result of Spergel et al. (2003). More importantly, as may be seen clearly in Figure 7, this independent constraint on the baryon density parameter, sampled some 400 kyr after BBN, is in excellent agreement with that from SBBN (see §5), providing strong support for the standard model of cosmology.

The independent determination of the baryon density parameter by the CBR reinforces the tension between SBBN and the relic abundances of ^{4}He and ^{7}Li inferred from the observational data (see §5). In the context of SBBN, the slightly higher best value of η from the WMAP data (compared to that from D plus SBBN) *increases* the expected primordial abundances of ^{4}He and ^{7}Li (see Figure 1), widening the gaps between the SBBN predictions and the data. Keeping in mind the observational and theoretical difficulties in deriving the primordial abundances from the data, it is nonetheless worthwhile to explore a class of nonstandard alternatives to the standard model of cosmology in which the early Universe expansion rate is modified ($S \neq 1$, $N_\nu \neq 3$).

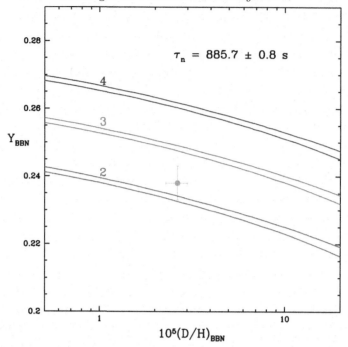

FIGURE 8. The BBN-predicted relation between the ^{4}He mass fraction Y_P and the deuterium abundance y_D for three, early-Universe expansion rates corresponding to $N_\nu = 2, 3, 4$. The filled circle with error bars is for the D and ^{4}He primordial abundances adopted here.

7. Nonstandard BBN: $S \neq 1$, $N_\nu \neq 3$

As outlined in §3, for fixed η as S increases the BBN-predicted abundances of D, ^{3}He, and ^{4}He increase (less time to destroy D and ^{3}He, more neutrons available for ^{4}He), while that of ^{7}Li decreases (less time to produce ^{7}Li). Since it is the ^{4}He mass fraction that is most sensitive to changes in the early Universe expansion rate and, since the SBBN-predicted value of Y_P is too large when compared to the data, $S < 1$ ($N_\nu < 3$) is required. For a *slower* than standard expansion rate the predicted abundances of D and ^{3}He *decrease* compared to their SBBN values (at fixed η) while that of ^{7}Li *increases*. Since the BBN-predicted abundance of D increases with decreasing baryon density, a decrease in S can be compensated for by a decrease in η. For $\eta_{10} \approx 6$ and $S - 1 \ll 1$, a good approximation (for fixed D) is $\Delta\eta_{10} \approx 6(S - 1)$ (Kneller & Steigman 2003). In Figure 8 are shown the ^{4}He–D (Y_P versus D/H) relations for three values of the expansion rate parameterized by N_ν. To first order, the combination of η and S that recovers the SBBN deuterium abundance will leave the ^{3}He abundance prediction unchanged as well, preserving its good agreement with the observational data. However, the consequences for ^{7}Li are not so favorable. The BBN abundance of ^{7}Li increases with decreasing S but decreases with a smaller η; the two effects nearly cancel, leaving essentially the same discrepancy as for SBBN. For ^{7}Li, a nonstandard expansion rate cannot relieve the tension between the BBN prediction and the observational data.

Setting aside ^{7}Li, it is of interest to consider the simultaneous constraints from BBN on the baryon density parameter and the expansion rate factor from the abundances of D and ^{4}He; it has already been noted that for this nonstandard case, D and ^{3}He will remain consistent. In Figure 9 are shown the 1σ, 2σ, and 3σ contours in the ΔN_ν–η plane derived from BBN and the D and ^{4}He relic abundances. As expected from the

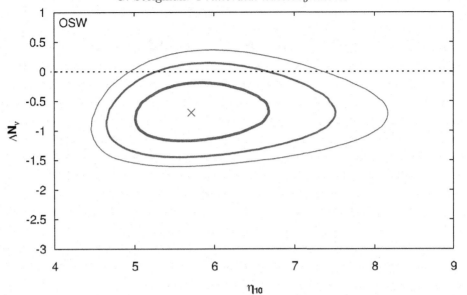

FIGURE 9. The 1σ, 2σ, and 3σ contours in the ΔN_ν – η_{10} plane from BBN and the relic D and ^{4}He abundances. The best fit values of ΔN_ν and η_{10} are marked by the cross.

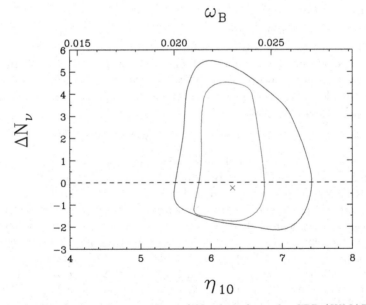

FIGURE 10. The 1σ and 2σ contours in the η–ΔN_ν plane from the CBR (*WMAP*) data. The best fit point ($\eta_{10} = 6.3$, $\Delta N_\nu = -0.25$) is indicated by the cross.

discussion above, the best fit value of η (the cross in Figure 9) has shifted downward to $\eta_{10} = 5.7$ ($\omega_B = 0.021$). While the best fit is for $\Delta N_\nu = -0.7$ ($S = 0.94$), it should be noted that the standard case of $N_\nu = 3$ is entirely compatible with the data at the $\sim 2\sigma$ level.

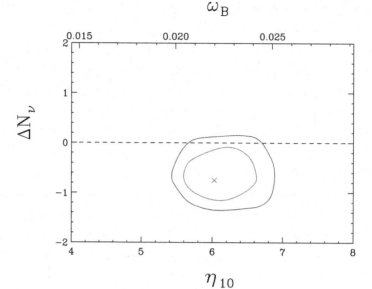

FIGURE 11. As for Figure 10, but for the *joint* BBN–CBR fit. The best fit point ($\eta_{10} = 6.0$, $\Delta N_\nu = -0.75$) is indicated by the cross.

8. Nonstandard CBR: $S \neq 1$, $N_\nu \neq 3$

The CBR temperature fluctuation anisotropy spectrum is sensitive to the early-Universe radiation density (ρ_R) as well as to the overall expansion rate. The early Universe is radiation dominated so that $\rho \approx \rho_R \propto 1 + 0.135\Delta N_\nu$ (see Eq. 2.6 and recall that $\rho \propto H^2$). The late Universe is matter dominated (MD; $\omega_M \equiv \Omega_M h^2$) and the crossover from RD to MD, important for the growth of fluctuations and for the age/size of the Universe at recombination, occurs for a redshift

$$z_{eq} = 2.4 \times 10^4 \omega_M (1 + 0.135\Delta N_\nu)^{-1} \; . \tag{8.1}$$

If the matter content is kept fixed while the radiation content is increased, corresponding to a faster than standard expansion rate, matter-radiation equality is delayed, modifying the growth of fluctuations prior to recombination and, also, the Universe is younger at recombination and has a smaller sound horizon, shifting the angular location of the acoustic peaks. The degeneracy between the radiation density (ΔN_ν or S) and ω_M is broken by the requirement that $\Omega_M + \Omega_\Lambda = 1$ and the HST Key Project determination of the Hubble parameter (see Barger et al. 2003a for details and further references). In Figure 10 are shown the 1σ and 2σ contours in the ΔN_ν –η plane from the CBR (WMAP) data; note the very different ΔN_ν scales and ranges in Figures 9 and 10. As is the case for BBN (see §7), the CBR favors a slightly slower than standard expansion. However, while the "best" fit value for the expansion rate factor is at $S < 1$ ($\Delta N_\nu < 0$), the CBR likelihood distribution of ΔN_ν values is very shallow and the WMAP data are fully consistent with $S = 1$ ($\Delta N_\nu = 0$).

Comparing Figures 9 and 10, it is clear that for this variant of the standard cosmology there is excellent overlap between the η–ΔN_ν confidence contours from BBN and those from the CBR (see Barger et al. 2003a). This variant of SBBN ($S \neq 1$) is consistent with the CBR. In Figure 11 (from Barger et al. 2003a) are shown the confidence contours in the η–ΔN_ν plane for a joint BBN–CBR fit. Again, while the best fit value for ΔN_ν is negative (driven largely by the adopted value for Y_P), $\Delta N_\nu = 0$ ($S = 1$) is quite acceptable.

9. Summary and conclusions

As cosmology deals with an abundance of precision data, redundancy will be the key to distinguishing systematic errors from evidence for new physics. BBN and the CBR provide complementary probes of the Universe at two epochs widely separated from each other and from the present. For the standard model assumptions ($N_\nu = 3$, $S = 1$) the SBBN-inferred baryon density is in excellent agreement with that derived from the CBR (with or without the extra constraints imposed by large scale structure considerations and/or the Lyman alpha forest). For this baryon density ($\eta_{10} \approx 6.1$, $\omega_B \approx 0.022$), the SBBN-predicted abundances of D and ^{3}He are in excellent agreement with the observational data. For ^{4}He the predicted relic mass fraction is $\sim 2\sigma$ higher than the primordial abundance inferred from current data, hinting at either new physics or the presence of unidentified systematic errors. For ^{7}Li too, the SBBN-predicted abundance is high compared to that derived from very metal-poor stars in the Galaxy. While the tension with ^{4}He can be relieved by invoking new physics in the form of a nonstandard (slower than expected) early-Universe expansion rate, this choice will not reconcile the BBN-predicted and observed abundances of ^{7}Li. When *both* the baryon density and the expansion rate factor are allowed to be free parameters, BBN (D, ^{3}He, and ^{4}He) and the CBR (WMAP) agree at 95% confidence for $5.5 \leqslant \eta_{10} \leqslant 6.8$ and $1.65 \leqslant N_\nu \leqslant 3.03$.

The engine powering the transformation of the study of cosmology from its youth to its current maturity has been the wealth of observational data accumulated in recent years. In this data-rich, precision era BBN, one of the pillars of modern cosmology, continues to play a key role. The spectacular agreement between the estimates of the baryon density derived from processes at widely separated epochs has confirmed the general assumptions of the standard models of cosmology and of particle physics. The tension with ^{4}He (and with ^{7}Li) provides a challenge, along with opportunities, to cosmology, to astrophysics, and to particle physics. Whether the resolution of these challenges is observational, theoretical, or a combination of both, the future is bright.

I am grateful to all my collaborators and I am happy to thank them for their various contributions to the material reviewed here. Many of the quantitative results (and figures) presented here are from recent collaborations or discusions with V. Barger, J. P. Kneller, H.-S. Lee, J. Linsky, D. Marfatia, K. A. Olive, R. J. Scherrer, V. V. Smith, and T. P. Walker. My research is supported at OSU by the DOE through grant DE-FG02-91ER40690. This manuscript was prepared while I was visiting the Instituto Astrônomico e Geofísico of the Universidade de São Paulo, and I thank them for their hospitality.

REFERENCES

ALLENDE-PRIETO, C., LAMBERT, D. L., & ASPLUND, M. 2001 *ApJ* **556**, L63.

BANIA, T., ROOD, R. T., & BALSER, D. 2002 *Nature* **415**, 54.

BARGER, V., KNELLER, J. P., LEE, H.-S., MARFATIA, D., & STEIGMAN, G. 2003a *Phys. Lett. B* **566**, 8.

BARGER, V., KNELLER, J. P., MARFATIA, D., LANGACKER, P., & STEIGMAN, G. 2003b *Phys. Lett. B* **569**, 123.

BENNETT, C. L., ET AL. 2003 *ApJS* **148**, 51.

BINETRUY, P., DEFAYYET, C., ELLWANGER, U., & LANGLOIS, D. 2000 *Phys. Lett. B* **477**, 285.

BURLES, S., NOLLETT, K. M., & TURNER, M. S. 2001 *Phys. Rev. D* **63**, 063512.

CLINE, J. M., GROJEAN, C., & SERVANT, G. 2000 *Phys. Rev. Lett.* **83**, 4245.

DEARBORN, D. S. P., STEIGMAN, G., & SCHRAMM, D. N. 1986 *ApJ* **203**, 35.

DEARBORN, D. S. P., STEIGMAN, G., & TOSI, M. 1996 *ApJ* **465**, 887.

GEISS, J. & GLOECKLER, G. 1998 *Space Sci. Rev.* **84**, 239.

GRUENWALD, R., STEIGMAN, G., & VIEGAS, S. M. 2002 *ApJ* **567**, 931.

IBEN, I., JR. 1967 *ApJ* **147**, 624.

IZOTOV, Y. I. & THUAN, T. X. 1998 *ApJ* **500**, 188.

IZOTOV, Y. I., THUAN T. X., & LIPOVETSKY V. A. 1997 *ApJS* **108**, 1.

KIRKMAN, D., TYTLER, D., SUZUKI, N., O'MEARA, J. M., & LUBIN, D. 2003 *ApJS* **149**, 1.

KNELLER, J. P. & STEIGMAN, G. 2003 BBN For Pedestrians, *in preparation.*

OLIVE, K. A. & SKILLMAN, E. 2001 *New Astron.* **6**, 119.

OLIVE, K. A., SKILLMAN, E., & STEIGMAN, G. 1997 *ApJ* **483**, 788.

OLIVE, K. A. & STEIGMAN, G. 1995 *ApJS* **97**, 49.

OLIVE, K. A., STEIGMAN, G., & WALKER, T. P. 2000 *Phys. Rep.* **333**, 389.

O'MEARA, J. M., TYTLER, D., KIRKMAN, D., SUZUKI, N., PROCHASKA, J. X., LUBIN, D., & WOLFE, A. M. 2001 *ApJ* **552**, 718.

PEIMBERT, A., PEIMBERT, M., & LURIDIANA, V. 2002 *ApJ* **565**, 668.

PINSONNEAULT, M. H., STEIGMAN, G., WALKER, T. P. & NARAYANAN, V. K. 2002 *ApJ* **574**, 398.

RANDALL, L. & SUNDRUM, R. 1999a *Phys. Rev. Lett.* **83**, 3370.

RANDALL, L. & SUNDRUM, R. 1999b *Phys. Rev. Lett.* **83**, 4690.

RYAN, S. G., BEERS, T. C., OLIVE, K. A., FIELDS, B. D., & NORRIS, J. E. 2000 *ApJ* **530**, L57.

SAUER, D. & JEDAMZIK, K. 2002 *A&A* **381**, 361.

SPERGEL, D. N. ET AL. 2003 *ApJS* **148**, 175.

STEIGMAN, G., SCHRAMM, D. N., & GUNN, J. E. 1977 *Phys. Lett. B* **66**, 202.

STEIGMAN, G., VIEGAS, S. M., & GRUENWALD, R. 1997 *ApJ* **490**, 187.

VASSILIADIS, E. & WOOD, P. R. 1993 *ApJ* **413**, 641.

VIEGAS, S. M., GRUENWALD, R., & STEIGMAN, G. 2000 *ApJ* **531**, 813.

Galactic structure

By ROSEMARY F. G. WYSE

Department of Physics & Astronomy, The Johns Hopkins University, Baltimore, MD 21218, USA

Our Milky Way Galaxy is a typical large spiral galaxy, representative of the most common morphological type in the local Universe. We can determine the properties of individual stars in unusual detail, and use the characteristics of the stellar populations of the Galaxy as templates in understanding more distant galaxies. The star formation history and merging history of the Galaxy is written in its stellar populations; these reveal that the Galaxy has evolved rather quietly over the last ∼10 Gyr. More detailed simulations of galaxy formation are needed, but this result apparently makes our Galaxy unusual if ΛCDM is indeed the correct cosmological paradigm for structure formation. While our Milky Way is only one galaxy, a theory in which its properties are very anomalous most probably needs to be revised. Happily, observational capabilities of next-generation facilities should, in the foreseeable future, allow the acquisition of detailed observations for all galaxies in the Local Group.

1. Introduction: The fossil record

The origins and evolution of galaxies such as our own Milky Way and of their associated dark matter haloes are among the major outstanding questions of astrophysics. Detailed study of the zero-redshift Universe provides complementary constraints on models of galaxy formation to those obtained from direct study of high-redshift objects. Stars of mass similar to that of the Sun live for essentially the present age of the Universe and nearby low-mass stars can be used to trace conditions in the high-redshift Universe when they formed, perhaps even the 'First Light' that ended the Cosmological Dark Ages. While these stars may well not have formed in the galaxy in which they now reside (especially if the CDM paradigm is valid), several important observable quantities are largely conserved over a star's lifetime—these include surface chemical elemental abundances (modulo effects associated with mass transfer in binaries) and orbital angular momentum (modulo the effects of torques and rapidly changing gravitational potentials). Excavating the fossil record of galaxy evolution from old stars nearby allows us to do Cosmology locally, and is possible to some extent throughout the Local Group, with the most detailed information available from the Milky Way Galaxy.

I here discuss our knowledge of the stellar populations of the Milky Way and the implications for models of galaxy formation. Complementary results for M31 are presented by Brown (this volume).

2. Large-scale structure of the stellar components of the Galaxy

There are four main stellar components of the Milky Way Galaxy.

• The thin stellar disk: This is the most massive stellar component of the Milky Way, and contains stars of a wide range of ages and is the site of on-going star formation. A defining quality of the thin disk is that the stars are on orbits of high angular momentum, close to circular orbits. The age distribution of thin-disk stars is not well determined even at the solar circle, but there are clearly old, ∼10 Gyr, thin-disk stars in the local vicinity.

• The thick stellar disk: This component, established some 20 years ago, has a scale-height around three times that of the thin disk. Again its properties are best established close to the solar Galactocentric distance; here the local thick disk consists of old stars, that are on average of lower metallicity than that of a typical old star in the thin disk, and are on orbits of lower angular momentum.

• The central bulge: This component is very centrally concentrated and mildly triaxial with rotational energy close to the expected value if it were an oblate, isotropic rotator. The dominant stellar population is old and metal-rich.

• The stellar halo: The bulk of the stars are old and metal-poor, on low angular-momentum orbits. A few percent of the stellar mass is in globular clusters.

2.1. *Large scale structure of the thin disk*

Our knowledge of the stellar populations in the thin disk is in fact rather poor, with only very limited data on age and metallicity distributions in either the inner or the outer disk. At the solar neighborhood, the gross characteristics of the metallicity distribution of the thin disk have been known for a long time—the narrow distribution (peaking somewhat below the solar metallicity) giving rise to the 'G-dwarf Problem,' or the deficit of metal-poor stars compared to the predictions of the Simple closed-box model of chemical evolution (e.g., van den Bergh 1962; Pagel & Patchett 1975). The favored solution to this 'problem' is to lift the 'closed-box' assumption, in particular to allow inflow of unenriched gas (cf. Larson 1972). Such inflow is rather natural in many models of disk formation and evolution (see Tosi's contribution to this volume).

The age distribution of stars in the thin disk is particularly important in setting the epoch of the onset of disk formation (assuming that the bulk of the old stars now in the thin disk were formed in the thin disk—see Steinmetz's contribution to this volume for an alternative view). In Cold-Dark-Matter–dominated cosmologies, the merging by which galaxies grow involves gravitational torques and dynamical friction, which result in significant angular momentum transport away from the central parts of dark matter haloes and their associated galaxies, and into the outer parts. This re-arrangement of angular momentum is particularly effective if the merging involves dense, non-dissipative (i.e., stellar and/or dark matter) substructure (cf. Zurek, Quinn & Salmon 1988; Zhang et al. 2002).

However, extended galactic disks as we observe them require detailed angular momentum *conservation* during the dissipative collapse and spin-up of proto-disk gas, within a dominant dark halo (cf. Fall & Efstathiou 1980). The angular momentum transport inherent in the merging process in a CDM universe results in disks that are too concentrated and have radial scale lengths that are too short (cf. Navarro, Frenk & White 1995; Navarro & Steinmetz 1997). A proposed solution to this problem delays the formation of stellar disks until after the epoch of most merging, with 'stellar feedback' as the proposed mechanism for the delay (cf. Cole et al. 2000). A delay in radiative cooling—the very first stage of stellar disk formation—until after a redshift of unity, or lookback times of ~ 8 Gyr, is apparently required (Eke, Efstathiou & Wright 2000). This has the obvious side effect that the extended disks that eventually form should contain few old stars. Further, the theoretical prediction is that disks form from the inside out, with lower angular-momentum gas settling to the (inner regions of the) disk plane earlier and only later settling of higher angular-momentum material, destined to form the outer disk. The solar circle is some two to three exponential scale lengths from the Galactic center, and thus forms later than the inner disk. Detailed predictions are lacking (and should be made), but allowing for one Gyr for cooling and star formation, the expectation from this delay in disk formation would then be that there should be very few stars in the local thin disk that are older than ~ 7 Gyr.

Distances and kinematics derived for nearby stars from parallaxes and proper motions from the *Hipparcos* satellite has allowed the determination of the color-absolute magnitude diagram for thin disk stars. Analysis of the locus in the CMD of subgiant stars gives a lower limit to the age of the oldest stars of 8 Gyr, obtained with an adopted upper limit to the metallicity of [Fe/H] = +0.3 (Jimenez, Flynn & Kotoneva 1998; Sandage, Lubin & Van den Berg 2003); the age limit increases approximately 1 Gyr for every 0.1 dex decrease in the adopted metallicity (to set the age scale, the Van den Berg isochrones used give ages of $\gtrsim$ 13 Gyr for metal-poor globular clusters). Analysis of the main sequence turn-off stars in the *Hipparcos* dataset provides a best-fit age for the oldest disk stars of $\gtrsim$ 11 Gyr (Binney, Dehnen & Bertelli 2000), adopting a metallicity distribution that peaks below the solar value and using isochrones that provide an age of $\gtrsim$ 12 Gyr for metal-poor globular clusters. It is clear that metallicity determinations for the *Hipparcos* sample are needed before a definitive value for the age of the oldest disk stars can be derived, but one should note that the available metallicity distributions, mostly for G/K dwarfs, all peak at [Fe/H] = −0.2 (e.g. Kotoneva et al. 2002), distinctly more metal-poor than the +0.3 dex that gave the lower limit in age of 8 Gyr for the oldest stars. An older age is then expected.

An alternative technique to derive ages uses the observed white dwarf (WD) luminosity function combined with theoretical models of white dwarf cooling. Hansen et al. (2002; see also Richer's contribution to this volume) analyzed the disk WD luminosity function of Leggett et al. (1998) together with their own data for the WD sequence in the globular cluster M4. They derived a ∼5 Gyr gap between the formation of M4 and the birth of the oldest disk stars, with ages of ∼13 Gyr and ∼8 Gyr respectively. However, completeness remains an issue for the disk WD luminosity function, and different determinations are available and provide older ages and less of a gap in age (e.g., Knox, Hawkins & Hambly 1999). Indeed, a recent re-analysis of the M4 data (De Marchi et al. 2003) has demonstrated that with a re-assessment of the errors, the derived WD luminosity function is, in fact, still rising at the last point and so only a lower limit in age, $\gtrsim$ 8 Gyr, can be derived. Clearly more and better data are needed, and should be available from the ACS on *HST* for M4 and from surveys such as SDSS for the faint, local disk WDs.

The star formation history of the local disk that is derived from the *Hipparcos* CMD (Hernandez, Valls-Gabaud & Gilmore 2000) has an amplitude that shows a slow overall decline, with quasi-periodic increases of a factor of a few on timescales of ∼1 Gyr. This result is consistent with other age indicators such as chromospheric activity (Rocha-Pinto et al. 2000), and with chemical evolution models.

The available data are then all consistent with a significant population of stars in the local disk with ages ∼8 Gyr, and perhaps as old as 11 Gyr. If these stars formed in the disk, then the formation of extended disks was *not* delayed until after a redshift of unity, as was proposed to 'solve' the disk angular-momentum problem in CDM models.

The outer disk of M31 also contains old stars (Ferguson & Johnson 2001; Guhathakurta this volume) and similar conclusions hold. Further, deep, high-resolution IR observations have revealed apparently relaxed disk galaxies at $z \gtrsim 1$ (Dickinson 2000), which presumably formed at least a few dynamical times earlier. Indeed, a candidate old disk has been identified at $z \sim 2.5$ (Stockton et al. 2003).

2.2. *Large scale structure of the thick disk*

The thick disk was defined through star counts 20 years ago (Gilmore & Reid 1983) and is now well established as a distinct component. Its origins remain the source of considerable debate. Locally, some ∼5% of stars are in the thick disk; the vertical scale-height is ∼1 kpc, and radial scale-length ∼3 kpc. Assuming a smooth double-exponential

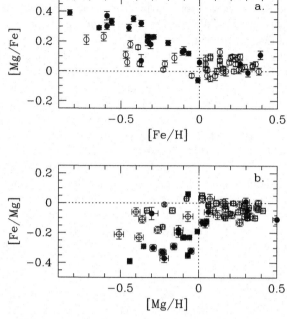

FIGURE 1. Taken from Feltzing et al. 2003, their Figure 2. Filled symbols represent stars whose kinematics are consistent with membership of the thick disk, while open symbols represent thin-disk stars. The uncertainties in Mg abundance are indicated by the error bars; uncertainties in Fe are smaller than the symbol sizes. At a given value of [Fe/H], the thick- and thin-disk stars are separated, with thick-disk stars having higher [Mg/Fe]. At the typical thick-disk metallicity, [Fe/H] ~ -0.5 dex, the value of [Mg/Fe] in thick-disk stars is equal to that seen in the stellar halo, and consistent with enrichment by Type II supernovae. More metal-rich thick-disk stars show some enrichment by iron-dominated ejecta from Type Ia supernovae.

spatial distribution with these parameter values, the stellar mass of the thick disk is 10–20% of that of the thin disk (the uncertainty allowing for the uncertainty in the structural parameters), or some $10^{10}\ M_\odot$.

Again the properties of the stellar populations in this component are rather poorly known far from the solar neighborhood. Locally, within a few kpc of the Sun, the typical thick-disk star is of intermediate metallicity, [Fe/H] ~ -0.6 dex, and old, with an age comparable to that of 47 Tuc, the globular cluster of the same metallicity, ~ 12 Gyr (see e.g., review of Wyse 2000). Detailed elemental abundances are now available for statistically significant sample sizes. These show that the pattern of elemental abundances differs between the thick and thin disks, with different values of the ratio $[\alpha/\mathrm{Fe}]$ at fixed [Fe/H], implying distinct star formation and enrichment histories for the thick and thin disks (Fuhrmann 1998, 2000; Prochaska et al. 2000; Feltzing, Bensby & Lundström 2003; Nissen 2003; see Figure 1). Such a difference argues against the model (Burkert, Truran & Hensler 1992) whereby the thick disk represents the earliest stages of disk star formation during continuous, self-regulated dissipational settling of gas to the thin disk.

Thick stellar disks can be formed from pre-existing thin stellar disks by heating, and a (minor) merger of a reasonably dense and massive satellite galaxy into a pre-existing thin-disk galaxy could be the heating mechanism (cf. Quinn, Hernquist & Fullagar 1993). In the merger, orbital energy is deposited in the internal degrees of freedom of both the thin disk and the satellite, and acts to disrupt the satellite and heat the disk. Depending on the orbit of the satellite, and on its density profile and mass (this last determines the dynamical friction timescale), tidal debris from the satellite will be distributed through the larger galaxy during the merger process. Thus the phase space structure of the debris from the satellite depends on many parameters, but in general, one expects that the final 'thick disk' will be a mix of heated thin-disk and satellite debris. The age and metallicity distributions of the thick disk can provide constraints on the mix.

Could the thick disk be dominated by the debris of tidally disrupted dwarf galaxies (cf. Abadi et al. 2003)? The removal of material from the satellite occurs essentially under the Roche relative-density criterion, so that one expects that the lower density, outer parts of accreted dwarfs will be tidally removed in the outer parts of the larger galaxy, with the inner, denser regions of the dwarf only being removed if the dwarf penetrates further inside the larger galaxy. As noted above, the local thick disk (within a few kpc of the Sun) is old and quite metal rich, with a mean iron abundance ~ -0.6 dex. Further, the bulk of these stars have enhanced, super-solar $[\alpha/Fe]$ abundances (Fuhrmann 1998, 2000; Prochaska et al. 2000; Feltzing et al. 2003; see Figure 1). Achieving such a high level of enrichment so long ago (the stellar age equals the age of 47 Tuc, at least 10 Gyr, as noted above), in a relatively short time—so that Type II supernovae dominate the enrichment, as evidenced by the enhanced levels of $[\alpha/Fe]$)—implies a high star-formation rate within a rather deep overall potential well. This does not favor dwarf galaxies.

Indeed, the inner disk of the LMC, our present most massive satellite galaxy, has a derived metallicity distribution (Cole, Smecker-Hane & Gallagher 2000) that is similar to that of the (local) thick disk, but, based on the color-magnitude diagram, these stars are of intermediate age. Thus the LMC apparently took until a few Gyr ago to self-enrich to an overall metallicity that equals that of the typical local thick-disk star in the Galaxy. Further, the abundances of the α-elements to iron in such metal-rich LMC stars are below the solar ratio (Smith et al. 2002), unlike the local thick-disk stars. This may be understood in terms of the different star-formation histories (cf. Gilmore & Wyse 1991). The LMC is not a good template for a putative dwarf to form the thick disk from its debris.

What about the Sagittarius dwarf, a galaxy that has clearly penetrated into the disk? From photometry the stars are, on average, quite enriched and of intermediate age (cf. the discovery paper of Ibata, Gilmore & Irwin 1994, where the member stars were clearly distinguished from the bulge-field stars; see also Layden & Saradejini 2000 and Cole 2001). The overall metallicity distribution of the Sagittarius dwarf spheroidal galaxy is not well defined, but it contains a significant population of stars with metallicity as high as the solar value (Bonifacio et al. 2000; Smecker-Hane & McWilliam 2003). One of its globular clusters, Terzian 7, has a metallicity equal to that of 47 Tuc, but an age several Gyr younger (Buonanno et al. 1995), and thus by inference, several Gyr younger than the thick-disk stars of the same metallicity. The derived age-metallicity relationship for the Sgr dSph, based on both the CMD (Saradejini & Layden 2000) and the spectroscopy of selected red giants (Smecker-Hane & McWilliam 2003), is consistent with stars more metal-rich than $[Fe/H] = -0.7$ dex being less than 8 Gyr old. Further, these stars have essentially solar values of the ratio of $[\alpha/Fe]$ (Bonifacio et al. 2000; Smecker-Hane & McWilliam 2003), and are thus different in several important properties from the local thick-disk stars. The Ursa Minor dSph is the sole satellite galaxy of the present retinue that contains only old stars, and thus has an age distribution similar to that of the local thick disk. However, these are exclusively metal poor, $[Fe/H] \sim -2$ dex, and again are not a good match to thick-disk stars.

Based on observations, there is no good analogue among the surviving dwarf galaxies for a possible progenitor of the thick disk. Theoretically, based on our (admittedly limited) understanding of supernova feedback, it seems very contrived to envisage a dwarf galaxy that had a deep enough potential well to self-enrich rapidly a long time ago, but that was sufficiently low density to be tidally disrupted to form the thick disk. One might argue that the satellites that were accreted earlier initiated star formation earlier (e.g. Bullock, Kravtsov & Weinberg 2000), and were typically more dense and able to self-enrich faster. However, the analyses of deep color-magnitude diagrams for the extant satellites of the Milky Way are consistent with *all* containing stars as old as the stellar halo of the Milky Way (e.g., Da Costa 1999), implying that the onset of star formation was co-eval and there are no (surviving) satellites that initiated star formation earlier.

In summary, it appears implausible that the bulk of the thick disk is the debris of accreted dwarf galaxies.

Heating of a pre-existing thin disk by a minor merger remains a viable mechanism for creating the bulk of the thick disk (e.g., Velazquez & White 1999). In this case, the old age of the thick disk, combined with the fairly continuous star formation in the thin disk, has two important consequences: (i) that there was an extended disk in place at a lookback time of greater than 10 Gyr, and (ii) that there has been no extraordinary heating—by mergers—of the thin disk since that time. Knowing the age distribution of stars in the thick disk—in both the observed thick disk and in predicted theoretical thick disks—is obviously crucial. Semi-analytic modeling of the heating of disks by merging of substructure in CDM cosmologies has shown that thin disks with reasonable scale-heights are produced at the present day (Benson et al. 2003). However, the presence or otherwise of thick disks has yet to be demonstrated in such simulations. Further, predictions need to be made for the age distribution of member stars of the thick and the thin disk, to be confronted with the observations.

All this being said, some fraction of the metal-poor stars assigned to the 'thick disk' on the basis of having orbital kinematics that are intermediate between those of the stellar halo and those of the thin disk, may well be debris from a satellite (perhaps the one that caused the disk heating), and we return to this point below, in Section 3.2.

2.3. *Large scale structure of the central bulge*

The metallicity distributions of low-mass stars in various low-reddening lines of sight towards the bulge (with projected Galactocentric distances of a few 100 pc to a few kpc) have been determined spectroscopically (e.g., McWilliam & Rich 1994; Ibata & Gilmore 1995; Sadler, Terndrup & Rich 1996) and photometrically (e.g., Zoccali et al. 2003), with the robust result that the peak metallicity is [Fe/H] ~ -0.3 dex, with a broad range and a tail to low abundances. (The distribution is well fit by the Simple closed-box model, unlike the solar neighborhood data). The available elemental abundances, limited to the brighter stars, show the enhanced [α/Fe] signatures of enrichment by predominantly Type II supernovae (McWilliam & Rich 1994, 2003), indicating rapid star formation. Indeed, the chemical abundances favor very rapid star formation and (self-)enrichment (Ferreras, Wyse & Silk 2003).

The age distributions derived from the analyses of deep *HST* and ISO color-magnitude diagrams—again over several degrees across the sky—are consistent with the dominant population being of old age $\gtrsim 10$ Gyr (Feltzing & Gilmore 2000: *HST*; van Loon et al. 2003: ISO; Zoccali et al. 2003: *HST*), confirming earlier conclusions from ground-based data (Ortolani et al. 1995). There is also a small intermediate-age component seen in the ISO data, and traced by OH/IR stars (Sevenster 1999), plus there is ongoing star formation in the plane. The interpretation of these younger stars in terms of the stellar populations in the bulge is complicated by the fact that the scale height of the thin disk is comparable to that of the central bulge, so that membership in either component is ambiguous. Indeed, the relation between the inner triaxial bulge/bar and the larger-scale bulge is as yet unclear (see Merrifield 2003 for a recent review). All that said, the dominant population in the bulge is clearly old and metal-rich.

In the hierarchical clustering scenario, bulges are built up during mergers, with several mechanisms contributing. The dense central regions of massive satellites may survive and sink to the center; the dynamical friction timescale for a satellite of mass M_{sat} orbiting in a more massive galaxy of mass M_{gal} is $t_{\mathrm{dyn\,fric}} \sim t_{\mathrm{cross}} M_{\mathrm{gal}}/M_{\mathrm{sat}}$, where t_{cross} is the crossing time of the more massive galaxy. With $t_{\mathrm{cross}} \sim 3 \times 10^8$ yr for a large galaxy, only the most massive satellites could contribute to the central bulge in a Hubble time. Gravitational torques during the merger process are also expected to drive disk gas to the central regions, and some fraction of stars in the disk will also be heated sufficiently to be 're-arranged' into a bulge (cf. Kauffmann 1996). The predicted age and metallicity distributions of the stars in the bulge are then dependent on the merger history; however, a uniform old population is not expected.

An alternative scenario for bulge formation appeals to an instability in the disk, either forming a bar which then buckles out of the plane to form a bulge (e.g., Raha et al. 1991), or is destroyed by the orbit-scattering effects of the accumulation of mass at its center (e.g., Hasan & Norman 1990). Again, one would expect a significant range of stellar ages in the bulge.

As noted above, the bulge is dominated by old, metal-rich stars. This favors neither of the two scenarios above, but rather points to formation of the bulge by an intense burst of star formation, *in situ*, a long time ago (cf. Elmegreen 1999; Ferreras, Wyse & Silk 2003). The inferred star formation rate is $\gtrsim 10\ M_\odot/$yr. A possible source of the gas is ejecta from star-forming regions in the halo; the rotation of the bulge is consistent with collapse and spin-up of halo material (cf. Wyse & Gilmore 1992; Ibata & Gilmore 1995), and the chemical abundances are also consistent with the mass ratios (see Carney, Latham & Laird 1990 and Wyse & Gilmore 1992).

2.4. *Large scale structure of the stellar halo*

The total stellar mass of the halo is $\sim 2 \times 10^9\ M_\odot$ (cf. Carney, Laird & Latham 1990), modulo uncertainties in the stellar halo density profile in each of the outer halo, where substructure may dominate, and the central regions, where the bulge dominates. Some $\sim 30\%$ of the stars in the halo are on orbits that take them through the solar neighborhood, to be identified by their 'high-velocity' with respect to the Sun. These stars form a rather uniform population—old and metal-poor, with enhanced values of the elemental abundance ratio $[\alpha/\mathrm{Fe}]$. The dominant signature of enrichment by Type II supernovae indicates a short duration of star formation, which could naturally arise due to star formation and self-enrichment occuring in low-mass star-forming regions that cannot sustain extended star formation.

In contrast, the typical star in a dwarf satellite galaxy now is of intermediate-age, and has solar values of $[\alpha/\mathrm{Fe}]$ (cf. Tolstoy et al. 2003). These differences in stellar populations between the field stellar halo and dwarf galaxies limit significant accretion ($\gtrsim 10\%$ by mass) into the stellar halo from satellite galaxies to have occurred at high redshift only, and at lookback times greater than ~ 8 Gyr (cf. Unavane, Wyse & Gilmore 1996). In contrast, typical CDM-models predict significant late accretion of sub-haloes, with $\sim 40\%$ of subhaloes that survive reionization falling into the host galaxy at redshifts less than $z = 0.5$, or a look-back time of less than 6 Gyr (Bullock, Kravtsov & Weinberg 2000). Again, the later accretion is preferentially to the outer parts, and to be consistent with the observations of the Milky Way, these sub-haloes must contain very few young stars, and not over-populate the outer galaxy with visible stars.

2.5. *Large-scale structure: Merging history*

The overall properties of the main stellar components of the Milky Way, as discussed above, can be understood if there was little merging or accretion of stars into the Milky Way for the last $\sim$10 Gyr (cf. Wyse 2001). How does this compare with 'merger trees' of N-body simulations? As an example, the publicly available Virgo GIF ΛCDM simulations (Jenkins et al. 1998) have 26 final haloes with mass similar to that of the Milky Way—taken to be 2×10^{12} $M_\odot$. Of these, only 7% have not merged with another halo of at least 20% by mass since a redshift of 2, a look-back time of $\sim$11 Gyr in this cosmology (L. Hebb, priv. comm.). A merger with these parameter values could produce a thick disk as observed in the Milky Way. None of these 'Milky Way' analogues pass a more stringent mass-ratio limit of no mergers more than 10% by mass (still capable of producing a thick disk, given appropriate orbit, etc.) since a redshift of 2. Reducing the epoch of last significant merger to unity (a look-back time of 8 Gyr) and adopting a maximum merging mass ratio of 20% makes the Milky Way more typical, with 35% of Milky Way analogues meeting these criteria. However, reducing the highest mass ratio to 10%, while maintaining this lower look-back time limit, results in only 4% of Milky Way analogue haloes passing these criteria. The Milky Way appears to be rather unusual in the ΛCDM cosmology.

Note that predictions of smooth, average 'universal' mass-assembly histories (e.g., Wechsler et al. 2002) are not useful for this comparison, since these curves suppress the detailed information necessary to predict the effect of the mass accretion. More useful is the detailed merging history as a function of radius (cf. Helmi et al. 2003b), since ideally one would like to know what fraction of mergers can penetrate into the realm of the baryonic Galaxy.

3. The small-scale structure of the stellar components of the Galaxy

While there is no evidence for recent, very significant, mergers into the Milky Way, mergers are clearly happening—as best evidenced by the Sagittarius dwarf galaxy (Ibata, Gilmore & Irwin 1994, 1995; Ibata et al. 1997; see Majewski's contribution to this volume). While the present and past mass of the Sagittarius dSph are rather uncertain, its assimilation into the Milky Way is best classed as a 'minor merger'—meaning mass ratio of less than 10%.

The small-scale structure in the Milky Way may reflect the minor-merger history—or may simply reflect inhomogeneities of different kinds.

3.1. *Small-scale structure in the thin disk*

The small-scale structure of the thin disk is rich and varied and includes stellar moving groups, the scatter in the age-metallicity relationship, spiral arms, the outer 'ring' structure and the central bar. All star formation appears to occur in clusters (e.g., Elmegreen 2002), which are then subject to both internally- and externally-driven dynamical processes that operate to disrupt them. Some clusters dissolve almost immediately once star formation is initiated and some remain gravitationally bound for many Gyr. The creation of phase-space structure is thus a natural part of the evolution of stellar disks.

The scatter in the age-metallicity relationship for stars at the solar neighborhood appears to be well established (Edvardsson et al. 1993), as is the offset between the metallicity of the Sun and of younger stars, and the interstellar medium in the solar neighborhood, with the Sun being more chemically enriched. These may have their origins in some combination of radial gradients and mixing (e.g., Francois & Matteucci 1993; Sellwood & Binney 2002) and infall of metal-poor gas, from the general intergalactic medium, or perhaps from satellite galaxies (e.g., Geiss et al. 2002). It should be noted that further motivation for appeal to accretion events from satellite galaxies had been found in the scatter in element ratios at a given iron abundance in the Edvardsson et al. data. However, more recent data has instead suggested that a more correct interpretation is that the elemental abundances of stars belonging to the thick disk are distinct from those belonging to the thin disk (e.g., Nissen 2003; see also Gilmore, Wyse & Kuijken 1989 and Figure 1 above), with no scatter in elemental ratios within a given component.

The low-latitude 'ring' seen in star counts (Newberg et al. 2002; Ibata et al. 2003; Bellazzini et al. 2004) in the anticenter direction at Galactocentric distances of ∼15 kpc may be structure in the outer disk, which has a well-established warp in the gas, and probably in stars (e.g., Carney & Seitzer 1993; Djorgovski & Sosin 1989). The recent detection of structure in H I—at just this distance and interpreted as a newly identified spiral arm—is intriguing (McClure-Griffiths et al. 2003). The ring may also be interpreted as resulting from the accretion of a satellite galaxy (e.g., Helmi et al. 2003a; Martin et al. 2004; Rocha-Pinto et al. 2003); a recent N-body hydrodynamic simulation within a ΛCDM cosmology has shown that it is possible for satellite galaxies to be accreted into a disk, provided they are massive enough for dynamical friction to circularize their orbit quickly enough (Abadi et al. 2003).

The available kinematics for 'ring' stars do not discriminate between the two possibilities of satellite or outer disk (Yanny et al. 2003; Crane et al. 2003). Given the complexity of the structure of outer disks, comprehensive color-magnitude data, plus metallicity distributions and kinematics will be needed to rule out the 'Occam's Razor' interpretation of the 'ring' as being a manifestation of structure in the outer disk.

The detailed structure of the thin disk will be revealed by large-scale spectroscopic surveys such as RAVE (Steinmetz 2003); the time is ripe to develop models that will distinguish between intrinsic structure due to the normal disk-star formation process and other effects (cf. Freeman & Bland-Hawthorn 2002).

3.2. *Small-scale structure in the thick disk*

As noted above in the minor-merger scenario for formation of the thick disk, one expects the 'thick disk' to be a mixture of heated thin disk plus satellite debris. An identification of satellite debris, made on the basis of distinct kinematics, was made by Gilmore, Wyse & Norris (2002). These authors obtained radial velocities for several thousand faint ($V \lesssim 19.5$) F/G dwarf stars selected by photometry to be unevolved stars in the thick disk/halo interface at several kpc from the Sun, in key intermediate-latitude lines of sight that probe orbital rotational velocity (particularly $\ell = 270°$). They found that the mean lag behind the Sun's azimuthal streaming velocity was significantly larger for the fainter stars than for the brighter stars (see Figure 2). Either there are (discontinuous?) steep kinematic gradients within the thick disk (cf. Majewski 1993), or a separate population exists. In the latter case, a viable explanation would be stars from a shredded satellite. Indeed, these stars have low metallicities, typically −1.5 dex (Norris et al., in prep.), a factor of ten below a typical thick-disk star (e.g., Gilmore, Wyse & Jones 1995), but typical of the old population in dwarf galaxies.

However, the values of critical defining parameters for the 'canonical' thick disk, probed locally, remain variable from study to study. For example, the 'accepted' value for the rotational lag is around 40 km sec^{-1} (e.g., Carney, Laird & Latham 1989), but values as low as 20 km sec^{-1} (Chiba & Beers 2000) and as high as 80 km sec^{-1} (Fuhrmann 2000) have been reported. Some of this variation is undoubtably due to the difficulty of deconvolving a complex mix of populations. The thin disk will dominate any local sample, and comparison with distant *in situ* surveys will help (cf. the technique of Wyse & Gilmore

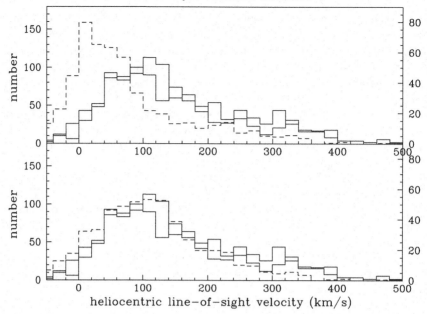

FIGURE 2. Modified from Gilmore, Wyse & Norris 2002. In each panel the solid histograms are observational data for faint $V \lesssim 19.5$ F/G stars in lines of sight where, at these distances, the line-of-sight velocity probes ~ 0.7–0.8 of the azimuthal streaming velocity. The dashed histogram is a model; in the upper panel the model is derived from standard local 'thick disk' kinematics which provide a good fit to the brighter stars, while the model in the lower panel has a significantly higher lag in $v_{\dots\dots}$ behind the Sun. This provides a significantly better fit to the data.

1995), as will using the discrimination inherent in the distinct elemental abundances of thick and thin disk stars (cf. Nissen 2003). Again, large statistically significant samples, so that tails of the distribution functions are well-defined, in key lines of sight, are needed.

3.3. *Small-scale structure of the bulge*

The bulge is clearly triaxial, but estimates of its three-dimensional structure are hindered by dust extinction, projection effects and the uncertainties in the structure of the disk along the line of sight (e.g., spiral arm pattern). The inner bulge, within ~ 1 kpc of the center, appears symmetric in deep infrared images taken with the *ISO* satellite (van Loon et al. 2003). The best fitting bar model (Bissantz & Gerhard 2002) to the *COBE* data has axial ratios 1:0.3–0.4:0.3 (i.e., barely triaxial) and a length of ~ 3.5 kpc. The effects of the bar potential may be the cause of the asymmetric stellar kinematics found by Parker, Humphreys & Beers (2003) in samples of stars on either side of the Galactic Center.

3.4. *Stellar halo small-scale structure*

Structure in coordinate space mixes and dissolves on dynamical timescales. The outer regions of the halo, say at Galactocentric distances of greater than 15 kpc where dynamical timescales are $\gtrsim 1$ Gyr, are thus most likely to host observable substructure. Indeed, as discussed more fully in Majewski's contribution to this volume, several streams are found in the outer halo, in both coordinate space and kinematics. The vast majority of the confirmed structure is due to a single system, the Sagittarius dwarf spheroidal (e.g., Ibata et al. 2001; Dohm-Palmer et al. 2001; Majewski et al. 2003; Newberg et al. 2002; Newberg et al. 2003). This contrasts with the predictions of many disrupted satellites

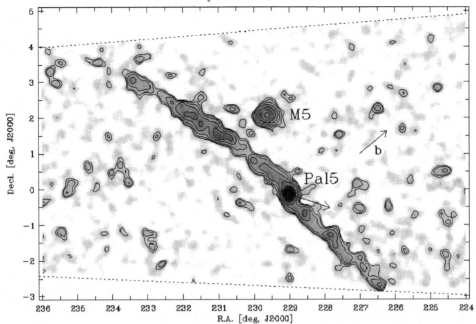

FIGURE 3. Taken from Odenkirchen et al. (2003), their Figure 3. The contours show the surface density of stars that are selected from their photometry to be members of Pal 5. There are clearly streams associated with this globular cluster. The arrow extending from the core of Pal 5 indicates the estimated direction of its orbit.

in CDM models (e.g., Bullock et al. 2000). The present mass of the Sagittarius dwarf is uncertain and model dependent, but most estimates are within a factor of three of $10^9 \, M_\odot$ (Ibata et al. 1997; Majewski et al. 2003). The mass lost by it to the halo is also model dependent; presently identified streams are perhaps 15% of the remaining bound mass. The evolutionary history of the Sagittarius dwarf is as yet unclear and much work remains to be done.

Tidal streams can be, and are, also associated with dynamically evolving globular clusters. The excellent photometry from the Sloan Digital Sky survey has allowed the tracing of extended, thin arms from the outer halo globular cluster Palomar 5 over 10 degrees across the sky (Odenkirchen et al. 2003; see Figure 3).

Streams are rare in the inner halo (which contains most of the stellar mass!). Simulations suggest that signatures in phase space, particularly if integrals of the motion can be estimated, can survive for about a Hubble time. A moving group has indeed be isolated (Helmi et al. 1999), but its mass is uncertain (see Chiba & Beers 2000), as is its origin—perhaps it is even associated with the Sagittarius dwarf (e.g., Majewski et al. 2003).

No structure is seen in coordinate space of the inner halo (within a few kpc of the Sun); the two point correlation function for main sequence stars brighter than $V = 19$ is flat (Gilmore, Reid & Hewett 1985; Lemon et al. 2003). This rules out significant recent accretion events that penetrate into the inner Galaxy, and ongoing disruption of inner globular clusters. Other tests for substructure show low-significance features consistent with known streams from the Sagittarius dwarf (Lemon et al. 2003), in agreement with results from blue horizontal branch stars (Sirko et al. 2003).

4. Concluding remarks

The properties of the stellar populations of the Milky Way contain much information about the star formation history and mass assembly history of the Galaxy. The Milky Way has merged with, is merging with, and will merge with, companion galaxies, which contribute stars, gas and dark matter. Debris from the Sagittarius dwarf galaxy dominates recent accretion into the outer Galaxy, while the data are consistent with little stellar accretion into the inner Galaxy, including the disk. Predominantly gaseous accretion is relatively unconstrained, and is favored by models of chemical evolution (cf. Tosi's contribution). Planned and ongoing large spectroscopic surveys will tightly constrain the existence and origins of stellar phase-space substructure. The relatively quiescent merging history of the Milky Way that is implied by the mean properties of the stellar components is rather atypical in ΛCDM cosmologies. What about the rest of the Local Group?

REFERENCES

ABADI, M., NAVARRO, J., STEINMETZ, M., & EKE, V. 2003 *ApJ* **597**, 21.

BELLAZZINI, M., IBATA, R., MONACO, L., MARTIN, L., IRWIN, M., & LEWIS, G. 2003 *MNRAS*, submitted (astro-ph/0311119).

BENSON, A., LACEY, C., FRENK, C., BAUGH, C., & COLE, S. 2004 *MNRAS* **351**, 1215.

BINNEY, J., DEHNEN, W., & BERTELLI, G. 2000 *MNRAS* **318**, 658.

BISSANTZ, N. & GERHARD, O. 2002 *MNRAS* **330**, 591.

BONIFACIO, P., HILL, V., MOLARI, P., PASQUINI, L., DiMARCANTONIO, P., & SANTINI, P. 2000 *A&A* **359**, 663.

BULLOCK, J., KRAVTSOV, A., & WEINBERG, D. 2000 *ApJ* **539**, 517.

BUONANNO, R., ET AL. 1995 *AJ* **109**, 663.

BURKERT, A., TRURAN, J., & HENSLER, G. 1992 *ApJ* **391**, 651.

CARNEY, B., LATHAM, D., & LAIRD, J. 1989 *AJ* **97**, 423.

CARNEY, B., LATHAM, D., & LAIRD, J. 1990 *AJ* **99**, 527.

CARNEY, B. & SEITZER, P. 1993 *AJ* **105**, 2127.

CHIBA, M. & BEERS, T. C. 2000 *AJ* **119**, 2843.

COLE, A. 2001 *ApJ* **559**, L17.

COLE, S., LACEY, C., BAUGH, C., & FRENK, C.S. 2000 *MNRAS* **319**, 168.

COLE, A., SMECKER-HANE, T., & GALLAGHER, J. 2000 *AJ* **120**, 1808.

DA COSTA, G. 1999. In *Third Stromlo Symposium: The Galactic Halo* (ed. B. Gibson, et al.). p. 153. Astron. Soc. Pacific.

CRANE, J., MAJEWSKI, S., ROCHA-PINTO, H., FRINCHABOY, P., SKRUTSKIE, M., & LAW, D. 2003 *ApJ* **594**, L119.

DE MARCHI, G., PARESCE, F., STRANIERO, O., & PRADA MORONI, P. 2003 *A&A* submitted; astro-ph/0310646.

DICKINSON, M. 2000. In *Building galaxies from the primordial Universe to the present*, XIXth Moriond meeting (eds. Hammer, et al.). p. 257. World Scientific Publishing.

DJORGOVSKI, G. & SOSIN, C. 1989 *ApJ* **341**, L13.

DOHM-PALMER, R., ET AL. 2001 *ApJ* **555**, L37.

EDVARDSSON, B., ANDERSEN, J., GUSTAFSSON, B., LAMBERT, D. L., NISSEN, P. E., & TOMKIN, J. 1993 *A&A* **275**, 101.

EKE, V., EFSTATHIOU, G., & WRIGHT, L. 2000 *MNRAS* **315**, L18.

ELMEGREEN, B. 1999 *ApJ* **517**, 103.

ELMEGREEN, B. 2002 *ApJ* **557**, 206.

FALL, S. M. & EFSTATHIOU, G. 1980 *MNRAS* **193**, 189.

FELTZING, S., BENSBY, T., & LUNDSTRÖM, I. 2003 *A&A* **397**, L1.

FELTZING, S. & GILMORE, G. 2000 *A&A* **355**, 949.

FERGUSON, A. M. N. & JOHNSON, R. 2001 *ApJ* **559**, L13.

FERRERAS, I., WYSE, R. F. G., & SILK, J. 2003 *MNRAS* **345**, 1381.

FRANCOIS, P. & MATTEUCCI, F. 1993 *A&A* **280**, 136.

FREEMAN, K. & BLAND-HAWTHORN, J. 2002 *ARAA* **40**, 487.

FUHRMANN, K. 1998 *A&A* **338**, 161.

FUHRMANN, K. 2000 http://www.xray.mpe.mpg.de/~fuhrmann

GEISS, J., GLOECKLER, G., & CHARBONNEL, C. 2002 *ApJ* **578**, 862.

GILMORE, G. & REID, I.N. 1983 *MNRAS* **202**, 1025.

GILMORE, G., REID, I. N., & HEWETT, P. 1985 *MNRAS* **213**, 257.

GILMORE, G. & WYSE, R. F. G. 1991 *ApJ* **367**, L55.

GILMORE, G., WYSE, R. F. G., & JONES, J. B. 1995 *AJ* **109**, 1095.

GILMORE, G., WYSE, R. F. G., & KUIJKEN, K. 1989 *ARAA* **27**, 555.

GILMORE, G., WYSE, R. F. G., & NORRIS, J. 2002 *ApJ* **574**, L39.

HANSEN, B., ET AL. 2002 *ApJ* **574**, L155.

HASAN, H. & NORMAN, C. 1990 *ApJ* **361**, 69.

HELMI, A., NAVARRO, J., MEZA, A., STEINMETZ, M., & EKE, V. 2003a *ApJ* **592**, L25.

HELMI, A., WHITE, S. D. M., DE ZEEUW, P. T., & ZHAO, H.-S. 1999 *Nature* **402**, 53.

HELMI, A., WHITE, S. D. M., & SPRINGEL, V. 2003b *MNRAS* **339**, 834.

HERNANDEZ, X., VALLS-GABAUD, D., & GILMORE, G. 2000 *MNRAS* **316**, 605.

IBATA, R. & GILMORE, G. 1995 *MNRAS* **275**, 605.

IBATA, R., GILMORE, G., & IRWIN, M. 1994 *Nature* **370**, 194.

IBATA, R., GILMORE, G., & IRWIN, M. 1995 *MNRAS* **277**, 781.

IBATA, R., IRWIN, M., LEWIS, G., FERGUSON, A., & TANVIR, N. 2003 *MNRAS* **340**, L21.

IBATA, R., LEWIS, G., IRWIN, M., TOTTEN, E., & QUINN, T. 2001 *ApJ* **551**, 294.

IBATA, R., WYSE, R. F. G., GILMORE, G., IRWIN, M., & SUNTZEFF, N. 1997 *AJ* **113**, 634.

JENKINS, A., ET AL. 1998 *ApJ* **499**, 20.

JIMENEZ, R., FLYNN, C., & KOVONTA, E. 1998 *MNRAS* **299**, 515.

KAUFFMANN, G. 1996 *MNRAS* **281**, 487.

KNOX, R., HAWKINS, M., & HAMBLY, N. 1999 *MNRAS* **306**, 736.

KOTONEVA, E., FLYNN, C., MATTEUCCI, F., & CHIAPPINI, C. 2002 *MNRAS* **336**, 879.

LARSON, R. B. 1972 *Nature Phys. Sci.* **236**, 7.

LAYDEN, A. & SARAJEDINI, A. 2000 *AJ* **119**, 1760.

LEGGETT, S., RUIZ, M., & BERGERON, P. 1998 *ApJ* **497**, 294.

LEMON, D., WYSE, R. F. G., LISKE, J., DRIVER, S., & HORNE, K. 2003 *MNRAS* in press; astro-ph/0308200.

MAJEWSKI, S. R. 1993 *ARAA* **31**, 575.

MAJEWSKI, S., SKRUTSKIE, M., WEINBERG, M., & OSTHEIMER, J. 2003 *ApJ* **599**, 1082.

MARTIN, N., IBATA, R., BELLAZZINI, M., IRWIN, M., LEWIS, G., & DEHNEN, W. 2003 *MNRAS* **348**, 12.

MCCLURE-GRIFFITHS, N., DICKEY, J., GAENSLER, B., & GREEN, A. 2004 *ApJ* **607** L127 (see http://waw.atnf.Cairo.au/news/press/spiralarm/).

MCWILLIAM, A. & RICH, R. M. 1994 *ApJS* **91**, 749.

MCWILLIAM, A. & RICH, R.M. 2003. In *Origin and evolution of the elements* eds. A. McWilliam & M. Rauch). Carnegie Observatories Astrophysics series, Vol. 4. Cambridge University Press; astro-ph/0312628.

MERRIFIELD, M. 2003. In *Milky Way surveys: the structure and evolution of our Galaxy,* (eds. D. Clemens, T. Brainerd & R. Shah), ASP Conf. Proc. ASP, in press; astro-ph/0308302.

NAVARRO, J., FRENK, C. S., & WHITE, S. D. M. 1995 *MNRAS* **275**, 56.

NAVARRO, J. & STEINMETZ, M. 1997 *ApJ* **478**, 13.

NEWBERG, H., ET AL. 2002 *ApJ* **569**, 245.

NEWBERG, H., ET AL. 2003 *ApJ* **596**, L191.

NISSEN, P. 2003. In *Origin and evolution of the elements* (eds. A. McWilliam & M. Rauch). Carnegie Observatories Astrophysics series, Vol. 4. Cambridge University Press, in press; astro-ph/0310326.

ODENKIRCHEN, M. ET AL. 2003 *AJ* **126**, 2385.

ORTOLANI, S., RENZINI, A., GILMOZZI, R., MARCONI, G., BARBUY, B., BICA, E., & RICH, R. M. 1995 *Nature* **377**, 701.

PAGEL, B. & PATCHETT, B. 1975 *MNRAS* **172**, 13.

PARKER, J., HUMPHREYS, R., & BEERS, T. 2004 *AJ*, in press; astro-ph/0312017.

PROCHASKA, J. X., NAUMOV, S., CARNEY, B., MCWILLIAM, A., & WOLFE, A. 2000 *AJ* **120**, 2513.

QUINN, P., HERNQUIST, L., & FULLAGAR, D. 1993 *ApJ* **403**, 74.

RAHA, N., SELLWOOD, J., JAMES, R., & KAHN, F. 1991 *Nature* **352**, 411.

ROCHA-PINTO, H. J., MAJEWSKI, S., SKRUTSKIE, M., & CRANE, J. 2003 *ApJ* **594**, L115.

ROCHA-PINTO, H. J., SCALO, J., MACIEL, W. J., & FLYNN, C. 2000 *A&A* **358**, 869.

SADLER, E., RICH, R. M. & TERNDRUP, D. 1996 *AJ* **112**, 171.

SANDAGE, A., LUBIN, L., & VAN DEN BERG, D. A. 2003 *PASP* **115**, 1187.

SELLWOOD, J. & BINNEY, J. 2002 *MNRAS* **336**, 785.

SEVENSTER, M. 1999 *MNRAS* **310**, 629.

SIRKO, E., ET AL. 2004 *AJ* **127**, 899.

SMECKER-HANE, T. & MCWILLIAM, A. 2003 *Astrophys. J.* accepted; astro-ph/0205411.

SMITH, V., ET AL. 2002 *AJ* **124**, 3241.

STEINMETZ, M. 2003. In *GAIA spectroscopy, science and technology* (ed. U. Munnari), ASP Conf. Proc. Vol. 298, p. 381. ASP.

STOCKTON, A., CANALIZO, G., & MAIHARA, T. 2004 *ApJ* **605**, 37.

TOLSTOY, E., VENN, K., SHETRONE, M., PRIMAS, F., HILL, V., KAUFER, A., & SZEIFERT, T. 2003 *AJ* **125**, 707.

UNAVANE, M., WYSE, R. F. G., & GILMORE, G. 1996 *MNRAS* **278**, 727.

VAN DEN BERGH, S. 1962 *AJ* **67**, 486.

VAN LOON, J., ET AL. 2003 *MNRAS* **338**, 857.

VELAZQUEZ, V. & WHITE, S. D. M. 1999 *MNRAS* **304**, 254.

WECHSLER, R., BULLOCK, J., PRIMACK, J., KRAVSOV, A., & DEKEL, A. 2002 *ApJ* **568**, 52.

WYSE, R. F. G. 2000 In *The Galactic Halo: From Globular Clusters to Field Stars* (eds. A. Noels, et al.). p. 305. Institut d'Astrophysique et de Geophysique de Liège.

WYSE, R. F. G. 2001 In *Galactic disks and disk galaxies* (eds. J. Funes & E. Corsini). ASP Conf. Series Vol. 230, p. 71. ASP.

WYSE, R. F. G. & GILMORE, G. 1992 *AJ* **104**, 114.

WYSE, R. F. G. & GILMORE, G. 1995 *AJ* **110**, 2771.

YANNY, B., ET AL. 2003 *ApJ* **588**, 824.

ZHANG, B., WYSE, R. F. G., STIAVELLI, M., & SILK, J. 2002 *MNRAS* **332**, 647.

ZOCCALI, M., ET AL. 2003 *A&A* **399**, 931.

ZUREK, W., QUINN, P. J., & SALMON, J. 1988 *ApJ* **330**, 519.

The Large Magellanic Cloud: Structure and kinematics

By ROELAND P. VAN DER MAREL

Space Telescope Science Institute, 3700 San Martin Drive, Baltimore, MD 21218, USA

I review our understanding of the structure and kinematics of the Large Magellanic Cloud (LMC), with a particular focus on recent results. This is an important topic, given the status of the LMC as a benchmark for studies of microlensing, tidal interactions, stellar populations, and the extragalactic distance scale. I address the observed morphology and kinematics of the LMC; the angles under which we view the LMC disk; its in-plane and vertical structure; the LMC self-lensing contribution to the total microlensing optical depth; the LMC orbit around the Milky Way; and the origin and interpretation of the Magellanic Stream. Our understanding of these topics is evolving rapidly, in particular due to the many large photometric and kinematic datasets that have become available in the last few years. It has now been established that: the LMC is considerably elongated in its disk plane; the LMC disk is thicker than previously believed; the LMC disk may have warps and twists; the LMC may have a pressure-supported halo; the inner regions of the LMC show unexpected complexities in their vertical structure; and precession and nutation of the LMC disk plane contribute measurably to the observed line-of-sight velocity field. However, many open questions remain and more work is needed before we can expect to converge on a fully coherent structural, dynamical and evolutionary picture that explains all observed features of the LMC.

1. Introduction

The Large Magellanic Cloud (LMC) is one of our closest neighbor galaxies at a distance of ~ 50 kpc. The Sagitarius dwarf is closer at ~ 24 kpc, but its contrast with respect to the Milky Way foreground stars is so low that it was discovered only about a decade ago. The LMC is therefore the closest, big, easily observable galaxy from our vantage point in the Milky Way. As such, it has become a benchmark for studies on various topics. It is of fundamental importance for studies of stellar populations and the interstellar medium (ISM), it is being used to study the presence of dark objects in the Galactic Halo through microlensing (e.g., Alcock et al. 2000a), and it plays a key role in determinations of the cosmological distance scale (e.g., Freedman et al. 2001). For all these applications it is important to have an understanding of the structure and kinematics of the LMC. This is the topic of the present review. For information on other aspects of the LMC, the reader is referred to the book by Westerlund (1997). The book by van den Bergh (2000) discusses more generally how the properties of the LMC compare to those of other galaxies in the Local Group.

The Small Magellanic Cloud (SMC) at a distance of ~ 62 kpc is a little further from us than the LMC, and is about five times less massive. Its structure is more irregular than that of the LMC, and is less well studied and understood. Recent studies of SMC structure and kinematics include the work by Hatzidimitriou et al. (1997), Udalski et al. (1998), Stanimirovic et al. (1999, 2004), Kunkel, Demers & Irwin (2000), Cioni, Habing & Israel (2000b), Zaritsky et al. (2000, 2002), Crowl et al. (2001) and Maragoudaki et al. (2001). However, our overall understanding of SMC structure and kinematics has not evolved much since the reviews by Westerlund and van den Bergh. The present review is therefore restricted to the LMC.

47

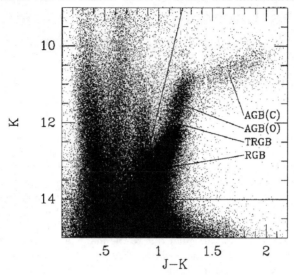

FIGURE 1. Near-IR $(J - K_s, K_s)$ CMD from 2MASS data for the LMC region of the sky. Only a quarter of the data is shown, to avoid saturation of the grey scale. The features due to the Red Giant Branch (RGB), the Tip of the RGB (TRGB), the oxygen-rich AGB stars [AGB(O)], and the carbon-rich AGB stars [AGB(C)] are indicated. The region enclosed by the solid lines was used to extract stars for creation of the LMC number density map in Figure 2.

2. Morphology

The LMC consists of an outer body that appears elliptical in projection on the sky, with a pronounced, off-center bar. The appearance in the optical wavelength regime is dominated by the bar, regions of strong star formation, and patchy dust absorption. The LMC is generally considered an irregular galaxy as a result of these characteristics. It is, in fact, the prototype of the class of galaxies called "Magellanic Irregulars" (de Vaucouleurs & Freeman 1973). Detailed studies of the morphological characteristics of the LMC have been performed using many different tracers, including optically detected starlight (Bothun & Thompson 1988; Schmidt-Kaler & Gochermann 1992), stellar clusters (Lynga & Westerlund 1963; Kontizas et al. 1990), planetary nebulae (Meatheringham et al. 1988) and non-thermal radio emission (Alvarez, Aparici, & May 1987). Recent progress has come primarily from studies of stars on the Red Giant Branch (RGB) and Asymptotic Giant Branch (AGB) and from studies of H I gas.

2.1. *Near-Infrared morphology*

Recently, two important near-IR surveys have become available for studies of the Magellanic Clouds, the Two Micron All Sky Survey (2MASS; e.g., Skrutskie 1998) and the Deep Near-Infrared Southern Sky Survey (DENIS; e.g., Epchtein et al. 1997). Cross-correlations of the data from these surveys and from other catalogs are now available as well (Delmotte et al. 2002). The surveys are perfect for a study of LMC morphology and structure. Near-IR data is quite insensitive to dust absorption, which is a major complicating factor in optical studies (Zaritsky, Harris, & Thompson 1997; Zaritsky 1999; Udalski et al. 2000; Alcock et al. 2000b). The surveys have superb statistics with of the order of a million stars. Also, the observational strategy with three near-IR bands (J, H and K_s in the 2MASS survey; I, J and K_s in the DENIS survey) allows clear separation of different stellar populations. In particular, the data are ideal for studies of evolved RGB and AGB stars, which emit much of their light in the near-IR. This is important for studies of

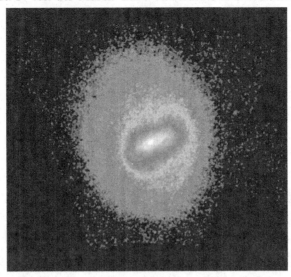

FIGURE 2. Surface number density distribution on the sky of RGB and AGB stars in the LMC from van der Marel (2001). North is to the top and east is to the left. The image covers an area of 23.55° × 21.55°. The Galactic foreground contribution was subtracted. A color version of the image is available at http://www.stsci.edu/~marel/lmc.html/.

LMC structure, because these intermediate-age and old stars are more likely to trace the underlying mass distribution of the LMC disk than younger populations that dominate the light in optical images.

Figure 1 shows the $(J - K_s, K_s)$ color-magnitude diagram (CMD) for the LMC region of the sky. Several finger-like features are visible, each due to different stellar populations in the LMC or in the foreground (Nikolaev & Weinberg 2000; Cioni et al. 2000a; Marigo, Girardi, & Chiosi 2003). The LMC features of primary interest in the present context are indicated in the figure, namely the Red Giant Branch (RGB), the Tip of the RGB (TRGB), the oxygen-rich AGB stars, and the carbon-rich AGB stars ("carbon stars"). Van der Marel (2001) used the color cut shown in the figure to extract a sample from the 2MASS and DENIS datasets that is dominated by RGB and AGB stars. These stars were used to make the number-density map of the LMC shown in Figure 2. Kontizas et al. (2001) showed the distribution on the sky of ~7000 carbon stars identified by eye from optical objective prism plates. Their map does not show all the rich detail visible in Figure 2, but it is otherwise in good agreement with it.

The near-IR map of the LMC is surprisingly smooth. The morphology is much less irregular than it is for the younger stellar populations that dominate the optical light. Apart from the central bar there is a hint of some spiral structure, as discussed previously by, e.g., de Vaucouleurs & Freeman (1973). However, the spiral features all have very low contrast with respect to their surroundings, and there is certainly no well organized spiral pattern in the LMC. Quantitative analysis can be performed on the basis of ellipse fits to the number density contours. This yields a surface number density profile that can be reasonably well fit by an exponential with a scale length of 1.4° (1.3 kpc). The radial profiles of the ellipticity ϵ (defined as $1 - q$, where q is the axial ratio) and the major axis position angle PA$_{maj}$ both show pronounced variations as function of distance from the LMC center, due to the presence of the central bar. However, at radii $r \gtrsim 4°$ the contour shapes converge to an approximately constant position angle PA$_{maj}$ = 189.3° ± 1.4° and ellipticity $\epsilon = 0.199 \pm 0.008$.

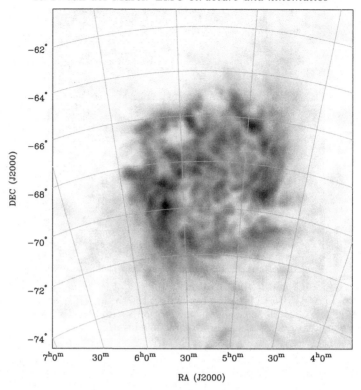

FIGURE 3. Peak brightness temperature image of H I in the LMC from Staveley-Smith et al. (2003). The data are from a Parkes multibeam survey and are sensitive to spatial structure in the range 200 pc to 10 kpc. The orientation is the same as in Figure 2, but the present image covers an area that is approximately 2.5 times smaller (i.e., 1.6 in each dimension).

2.2. *H I morphology*

To study the distribution of H I in the LMC on small scales requires many pointings to cover the LMC at high angular resolution. Kim et al. (1998) used the Australia Telescope Compact Array (ATCA) to obtain a map that is sensitive on scales of 15–500 pc. On these scales, the morphology is dominated by H I filaments with numerous shells and holes. The turbulent and fractal nature of the ISM on these scales is the result of dynamical feedback into the ISM from star formation processes. In the context of the present review we are more interested in the large scale distribution of H I gas. This issue has been studied for decades, including work by McGee & Milton (1966), Rohlfs et al. (1984) and Luks & Rohlfs (1992). Most recently, Staveley-Smith et al. (2003) obtained the map shown in Figure 3, using a multibeam survey with the Parkes telescope. This map is sensitive to spatial structure in the range 200 pc to 10 kpc. The H I distribution is very patchy (see also Kim et al. 2003, who combined the data of Kim et al. 1998 and Staveley-Smith et al. 2003 into a single map that is sensitive to structures on all scales down to 15 pc). The H I in the LMC is not centrally concentrated and the brightest areas are several degrees from the center. The overall distribution is approximately circular and there is no sign of a bar. All of these characteristics are strikingly different from the stellar distribution shown in Figure 2.

Figure 4 shows contours of the H I distribution on a somewhat larger scale, from the HIPASS data presented in Putman et al. (2003), overlaid on the LMC near-IR star count map from Figure 2. This comparison suggests that collisionless tracers such as RGB and

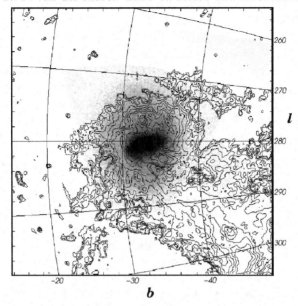

FIGURE 4. Contours of H I column density from Putman et al. (2003) overlaid on the LMC near-IR star count map from Figure 2. This figure is in Galactic coordinates, but (l, b) are shown so as to allow easy comparison with Figures 2 and 3. North is 9° clockwise from the vertical direction on the page. The image covers an area that is approximately 2.5 times larger than Figure 2 (i.e., 1.6 times in each dimension). The figure was kindly provided by M. Putman.

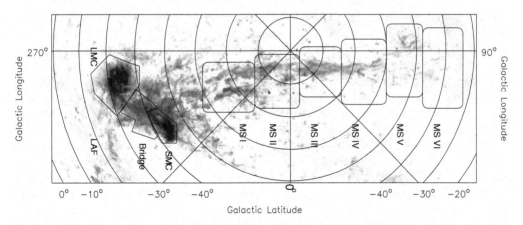

FIGURE 5. H I column density map of the Magellanic System in galactic coordinates from Putman et al. (2003). The LMC and SMC are indicated, as well as the Magellanic Bridge between them, several individual gas clumps in the Magellanic Stream (MS#), and the start of the Leading Arm Feature (LAF). The orientation of the LMC on the page is similar to that in Figure 4.

AGB stars are best suited to study the structure and mass distribution of the LMC disk, whereas H I gas may be better suited to study the effects of tidal interactions. The stream of gas towards the right of Figure 4 is the start of the Magellanic Bridge towards the SMC, which provides one of the many pieces of evidence for strong tidal interactions between the LMC, the Milky Way and the SMC.

3. Magellanic Stream

Figure 5 shows a wide-area H$\,$I map of the Magellanic System. The most prominent feature is the Magellanic Stream, a $10°$ wide filament with $\sim 2 \times 10^8 \; M_\odot$ of neutral hydrogen that spans more than $100°$ across the sky (e.g., Westerlund 1997; Putman et al. 2003). It consists of gas that trails the Magellanic Clouds as they orbit the Milky Way. A less prominent leading gas component was recently discovered as well (Lu et al. 1998; Putman et al. 1998), the start of which is seen at the left in Figure 5.

Many detailed theoretical models have been constructed for the Magellanic Stream (e.g., Murai & Fujimoto 1980; Lin & Lynden-Bell 1982; Shuter 1992; Liu 1992; Heller & Rohlfs 1994; Gardiner, Sawa & Fujimoto 1994; Moore & Davis 1994; Lin, Jones & Klemola 1995; Gardiner & Noguchi 1996; Yoshizawa & Noguchi 2003; Mastropietro et al. 2004; Conners et al. 2004). Models in which tidal stripping is the dominating process have been particularly successful. The most sophisticated recent calculations in this class are those by Yoshizawa & Noguchi (2003) and Conners et al. (2004), both of which build on earlier work by Gardiner & Noguchi (1996). In these models the LMC and SMC form a gravitationally bound system that orbits the Milky Way. The Magellanic Stream and the Leading Arm represent material that was stripped from the SMC ~ 1.5 Gyr ago. This was the time of the previous perigalactic passage, which coincided with a close encounter between the Clouds. The models successfully reproduce many properties of the Magellanic Stream, including its position, morphology, width variation, and the velocity profile along the Stream. The models also explain the presence of the Leading Arm, and why it is less prominent than the trailing Stream.

Given the successes of tidal models, it has always been surprising that no population of stars associated with the Stream has ever been found (e.g., Irwin 1991; Guhathakurta & Reitzel 1998; Majewski et al. 2003). In a tidal model where the stripping is dominated by gravity, one might naively expect that both stars and gas are stripped equally. However, galaxies generally have H$\,$I gas disks that are more extended than the stellar distribution. Since material is preferentially stripped from the outskirts of a galaxy, this can explain why there may not be any stars associated with the Stream (Yoshizawa & Noguchi 2003). Alternatively, it has been argued that the lack of stars in the Stream may point to important contributions from other physical processes than tidal effects. For example, Moore & Davis (1994) and Mastropietro et al. (2004) suggest that the Stream consists of material which was ram-pressure stripped from the LMC during its last passage through a hot ($\sim 10^6$ K) ionized halo component of the Milky Way. Another alternative was proposed by Heller & Rohlfs (1994), who suggested that hydrodynamical forces (rather than gravitational/tidal forces) during a recent LMC-SMC interaction are responsible for the existence of the Stream.

Models of the Magellanic Stream have traditionally used the properties of the Stream to estimate the orbit of the LMC, rather than to base the calculations on estimates of the LMC proper motion. An important characteristic of the orbit is the present-day tangential velocity in Galactocentric coordinates, for which values have been inferred that include $v_{\rm LMC,tan} = 369 \; \mathrm{km\,s}^{-1}$ (Lin & Lynden-Bell 1982), $355 \; \mathrm{km\,s}^{-1}$ (Shuter 1992), $352 \; \mathrm{km\,s}^{-1}$ (Heller & Rohlfs 1994), $339 \; \mathrm{km\,s}^{-1}$ (Murai & Fujimoto 1980), $320 \; \mathrm{km\,s}^{-1}$ (Liu 1992) and $285 \; \mathrm{km\,s}^{-1}$ (Gardiner et al. 1994; Gardiner & Noguchi 1996), respectively. Proper motion measurements have improved significantly over time, and it is now known that $v_{\rm LMC,tan} = 281 \pm 41 \; \mathrm{km\,s}^{-1}$ (as discussed in detail in Section 4). This is consistent with the lower range of the values predicted by models for the Stream. The data are most consistent with the models of Gardiner & Noguchi (1996) and their follow-ups, which also provide some of the best fits to many other properties of the Magellanic

Stream. The observational error on $v_{\text{LMC,tan}}$ is almost small enough to start ruling out some of the other models. It is possible that future proper motion measurements may yield a much more accurate determination of the LMC orbit. Models of the Magellanic Stream then hold the promise of providing important constraints on the mass, shape, and radial density profile of the Milky Way's dark halo (e.g., Lin et al. 1995).

4. Orbit

To understand the tidal interactions in the Magellanic System it is important to know the orbit of the LMC around the Milky Way. This requires knowledge of all three of the velocity components of the LMC center of mass. The line-of-sight velocity can be accurately determined from the Doppler velocities of tracers (see Section 5). By contrast, determination of the velocity in the plane of the sky is much more difficult. For the LMC, proper motion determinations are available from the following sources: Kroupa et al. (1994), using stars from the PPM Catalogue; Jones, Klemola, & Lin (1994), using photographic plates with a 14-year epoch span; Kroupa & Bastian (1997), using *Hipparcos* data; Drake et al. (2002), using data from the MACHO project; and Anguita, Loyola, & Pedreros (2000) and Pedreros, Anguita, & Maza (2002), using CCD frames with an 11-year epoch span. Some of the measurements pertain to fields in the outer parts in the LMC disk, and require corrections for the orientation and rotation of the LMC disk. The measurements are all more-or-less consistent with each other to within the error bars. The exception to this is the Anguita et al. result, which almost certainly suffers from an unidentified systematic error. When this latter result is ignored, the weighted average of the remaining measurements yields proper motions towards the West and North of (van der Marel et al. 2002)

$$\mu_W = -1.68 \pm 0.16 \text{ mas yr}^{-1} \ , \qquad \mu_N = 0.34 \pm 0.16 \text{ mas yr}^{-1} \ . \tag{4.1}$$

Transformation of the proper motion to a space velocity in km s^{-1} requires knowledge of the LMC distance D_0. Many techniques have been used over the years to estimate the LMC distance, but unfortunately, there continue to be systematic differences between the results from different techniques that far exceed the formal errors. It is beyond the scope of the present review to address this topic in detail. Instead, the reader is referred to the recent reviews by, e.g., Westerlund (1997), Gibson et al. (2000), Freedman et al. (2001) and Alves (2004b). Freedman et al. adopt a distance modulus $m - M = 18.50 \pm 0.10$ on the basis of a review of all published work. This corresponds to $D_0 = 50.1 \pm 2.5$ kpc. At this distance, a proper motion of 1 mas yr^{-1} corresponds to $238 \pm 12 \text{ km s}^{-1}$ and 1 degree on the sky corresponds to 0.875 ± 0.044 kpc.

Combination of the distance and proper motion of the LMC yields velocities $v_W = -399 \text{ km s}^{-1}$ and $v_N = 80 \text{ km s}^{-1}$ towards the West and North, respectively, with errors of $\sim 40 \text{ km s}^{-1}$ in each direction. This can be combined with the observed line-of-sight velocity, $v_{\text{sys}} = 262.2 \pm 3.4 \text{ km s}^{-1}$ (see Section 5.2), to obtain the three-dimensional motion of the LMC with respect to the Milky Way. It is usual to adopt a Cartesian coordinate system (X, Y, Z), with the origin at the Galactic Center, the Z-axis pointing towards the Galactic North Pole, the X-axis pointing in the direction from the sun to the Galactic Center, and the Y-axis pointing in the direction of the sun's Galactic Rotation. The observed LMC velocities must be corrected for the reflex motion of the sun, which is easily done with use of standard estimates for the position and velocity of the sun with respect to the Galactic center. The (X, Y, Z) position of the LMC is then found to be

(van der Marel et al. 2002)

$$\vec{r}_{\mathrm{LMC}} = (-0.78, -41.55, -26.95) \text{ kpc} \quad, \tag{4.2}$$

and its three-dimensional space velocity is

$$\vec{v}_{\mathrm{LMC}} = (-56 \pm 36, -219 \pm 23, 186 \pm 35) \text{ km s}^{-1} \quad. \tag{4.3}$$

This corresponds to a distance of 49.53 kpc from the Galactic center, and a total velocity of 293 ± 39 km s^{-1} in the Galactocentric rest frame. The motion has a radial component of 84 ± 7 km s^{-1} pointing away from the Galactic center, and a tangential component of 281 ± 41 km s^{-1}. The proper motion of the SMC is known only with errors of ~ 0.8 mas yr^{-1} in each coordinate (Kroupa & Bastian 1997), which is five times less accurate than for the LMC. However, to within these errors the SMC is known to have a galactocentric velocity vector that agrees with that of the LMC.

The combination of a small but positive radial velocity and a tangential velocity that exceeds the circular velocity of the Milky Way halo implies that the LMC must be just past pericenter in its orbit. The calculation of an actual orbit requires knowledge of the three-dimensional shape and the radial profile of the gravitational potential of the Milky Way dark halo. Gardiner et al. (1994) and Gardiner & Noguchi (1996) calculated orbits in a spherical Milky Way halo potential with a rotation curve that stays flat at 220 km s^{-1} out to a galactocentric distance of at least 200 kpc. Such a potential is consistent with our present understanding of the Milky Way dark halo (Kochanek 1996; Wilkinson & Evans 1999; Ibata et al. 2001). The calculations properly take into account that the LMC and SMC orbit each other, while their center of mass orbits the Milky Way. However, since the LMC is more massive than the SMC, its motion is not too different from that of the LMC-SMC center of mass. For an assumed present-day LMC velocity $\vec{v}_{\mathrm{LMC}}$ consistent with that given in equation (4.3) the apocenter to pericenter ratio is inferred to be $\sim 2.5 : 1$. The perigalactic distance is ~ 45 kpc and the orbital period around the Milky Way is ~ 1.5 Gyr.

5. Kinematics

The kinematical properties of the LMC provide important clues to its structure. Observations have therefore been obtained for many tracers. The kinematics of gas in the LMC have been studied primarily using H I (e.g., Rohlfs et al. 1984; Luks & Rohlfs 1992; Kim et al. 1998). Discrete LMC tracers which have been studied kinematically include star clusters (Freeman, Illingworth, & Oemler 1983; Schommer et al. 1992), planetary nebulae (Meatheringham et al. 1988), H II regions and supergiants (Feitzinger, Schmidt-Kaler, & Isserstedt 1977), and carbon stars (Kunkel et al. 1997b; Graff et al. 2000; Alves & Nelson 2000; van der Marel et al. 2002). A common result from all these studies is that the line-of-sight velocity dispersion of the tracers is generally at least a factor ~ 2 smaller than their rotation velocity. This implies that the LMC is kinematically cold, and must therefore to lowest approximation be a disk system.

5.1. *General expressions*

To understand the kinematics of an LMC tracer population, it is necessary to have a general model for the line-of-sight velocity field that can be fit to the data. All studies thus far have been based on the assumption that the mean streaming (i.e., the rotation) in the disk plane can be approximated to be circular. However, even with this simplifying assumption, it is not straightforward to model the kinematics of the LMC, because it is so near to us. Its main body spans more than 20° on the sky and one therefore cannot make

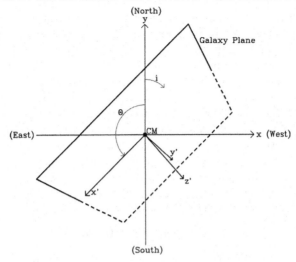

FIGURE 6. Schematic illustration of the observer's view of the LMC disk. The plane of the disk is titled diagonally out of the paper. The inclination i is the angle between the (x, y) plane of the sky, and the (x', y') plane of the galaxy disk. The x'-axis is the line of nodes, defined as the intersection of the (x, y) plane of the sky and the (x', y') plane of the galaxy disk. The angle Θ is the position angle of the line of nodes in the plane of the sky.

the usual approximation that "the sky is flat" over the area of the galaxy. Spherical trigonometry must be used, which yields the general expression (van der Marel et al. 2002):

$$v_{los}(\rho, \Phi) = s\,V(R')f \sin i \cos(\Phi - \Theta)$$
$$+ v_{sys} \cos \rho$$
$$+ v_t \sin \rho \cos(\Phi - \Theta_t)$$
$$+ D_0(di/dt) \sin \rho \sin(\Phi - \Theta) \quad , \tag{5.1}$$

with

$$R' = D_0 \sin \rho / f, \qquad f \equiv \frac{\cos i \cos \rho - \sin i \sin \rho \sin(\Phi - \Theta)}{[\cos^2 i \cos^2(\Phi - \Theta) + \sin^2(\Phi - \Theta)]^{1/2}} \quad . \tag{5.2}$$

In this equation, v_{los} is the observed component of the velocity along the line of sight. The quantities (ρ, Φ) identify the position on the sky: ρ is the angular distance from the center and Φ is the position angle with respect to the center (measured from North over East). The kinematical center is at the center of mass (CM) of the galaxy. The quantities (v_{sys}, v_t, Θ_t) describe the velocity of the CM in an inertial frame in which the sun is at rest: v_{sys} is the systemic velocity along the line of sight, v_t is the transverse velocity, and Θ_t is the position angle of the transverse velocity on the sky. The angles (i, Θ) describe the direction from which the plane of the galaxy is viewed: i is the inclination angle ($i = 0$ for a face-on disk), and Θ is the position angle of the line of nodes, as illustrated in Figure 6. The line-of-nodes is the intersection of the galaxy plane and the sky plane. The velocity $V(R')$ is the rotation velocity at cylindrical radius R' in the disk plane. D_0 is the distance to the CM, and f is a geometrical factor. The quantity $s = \pm 1$ is the 'spin sign' that determines in which of the two possible directions the disk rotates.

The first term in equation (5.1) corresponds to the internal rotation of the LMC. The second term is the part of the line-of-sight velocity of the CM that is seen along the line of sight, and the third term is the part of the transverse velocity of the CM that is seen along the line of sight. For a galaxy that spans a small area on the sky (very small ρ), the second term is simply v_{sys} and the third term is zero. However, the LMC does not have a small angular extent and the inclusion of the third term is particularly important. It corresponds to a solid-body rotation component. Given the LMC transverse velocity implied by equation (4.1), it rises to an amplitude of 71 $km\,s^{-1}$ at $\rho = 10°$, which significantly exceeds the amplitude of the intrinsic rotation contribution (first term of eq. [5.1]) at that radius. The fourth term in equation (5.1) describes the line-of-sight component due to changes in the inclination of the disk with time, as are expected due to precession and nutation of the LMC disk plane as it orbits the Milky Way (Weinberg 2000). This term also corresponds to a solid-body rotation component.

The general expression in equation (5.1) appears complicated, but it is possible to gain some intuitive insight by considering some special cases. Along the line of nodes one has that $\sin(\Phi - \Theta) = 0$ and $\cos(\Phi - \Theta) = \pm 1$, so that

$$\hat{v}_{los}(\text{along}) = \pm[v_{tc} \sin\rho - V(D_0 \tan\rho) \sin i \cos\rho] \quad . \tag{5.3}$$

Here it has been defined that $\hat{v}_{los} \equiv v_{los} - v_{sys} \cos\rho \approx v_{los} - v_{sys}$. The quantity $v_{tc} \equiv v_t \cos(\Theta_t - \Theta)$ is the component of the transverse velocity vector in the plane of the sky that lies along the line of nodes; similarly, $v_{ts} \equiv v_t \sin(\Theta_t - \Theta)$ is the component perpendicular to the line of nodes. Perpendicular to the line of nodes one has that $\cos(\Phi - \Theta) = 0$ and $\sin(\Phi - \Theta) = \pm 1$, and therefore

$$\hat{v}_{los}(\text{perpendicular}) = \pm w_{ts} \sin\rho \quad . \tag{5.4}$$

Here it has been defined that $w_{ts} = v_{ts} + D_0(di/dt)$. This implies that perpendicular to the line of nodes $\hat{v}_{los}$ is linearly proportional to $\sin\rho$. By contrast, along the line of nodes this is true only if $V(R')$ is a linear function of R'. This is not expected to be the case, because galaxies do not generally have solid-body rotation curves; disk galaxies tend to have flat rotation curves, at least outside the very center. This implies that, at least in principle, both the position angle Θ of the line of nodes and the quantity w_{ts} are uniquely determined by the observed velocity field: Θ is the angle along which the observed $\hat{v}_{los}$ are best fit by a linear proportionality with $\sin\rho$, and w_{ts} is the proportionality constant.

5.2. *Carbon star kinematics*

Among the discrete tracers in the LMC that have been studied kinematically, carbon stars have yielded some of the largest and most useful datasets in recent years. Van der Marel et al. (2002) fitted the general velocity field expression in equation (5.1) to the data for 1041 carbon stars, obtained from the work of Kunkel et al. (1997a) and Hardy, Schommer & Suntzeff (unpublished). The combined dataset samples both the inner and the outer parts of the LMC, although with a discontinuous distribution in radius and position angle. Figure 7 shows the data, with the best model fit overplotted. Overall, the model provides a good fit to the data.

As discussed in Section 5.1, the line-of-nodes position angle is uniquely determined by the data; the model yields $\Theta = 129.9° \pm 6.0°$. The LMC inclination cannot be determined kinematically, but it is known reasonably well from other considerations (see Section 6). With both viewing angles and the LMC proper motion known, the rotation curve $V(R')$ follows from equation (5.3). The result is shown in Figure 8. The inferred rotation curve $V(R')$ rises linearly in the central region and is roughly flat at $V \approx 50$ $km\,s^{-1}$ for $R' \gtrsim 4$ kpc. The negative value at the innermost radius $R' \approx 0.5$ kpc has limited significance; it is probably affected by the sparse coverage of the data at these radii (see Figure 7), as well as potential non-circular streaming motions in the region of the bar. The error bars on $V(R')$ in Figure 8 reflect the random errors due to the sampling of the

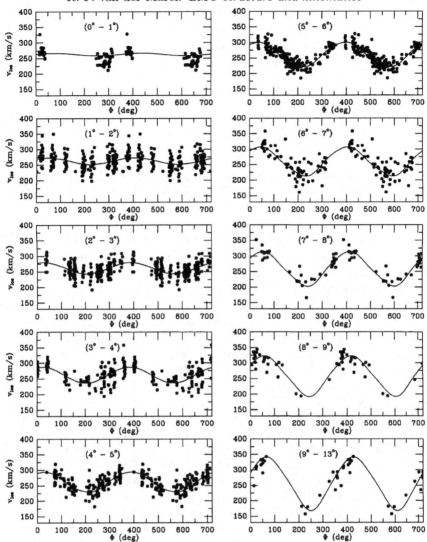

FIGURE 7. Carbon star line-of-sight velocity data from Kunkel, Irwin & Demers (1997a) and Hardy et al. (unpublished), as function of position angle Φ on the sky. The displayed range of the angle Φ is $0°$–$720°$, so each star is plotted twice. Each panel corresponds to a different range of angular distances ρ from the LMC center, as indicated. The curves show the predictions of the best-fitting circularly-rotating disk model from van der Marel et al. (2002).

data. However, there are also contributions from the errors in the line-of-nodes position angle Θ, the LMC proper motion and the inclination i. In particular, $V(R')$ scales as $1/\sin i$; and if the component v_{tc} of the transverse velocity vector along the line of nodes is larger than given in equation (4.1), then $V(R')$ will go up. When all uncertainties are properly accounted for, the amplitude of the flat part of the rotation curve and its formal error become $V = 49.8 \pm 15.9$ km s^{-1}.

When asymmetric drift is corrected for, the circular velocity in the disk plane can be calculated to be $V_{\mathrm{circ}} = 64.8 \pm 15.9$ km s^{-1}. The implied total mass of the LMC inside the last measured data point is therefore $M_{\mathrm{LMC}}(8.9\ \mathrm{kpc}) = (8.7 \pm 4.3) \times 10^9\ M_\odot$. By contrast, the total stellar mass of the LMC disk is $\sim 2.7 \times 10^9\ M_\odot$. and the mass of the

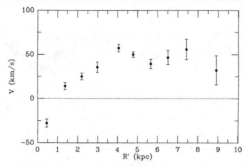

FIGURE 8. Rotation velocity V of carbon stars in the plane of the LMC disk as function of the cylindrical radius R' in kpc from the carbon star fits of van der Marel et al. (2002). For reference, the exponential disk scale length of the LMC is approximately 1.4 kpc.

neutral gas in the LMC is $\sim 0.5 \times 10^9\ M_\odot$ (Kim et al. 1998). The combined mass of the visible material in the LMC is therefore insufficient to explain the dynamically inferred mass. The LMC must therefore be embedded in a dark halo. This is consistent with the fact that the observed rotation curve amplitude is relatively flat as a function of radius. The LMC tidal radius can be calculated to be $r_t = 15.0 \pm 4.5$ kpc, which corresponds to an angle on the sky of $17.1° \pm 5.1°$. The uncertainty in the tidal radius is due primarily to our ignorance of how far the LMC dark halo extends. Either way, the tidal radius extends beyond the region for which most observations of the main body of the LMC are available. However, it should be kept in mind that the tidal radius marks the position beyond which material becomes unbound. The structure of a galaxy can be altered well inside of this radius (Weinberg 2000).

As discussed in Section 5.1, the line-of-sight velocity field constrains the value of $w_{ts} = v_{ts} + D_0(di/dt)$. The carbon stars yield $w_{ts} = -402.9 \pm 13.0$ km s^{-1}. Given our knowledge of the proper motion and distance of the LMC (see Section 4) this implies that $di/dt = -0.37 \pm 0.22$ mas yr$^{-1} = -103 \pm 61$ degrees/ Gyr. The LMC is the first galaxy for which it has been possible to measure di/dt. The N-body simulations by Weinberg (2000) show that the tidal torques from the Milky Way are expected to induce precession and nutation in the symmetry axis of the LMC disk plane. Although a detailed data-model comparison is not possible at the present time, it is comforting to note that the observed di/dt is of the same order as the rate of change of the disk orientation seen in the simulations.

5.3. *H I kinematics*

H I gas provides another powerful method to study the kinematics of the LMC. High quality data is available from, e.g., Kim et al. (1998). Unfortunately, the kinematical analysis presented by Kim et al. was not as general as that discussed above for carbon stars. They did not leave w_{ts} as a free parameter in the fit. Instead, they corrected their data at the outset for the transverse motion of the LMC using the proper motion measured by Jones et al. (1994), $(\mu_W, \mu_N) = (-1.37 \pm 0.28, -0.18 \pm 0.27)$ mas yr^{-1}, and assumed that $di/dt = 0$. This fixes $w_{ts} = -175 \pm 72$ km s^{-1}, which is inconsistent with the value inferred from the carbon star velocity field. Kim et al. obtained a kinematic line of nodes that is both twisting with radius and inconsistent with the value determined from the carbon stars. In addition, they inferred a rotation curve for which the amplitude exceeds that in Figure 8 by $\sim 40\%$. It is likely that these results are affected by the imposed value of w_{ts}. A more general analysis of the H I kinematics is therefore desirable, but unfortunately, is not currently available. The same limitations apply to many of the other published studies of LMC tracer kinematics cited at the start of Section 5.

Independent of how the data are analyzed, it is likely that collisionless tracers provide a more appropriate means to study the structure of the LMC using equilibrium models than does H I gas. The dynamical center of the carbon star velocity field is found to be consistent (to within the $\sim 0.4°$ per-coordinate errors) with both the center of the bar and the center of the outer stellar contours. However, it has long been known that the center of the H I rotation velocity field does not coincide with the center of the bar, and it also doesn't coincide with the center of the outer contours of the stellar distribution. It is offset from both by ~ 1 kpc. This indicates that the H I gas may not be in equilibrium in the the the gravitational potential. Added evidence for this comes from the fact that the LMC and the SMC are enshrouded in a common H I envelope, and that they are connected by a bridge of H I gas (see Figure 5). Even at small radii the LMC gas disk appears to be subject to tidal disturbances (see Figure 4) that may well affect the velocity field.

6. Viewing angles

A disk that is intrinsically circular will appear elliptical in projection on the sky. The viewing angles of the disk (see Figure 6) are then easily determined: the inclination is $i = \arccos(1 - \epsilon)$, where ϵ is the apparent ellipticity on the sky, and the line-of-nodes position angle Θ is equal to the major axis position angle PA_{maj} of the projected body. The viewing angles of the LMC have often been estimated under this assumption, using the projected contours for many different types of tracers (de Vaucouleurs & Freeman 1973; Bothun & Thompson 1988; Schmidt-Kaler & Gochermann 1992; Weinberg & Nikolaev 2001; Lynga & Westerlund 1963; Kontizas et al. 1990; Feitzinger et al. 1977; Kim et al. 1998; Alvarez et al. 1987). However, it now appears that this was incorrect. The kinematics of carbon stars imply $\Theta = 129.9° \pm 6.0°$ (see Section 5.2), whereas the near-IR morphology of the LMC implies $PA_{maj} = 189.3° \pm 1.4°$ (see Section 2). The result that $\Theta \neq PA_{maj}$ implies that the LMC cannot be intrinsically circular. The value of PA_{maj} is quite robust; studies of other tracers have yielded very similar results, although often with larger error bars. The result that $\Theta \neq PA_{maj}$ therefore hinges primarily on our confidence in the inferred value of Θ. There have been other kinematical studies of the line of nodes, in addition to that described in Section 5.2. These have generally yielded values of Θ that are both larger and twisting with radius (e.g., Kim et al. 1998; Alves & Nelson 2000). However, the accuracy of these results is suspect because of the important simplifying assumptions that were made in the analyses (see Section 5.3). No allowance was made for a potential solid-body rotation component in the velocity field due to precession and nutation of the LMC disk, which is both predicted theoretically (Weinberg 2000) and implied observationally by the carbon star data (see Section 5.2).

Arguably the most robust way to determine the LMC viewing angles is to use geometrical considerations, rather than kinematical ones. For an inclined disk, one side will be closer to us than the other. Tracers on that one side will appear brighter than similar tracers on the other side. This method does not rely on absolute distances or magnitudes, which are notoriously difficult to estimate, but only on relative distances or magnitudes. To lowest order, the difference in magnitude between a tracer at the galaxy center and a similar tracer at a position (ρ, Φ) in the disk (as defined in Section 5.1) is

$$\mu = \left(\frac{5\pi}{180 \ln 10}\right) \rho \tan i \sin(\Phi - \Theta) , \qquad (6.1)$$

where the angular distance ρ is expressed in degrees. The constant in the equation is $(5\pi)/(180 \ln 10) = 0.038$ magnitudes. Hence, when following a circle on the sky around

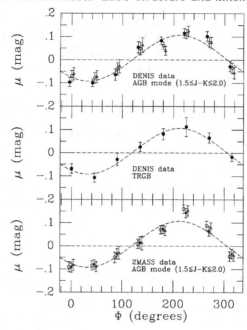

FIGURE 9. Variations in the magnitude of tracers as function of position angle Φ from van der Marel & Cioni (2001). *(a)* Modal magnitude of AGB stars in the DENIS data with colors $1.5 \leqslant J - K_s \leqslant 2.0$. *(b)* TRGB magnitude from DENIS data. *(c)* As (a), but using data from the 2MASS Point Source Catalog. All panels refer to an annulus of radius $2.5° \leqslant \rho \leqslant 6.7°$ around the LMC center. Filled circles, open circles, four-pointed stars and open triangles refer to the I, J, H and K_s-band, respectively. Results in different bands are plotted with small horizontal offsets to avoid confusion. The dashed curve shows the predictions for an inclined disk with viewing angles $i = 34.7°$ and $\Theta = 122.5°$.

the galaxy center one expects a sinusoidal variation in the magnitudes of tracers. The amplitude and phase of the variation yield estimates of the viewing angles (i, Θ).

Van der Marel & Cioni (2001) used a polar grid on the sky to divide the LMC area into several rings, each consisting of a number of azimuthal segments. The data from the DENIS and 2MASS surveys were used for each segment to construct near-IR CMDs similar to that shown in Figure 1. For each segment the modal magnitude (magnitude where the luminosity function peaks) was determined for carbon-rich AGB stars selected by color, as had been suggested by Weinberg & Nikolaev (2001). Figure 9 shows the inferred variation in magnitude as function of position angle Φ for the radial range $2.5° \leqslant \rho \leqslant 6.7°$. The expected sinusoidal variations are confidently detected. The top panel shows the results for stars selected from the DENIS survey with the color selection criterion $1.5 \leqslant J - K \leqslant 2.0$. The bottom panel shows the results from the 2MASS survey with the same color selection. The same sinusoidal variations are seen, indicating that there are no relative calibration problems between the surveys. Also, the same variations are seen in the I, J, H and K_s bands, which implies that the results are not influenced significantly by dust absorption. The middle panel shows the variations in the TRGB magnitudes as a function of position angle, from the DENIS data. RGB stars show the same variations as the AGB stars, suggesting that the results are not influenced significantly by potential peculiarities associated with either of these stellar populations. The observed variations can therefore be confidently interpreted as a purely geometrical effect. The implied viewing angles are $i = 34.7° \pm 6.2°$ and $\Theta = 122.5° \pm 8.3°$. The Θ value thus inferred geometrically is entirely consistent with the value inferred kinematically (see

Section 5.2). Moreover, there is an observed drift in the center of the LMC isophotes at large radii which is consistent with both estimates, when interpreted as a result of viewing perspective (van der Marel 2001).

The aforementioned analyses are sensitive primarily to the structure of the outer parts of the LMC. Several other studies of the viewing angles have focused mostly on the region of the bar, which samples only the central few degrees. Many of these studies have been based on Cepheids. Their period-luminosity relation allows calculation of the distance to each individual Cepheid from a light curve. The relative distances of the Cepheids in the sample can then be analyzed in similar fashion as discussed above to yield the LMC viewing angles. Cepheid studies in the 1980s didn't have many stars to work with. Caldwell & Coulson (1986) analyzed optical data for 73 Cepheids and obtained $i = 29° \pm 6°$ and $\Theta = 142° \pm 8°$. Laney & Stobie (1986) obtained $i = 45° \pm 7°$ and $\Theta = 145° \pm 17°$ from 14 Cepheids, and Welch et al. (1987) obtained $i = 37° \pm 16°$ and $\Theta = 167° \pm 42°$ from 23 Cepheids, both using near-IR data. The early Cepheid studies have now all been superseded by the work of Nikolaev et al. (2004). They analyzed a sample of more than 2000 Cepheids with lightcurves from MACHO data. Through use of photometry in five different bands, including optical MACHO data and near-IR 2MASS data, each star could be individually corrected for dust extinction. From a planar fit to the data they obtained $i = 30.7° \pm 1.1°$ and $\Theta = 151.0° \pm 2.4°$. Other recent work has used the magnitude of the Red Clump to analyze the relative distances of different parts of the LMC. Olsen & Salyk (2002) obtained $i = 35.8° \pm 2.4°$ and $\Theta = 145° \pm 4°$, also from an analysis that was restricted mostly to the the inner parts of the LMC.

There is one caveat associated with all viewing angle results for the central few degrees of the LMC. Namely, it appears that the stars in this region are not distributed symmetrically around a single well-defined plane, as discussed in detail in Section 8.4. In the present context we are mainly concerned with the influence of this on the inferred viewing angles. Olsen & Salyk (2002) perform their viewing angle fit by ignoring fields south-west of the bar, which do not seem to agree with the planar solution implied by their remaining fields. By contrast, Nikolaev et al. (2004) fit all the stars in their sample, independent of whether or not they appear to be part of the main disk plane. Clearly, the (i, Θ) results of Olsen & Salyk and Nikolaev et al. are the best-fitting parameters of well-posed problems. However, it is somewhat unclear whether they can be assumed to be unbiased estimates of the actual LMC viewing angles. For a proper understanding of this issue one would need to have both an empirical and a dynamical understanding of the nature of the extra-planar structures in the central region of the LMC. Only then is it possible to decide whether the concept of a single disk plane is at all meaningful in this region, and which data should be included or excluded in determining its parameters. This is probably not an issue for the outer parts of the LMC, given that the AGB star results of van der Marel & Cioni (2001) provide no evidence for extra-planar structures at radii $\rho \geqslant 2.5°$.

In summary, all studies agree that i is in the range $30°–40°$. At large radii, Θ appears to be in the range $115°–135°$. By contrast, at small radii all studies indicate that Θ is in the range $140°–155°$. As mentioned, it is possible that the results at small radii are systematically in error due to the presence of out-of-plane structures. Alternatively, it is quite well possible that there are true radial variations in the LMC viewing angles due to warps and twists of the disk plane. Many authors have suggested this as a plausible interpretation of various features seen in LMC datasets (van der Marel & Cioni 2001; Olsen & Salyk 2002; Subramaniam 2003; Nikolaev et al. 2004). Moreover, numerical simulations have shown that Milky Way tidal effects can drive strong warps in the LMC disk plane (Mastropietro et al. 2004).

7. Ellipticity

The inferred LMC viewing angles can be used to deproject the observed morphology that is seen in projection on the sky (van der Marel 2001). This yields an in-plane ellipticity ϵ in the range ~ 0.2–0.3, depending somewhat on the adopted viewing angles; e.g., the Nikolaev et al. (2004) angles give $\epsilon = 0.21$ and the van der Marel & Cioni (2001) angles give $\epsilon = 0.31$. The conclusion that the LMC is elongated is in itself not surprising. The dark matter halos predicted by cosmological simulations are generally triaxial (e.g., Dubinski & Carlberg 1991), and the gravitational potential in the equatorial plane of such halos does not have circular symmetry. So it is generally expected that disk galaxies are elongated rather than circular. Furthermore, it is possible to construct self-consistent dynamical models for elliptical disks (e.g., Teuben 1987). What is surprising is that the LMC ellipticity is fairly large. Studies of the apparent axis ratio distribution of spiral galaxy disks (Binney & de Vaucouleurs 1981; Lambas, Maddox & Loveday 1992) of the structure of individual spiral galaxies (Rix & Zaritsky 1995; Schoenmakers, Franx & de Zeeuw 1997; Kornreich, Haynes & Lovelace 1998; Andersen et al. 2001) and of the scatter in the Tully-Fisher relation (Franx & de Zeeuw 1992) indicate that the average (deprojected) ellipticity of spiral galaxies is only 5–10%. So while spiral galaxies are generally elongated, their elongation is usually smaller than inferred here for the LMC. Of course, galaxies of type Sm and Im are (by definition) more irregular and lopsided than spirals. So it is not *a priori* clear whether or not the LMC is atypically elongated for its Hubble type.

It is interesting to ask what may be the cause of the large in-plane ellipticity of the LMC. The prime candidate is distortion by the Milky Way tidal field. The present-day tidal force on the LMC by the Milky Way exceeds that from the SMC. Moreover, the Milky Way tidal field is responsible for other well-known features of the Magellanic system, such as the Magellanic Stream (see Section 3). *N*-body simulations have shown that the structure of the LMC can be altered significantly by the Milky Way tidal force (Weinberg 2000) and the simulations of Mastropietro et al. (2004) indeed predict a considerable in-plane elongation for the LMC. Also, the LMC elongation in projection on the sky points approximately towards the Galactic Center and is perpendicular to the Magellanic Stream (van der Marel 2001), as predicted naturally by simulations of tidal effects (Mastropietro et al. 2004). However, a very detailed data-model comparison is not possible at the present time. That would require accurate knowledge of the past history as a function of time of the LMC orbit, of the disk-plane orientation due to precession and nutation, and of the LMC-SMC distance. Such knowledge is not available at the present time.

A consequence of the ellipticity of the LMC disk is that one cannot expect the streamlines of tracers in the disk to be perfectly circular, by contrast to what has been assumed in all kinematical studies to date. The effect of this is probably not large, because the gravitational potential of a mass distribution is always rounder than the mass distribution itself. One effect of ellipticity is an apparent offset between the kinematical line of nodes and the true line of nodes (Schoenmakers, Franx & de Zeeuw 1997). This effect is presently not at observable levels, given that kinematical analysis of carbon stars (see Section 5) yields a line-of-nodes position angle Θ that is in adequate agreement with geometrical determinations (see Section 6). However, it should be kept in mind that, as the sophistication of the studies of LMC kinematics increases, it might become necessary to account for the effect of non-circularity on the observed kinematics.

8. Vertical structure and microlensing

8.1. Scale height

The scale height of the LMC disk can be estimated from the observed line-of-sight velocity dispersion σ. For carbon stars, the scatter of the velocity measurements around the best-fitting rotating disk model (Figure 7) yields $\sigma = 20.2 \pm 0.5$ km s^{-1}. The ratio of rotation velocity to velocity dispersion is therefore $V/\sigma = 2.9 \pm 0.9$. For comparison, the thin disk of the Milky Way has $V/\sigma \approx 9.8$ and its thick disk has $V/\sigma \approx 3.9$. In a relative sense one might therefore expect the LMC disk to be similar, but somewhat thicker than the Milky Way thick disk. Weinberg (2000) argued from N-body simulations that such considerable thickness could be the result of Milky Way tidal effects on the LMC.

The radial profile of the velocity dispersion contains information on the LMC scale height as function of radius. The carbon star velocity dispersion is close to constant as function of radius, and this is not what is expected for a disk with a constant scale height. To fit this behavior one must assume that the scale height increases with radius in the disk (Alves & Nelson 2000). This can arise naturally as a result of tidal forces from the Milky Way, which become relatively more important (compared to the LMC self-gravity) as one moves to larger radii. Alves & Nelson considered an isothermal disk with a vertical density profile proportional to sech$^2(z/z_0)$, where z_0 can vary with disk radius. Application to the carbon star data of van der Marel et al. (2002) yields $z_0 = 0.27$ kpc at the LMC center, rising to $z_0 = 1.5$ kpc at a radius of 5.5 kpc.

The LMC carbon stars are part of the intermediate-age population which is believed to be fairly representative for the bulk of the mass in the LMC. In this sense, the results inferred for the carbon star population are believed to be characteristic for the LMC as a whole. However, it is certainly not true that all populations have the same kinematics. As in the Milky Way, younger populations have a smaller velocity dispersion (and hence a smaller scale height) than older populations. A summary of measurements for various populations is given by Gyuk, Dalal, & Griest (2000). The youngest populations (e.g., supergiants, H II regions, H I gas) have dispersions of only $\sigma \approx 6$ km s^{-1}. Old long-period variables have dispersions $\sigma \approx 30$ km s^{-1} (Bessell, Freeman, & Wood 1986) and so do the oldest star clusters (Freeman et al. 1983; Schommer et al. 1992). These values are considerably below the LMC circular velocity (see Section 5.2). This has generally been interpreted to mean that the LMC does not have an old pressure supported halo similar to that of the Milky Way.

The first possible evidence for the presence of a pressure supported halo was presented recently by Minniti et al. (2003). They measured a dispersion $\sigma \approx 53 \pm 10$ km s^{-1} for a sample of 43 RR Lyrae stars within 1.5° from the LMC center. This value is consistent with what would be expected for a pressure supported halo in equilibrium in the gravitational potential implied by the circular velocity (Alves 2004a). The RR Lyrae stars make up $\sim 2\%$ of the visible mass of the LMC, similar to the value for the Milky Way halo. However, it is surprising that the surface density distribution of the LMC RR Lyrae stars is well fit by an exponential with the same scale length as inferred for other tracers known to reside in the disk (Alves 2004a). This is very different from the situation for the Milky Way halo, where RR Lyrae stars follow a power-law density profile. This suggests that maybe the RR Lyrae stars in the LMC did form in the disk, instead of in a halo. In this scenario they might simply have attained their large dispersions by a combination of disk heating and Milky Way tidal forces (Weinberg 2000). To discriminate between halo and disk origins it will be important to determine whether the velocity field of the RR Lyrae stars has a rotation component. This will require observations at larger galactocentric distances.

8.2. *Microlensing optical depth*

One of the most important reasons for seeking to understand the vertical structure of the LMC is to understand the results from microlensing surveys. The observed microlensing optical depth towards the LMC is $\tau_{obs} = 12^{+4}_{-3} \times 10^{-8}$ with an additional 20–30% of systematic error (Alcock et al. 2000a). It has generally been found that equilibrium models for the LMC do not predict enough LMC self-lensing events to account for the observed optical depth (Gyuk, Dalal, & Griest 2000; Jetzer, Mancini, & Scarpetta 2002). For the most favored set of LMC model parameters in the Gyuk et al. study the predicted self-lensing optical depth is $\tau_{self} = 2.2 \times 10^{-8}$, a factor of 5.5 less than the observed value. To account for the lenses it is therefore necessary to assume that some $\sim 20\%$ of the Milky Way dark halo is made up of lenses of mass 0.15–0.9 $M_{\odot}$ (Alcock et al. 2000a). However, it is a mystery what the composition of this lensing component could be. A large population of old white dwarfs has been suggested, but this interpretation is not without problems (e.g., Fields, Freese, & Graff 2000; Flynn, Holopainen, & Holmberg 2003). It is therefore worthwhile to investigate whether the models that have been used to estimate the LMC self-lensing might have been oversimplified. This is particularly important since there is evidence from the observed microlensing events themselves that many of the lenses may reside in the LMC (Sahu 2003).

To lowest order, the self-lensing optical depth in simple disk models of the LMC depends exclusively on the observed velocity dispersion and not separately on either the galaxy mass or scale height (Gould 1995). One way to increase the self-lensing predicted by LMC models is therefore to assume that a much larger fraction of the LMC mass resides in high velocity dispersion populations than has previously been believed. Salati et al. (1999) showed that the observed optical depth can be reproduced if one assumes that 70% of the mass in the LMC disk consists of objects with σ ranging from 25 km s^{-1} to 60 km s^{-1}. However, this would seem difficult to explain on the basis of present data. Although RR Lyrae stars have now been observed to have a high velocity dispersion, these old stars make up only 2% of the visible mass. So their influence on self-lensing predictions is negligible. A better way to account for the observed optical depth might be to assume that the vertical structure of the LMC is more complicated than for normal disk galaxies. The self-lensing optical depth might then have been considerably underestimated.

8.3. *Foreground and background populations*

One way to explain the observed microlensing optical depth is to assume that there might be stellar populations outside of the main LMC disk plane. For example, there might be a population of stars in front of or behind the LMC that was pulled from the Magellanic Clouds due to Milky Way tidal forces (Zhao 1998) or there might be a non-virialized shroud of stars at considerable distances above the LMC plane due to Milky Way tidal heating (Weinberg 2000; Evans & Kerins 2000). If microlensing source stars belong to such populations, then this would yield observable signatures in their characteristics. In particular, if microlensing source stars are behind the LMC, then they should be systematically fainter (Zhao, Graff, & Guhathakurta 2000) and redder (Zhao 1999, 2000) than LMC disk stars. *HST*/WFPC2 CMDs of fields surrounding microlensing events show no evidence for this, although the statistics are insufficient to rule out specific models with strong confidence (Alcock et al. 2001). Other tests of the spatial distribution and properties of the LMC microlensing events also do not (yet) discriminate strongly between different models for the location and nature of either the lenses or the sources (Alcock et al. 2000a; Gyuk, Dalal, & Griest 2000; Jetzer, Mancini, & Scarpetta 2002). This is primarily because only 13–17 LMC microlensing events are known. So to assess whether the LMC has out-of-plane structures it is best to focus on other types of tracers that might yield better statistics.

Zaritsky & Lin (1997) studied the optical CMD of a field $\sim 2°$ north-west of the LMC center and found a vertical extension at the bright end of the Red Clump. They suggested that this feature is due to stars 15–17 kpc above the LMC disk. However, this interpretation has been challenged for many different reasons (e.g., Bennett 1998; Gould 1998, 1999). Most seriously, Beaulieu & Sackett (1998) pointed out that a Vertical Red Clump (VRC) extension is naturally expected from stellar evolution due to young helium core burning stars. The same feature is seen in the Fornax and Sextans A dwarfs (Gallart 1998). It remains somewhat open to discussion whether the number of stars in the LMC VRC is larger than expected given our understanding of the LMC star formation history, so a foreground population is not strictly ruled out (Zaritsky et al. 1999). However, it certainly doesn't appear to be a favored interpretation of the data. Also, the kinematics of the VRC stars is indistinguishable from that of LMC Red Clump stars (Ibata, Lewis, & Beaulieu 1998). More recently, Zhao et al. (2003) performed a detailed radial velocity survey of 1300 stars of various types within $\sim 2°$ from the LMC center. They found no evidence for stars with kinematics that differ significantly from that of the main LMC disk. This rules out a significant foreground or background population of stars that reside in a tidal stream that is seen superposed onto the LMC by chance (Zhao 1999). Any foreground or background population must be physically associated with the LMC, and share its kinematics. Sub-populations within the LMC with subtly different kinematics have been suggested (Graff et al. 2000), but only at low statistical significance.

Other arguments for stars at large distances from the LMC disk also have not been convincing. Kunkel et al. (1997b) argued on the basis of carbon star velocities that the LMC has a polar ring. However, the carbon star velocity field shown in Figure 7 seems to be well fit by a single rotation disk model. It is possible that the Kunkel et al. study was affected by the use of an LMC transverse velocity value of only 240 $km\,s^{-1}$, which is considerably below the value indicated by presently available proper motion data (406 ± 44 $km\,s^{-1}$; see Section 4). Weinberg & Nikolaev (2001) found a tail of relatively faint stars in luminosity functions of AGB stars selected by $J - K$ color from 2MASS data. They suggested that this is not due to dust extinction but might indicate stars behind the LMC. On the other hand, the distribution of these stars on the sky (van der Marel, unpublished) bears a strong resemblance to the far infrared IRAS map of LMC dust emission (Schwering 1989). This casts doubt on the interpretation that the brightnesses of these stars have not been affected by dust. Weinberg & Nikolaev (2001) also found a slightly non-Gaussian tail in their AGB star luminosity functions towards brighter magnitudes. However, this need not indicate stars in the foreground, given that there is no *a priori* reason why the AGB star luminosity function would have to be Gaussian. In another study, Alcock et al. (1997) found no evidence for unexpected numbers of RR Lyrae stars at distances beyond ~ 15 kpc from the LMC disk plane.

8.4. *Non-planar structure in the LMC disk*

The studies discussed in Sections 5 and 6 indicate that the overall structure of the LMC is that of a (thick) disk. From Section 8.3 it follows that there is no strong evidence for unexpected material far from the disk plane. However, this does not mean that there may not be non-planar structures in the disk itself. For example, it was already discussed in Section 6 that there might be warps and twists in the disk plane.

Early evidence for non-planar material came from H I observations. One of the most prominent optical features of the LMC is the star forming 30 Doradus complex, located just north of the eastern tip of the bar. This region is a very strong source of UV radiation,

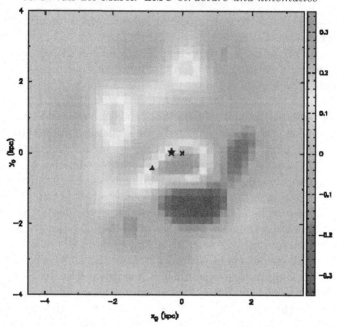

FIGURE 10. Map of the average vertical distances of Cepheids (in kpc) from a best-fitting plane solution as function of in-plane coordinates $(x., y.)$ from Nikolaev et al. (2004). The orientation of the bar in this representation is similar as in Figure 2. Negative (positive) distances denote material behind (in front of) the fitted plane. The cross indicates the H I rotation center according to Kim et al. (1998), the star shows the geometric center of the Cepheid sample, and the triangle gives the center of the outer carbon star isophotes from van der Marel & Cioni (2001). A color version of the image is available in the Nikolaev et al. paper.

yet the H I gas disk of the LMC does not show a void in this part of the sky. This indicates that the 30 Doradus complex cannot be in the plane of the LMC disk, but must be at least 250–400 kpc away from it (Luks & Rohlfs 1992). H I channel maps show that there is a separate H I component, called the "L-component," that is distinct from the main LMC disk. It has lower line-of-sight velocities than the main disk by 20–30 km s^{-1}, contains some 19% of the H I gas in the LMC, and does not extend beyond 2°–3° from the LMC center (Luks & Rohlfs 1992). Absorption studies indicate that this component is behind the LMC disk (Dickey et al. 1994). The 30 Doradus complex is spatially located at the center of the L-component. It is probably directly associated with it, given that the L-component shows a hole of H I emission at the 30 Doradus position as expected from ionization. These results suggest that in the central few degrees of the LMC an important fraction of the gas and stars may not reside in the main disk.

    It has now proven possible to test some of these ideas more directly by using large stellar databases. Section 6 already mentioned the study of more than 2000 Cepheids by Nikolaev et al. (2004). They obtained accurate reddening-corrected (relative) distances to each of the Cepheids. The distance residuals with respect to the best-fitting plane do not follow a random Gaussian distribution. Instead they show considerable structure, as shown in Figure 10. The two-dimensional distribution of the residuals on the sky was interpreted as a result of two effects, namely a symmetric warp in the disk, and the fact that the bar is located ~ 0.5 kpc in front of the main disk. Interestingly, an offset between the LMC bar and disk had previously been suggested by Zhao & Evans (2000) as an explanation for the observed LMC microlensing optical depth. If the disk and the bar of the LMC are not dynamically connected, then this may also explain why in projection the LMC bar appears offset from the center of the outer isophotes.

The variation of the reddening-corrected Red Clump magnitude over the face of the LMC has also been used to study vertical structures in the LMC disk. Olsen & Salyk observed 50 randomly selected fields in the central 4° of the LMC at CTIO. They found that fields between 2°–4° south-west of the LMC center are 0.1 mag brighter than expected from the best plane fit. They interpreted this to indicate that stars in these fields lie some 2 kpc above the LMC disk, and argued for warps and twists in the LMC disk plane. Subramaniam (2003) used Red Clump magnitudes determined from stars in the OGLE database to address the same issue. This yielded a map of the Red Clump magnitude along the length and width of the LMC bar. This map shows considerable structure and clearly cannot be fit as a single plane. Subramaniam (2004) suggested that the residuals can be interpreted as the result of a misaligned secondary bar inside the primary bar.

Eclipsing binaries have also provided interesting information on this subject. These binaries can be modeled in detail to yield a fairly accurate distance. The distances of sources studied so far seem to indicate a slightly lower LMC distance modulus of $m - M \approx 18.4$ (Ribas et al. 2002) than has been inferred from other tracers (see Section 4). However, it is possible to obtain somewhat higher values with a slightly different analysis (Groenewegen & Salaris 2001; Clausen et al. 2003) so this is no great cause for concern (Alves 2004b). What is interesting though is that four eclipsing binaries analyzed by the same team with the same method show a considerable spread in distance. This has been interpreted to mean than one of the binaries lies ~ 3 kpc behind the LMC disk plane (Ribas et al. 2002) and that another one lies ~ 4 kpc in front of it (Fitzpatrick et al. 2003).

The above studies indicate that there is considerable and complicated vertical structure in the central few degrees of the LMC disk. However, this is a rapidly developing field, and many important questions remain open. In particular, it is unclear whether the features reported in the various studies are actually the same or not. Qualitative comparison is not straightforward. Different authors study different areas of the LMC, they plot residuals with respect to different planes, and they present their results in figures that plot different types of quantities. However, cursory inspection of the various papers shows very few features that are obviously in common between the studies. Quantitative comparison is therefore needed. Direct comparison to the results for the outer parts of the LMC, where there is little evidence for extra-planar structures (van der Marel & Cioni 2001), is also important. If discrepancies emerge from such comparisons, then this can mean two things. Either some studies are in error (e.g., due to use of inaccurate dust corrections, or by incorrect interpretation of stellar evolutionary variations as distance variations) or different tracers do not trace the same structure. The latter might well the case. Cepheids are young stars with ages less than a few times 10^8 years whereas stars on the RGB and AGB are typically older than one Gyr. Since the structure of the LMC and its gaseous component vary with time as a result of tidal interactions with the Milky Way and the SMC, one wouldn't necessarily expect stars formed at different epochs to trace identical structures.

Another open question is what the physical and dynamical interpretation is of the extra-planar structures that are being detected. Many authors have used the term "warp." However, the residuals that have been reported do not much resemble the smoothly varying residuals that are expected from a single plane with a global warp. The structures appear to be both different and more complicated than a single warp. If a decomposition into different components is attempted, then such a decomposition must use entities that can be understood dynamically (disk, bar, bulge, warp, etc.) and that make sense in the context of the overall evolutionary history of the Magellanic System. Components with holes or sharp edges (e.g., Subramaniam 2004) are not particularly realistic from a dynamical viewpoint. Phase mixing of stellar orbits quickly removes such sharp discontinuities. Also, if the vertical structures detected in the inner region of the LMC disk are due to components that are not connected to the main disk plane, then why does the projected image of the LMC (e.g., Figure 2) look so smooth? And why is there so little evidence from stellar kinematics for components with decoupled kinematics (e.g., Figure 7; Zhao et al. 2003)? Clearly, many questions remain to be answered before we can come to a full understanding of the vertical structure of the LMC.

9. Concluding remarks

The structure and kinematics of the LMC continue to be active areas of research. As outlined in this review, much progress has been made recently. Improved datasets have played a key role in this, most notably the advent of large stellar datasets of magnitudes in many bands, lightcurves, and line-of-sight kinematics, and also the availability of sensitive H I observations over large areas. As a result we now have a fairly good understanding of the LMC morphology and kinematics. The proper motion of the LMC is reasonably well measured and the global properties of the LMC orbit around the Milky Way are understood. The angles that determine how we view the LMC are now known much more accurately than before and this has led to the realization that the LMC is quite elliptical in its disk plane. We are starting to delineate the vertical structure of the LMC and are finding complexities that were not previously expected.

Despite the excellent progress, many questions on LMC structure still remain open. Why is the bar offset from the center of the outer isophotes of the LMC? Why is the dynamical center of the H I offset from the center of the bar, from the center of the outer isophotes, and from the dynamical center of the carbon stars? Why do studies of the inner and outer regions of the LMC yield differences in line-of-nodes position angle of up to 30°? Does the LMC have a pressure supported halo? Are there populations of stars at large distances from the LMC plane? What is the origin and dynamical nature of the non-planar structures detected in the inner regions of the LMC? Do different tracers outline the same non-planar structures?

It might be necessary to answer all of these open questions before we can convince ourselves that the optical depth for LMC self-lensing has been correctly estimated. This seems to be the most critical step in establishing whether or not the Milky Way halo contains hitherto unknown compact lensing objects (MACHOs). The open questions about LMC structure are important also in their own right. The tidal interaction between the Magellanic Clouds and the Milky Way provides one of our best laboratories for studying the processes of tidal disruption and hierarchical merging by which all galaxies are believed to grow. A better understanding of LMC structure may also provide new insight into the origin of the Magellanic Stream, which continues to be debated. And with improved proper motion measurements of the Magellanic Clouds, the Stream may become a unique tool to constrain the shape and radial density distribution of the Milky Way halo at radii inaccessible using other tracers.

REFERENCES

ALCOCK, C., ET AL. 1997 *ApJ* **490**, L59.

ALCOCK, C., ET AL. 2000a *ApJ* **542**, 281.

ALCOCK, C., ET AL. 2000b *AJ* **119**, 2194.

ALCOCK, C., ET AL. 2001 *ApJ* **552**, 582.

ALVAREZ, H., APARICI, J., & MAY, J. 1987 *A&A* **176**, 25.

ALVES, D. R. 2004a *ApJ* **601**, L151.

ALVES, D. R. 2004b in *Highlights of Astronomy*, Vol. 13, (ed. O. Engvold). Elsevier, in press; astro-ph/0310673.

ALVES, D. R. & NELSON, C. A. 2000 *ApJ* **542**, 789.

ANDERSEN, D. R., BERSHADY, M. A., SPARKE, L. S., GALLAGHER III, J. S., & WILCOTS, E. M. 2001 *ApJ* **551**, L131.

ANGUITA, C., LOYOLA, P., & PEDREROS, M. H. 2000 *AJ* **120**, 845.

BEAULIEU, J.-P. & SACKETT, P. D. 1998 *AJ* **116**, 209.

BENNETT, D. P. 1998 *ApJ* **493**, L79.

BESSELL, M. S., FREEMAN, K. C., & WOOD, P. R. 1986 *ApJ* **310**, 710.

BINNEY, J. J. & DE VAUCOULEURS, G. 1981 *MNRAS* **194**, 679.

BOTHUN, G. D. & THOMPSON, I. B. 1988 *AJ* **96**, 877.

CALDWELL, J. A. R. & COULSON, I. M. 1986 *MNRAS* **218**, 223.

CIONI, M. R., ET AL. 2000a *A&AS* **144**, 235.

CIONI, M.-R. L., HABING, H. J., & ISRAEL, F. P. 2000b *A&A* **358**, L9.

CLAUSEN, J. V., STORM, J., LARSEN, S. S., & GIMENEZ, A. 2003 *A&A* **402**, 509.

CONNERS, T. W., KAWATA, D., MADDISON, S. T., & GIBSON, B. K. 2004, *PASA*, in press; astro-ph/0402187.

CROWL, H. H., SARAJEDINI, A., PIATTI, A. E., GEISLER, D., BICA, E., CLARIA, J. J., & SANTOS, J. F. C., JR. 2001 *AJ* **122**, 220.

DELMOTTE, N., LOUP, C., EGRET, D., CIONI, M.-R., & PIERFEDERICI, F. 2002 *A&A* **396**, 143.

DE VAUCOULEURS, G. & FREEMAN, K. C. 1973 *Vistas Astron.* **14**, 163.

DICKEY, J. M., MEBOLD, U., MARX, M., AMY, S., HAYNES, R. F., & WILSON, W. 1994 *A&A* **289**, 357.

DRAKE, A., ET AL. 2002. In 199th *Meeting of the American Astronomical Society*, 52.05.

DUBINSKI, J. & CARLBERG, R. G. 1991 *ApJ* **378**, 496.

EPCHTEIN, N., ET AL. 1997 *Messenger* **87**, 27.

EVANS, N. W. & KERINS, E. 2000 *ApJ* **529**, 917.

FEITZINGER, J. V., SCHMIDT-KALER, T., & ISSERSTEDT, J. 1977 *A&A* **57**, 265.

FIELDS, B., FREESE, K., & GRAFF, D. S. 2000 *ApJ* **534**, 265.

FITZPATRICK, E. L., RIBAS, I., GUINAN, E. F., MALONEY, F. P., & CLARET, A. 2003 *ApJ* **587**, 685.

FLYNN, C., HOLOPAINEN, J., & HOLMBERG, J. 2003 *MNRAS* **339**, 817.

FRANX, M. & DE ZEEUW, P. T. 1992 *ApJ* **392**, L47.

FREEDMAN, W. L., ET AL. 2001 *ApJ* **553**, 47.

FREEMAN, K. C., ILLINGWORTH, G., & OEMLER, A. 1983 *ApJ* **272**, 488.

GALLART, C. 1998 *ApJ* **495**, L43.

GARDINER, L. T. & NOGUCHI, M. 1996 *MNRAS* **278**, 191.

GARDINER, L. T., SAWA, T., & FUJIMOTO, M. 1994 *MNRAS* **266**, 567.

GIBSON, B. K. 2000 *Mem. Soc. Astron. Italiana* **71**, 693.

GOULD, A. 1995 *ApJ* **441**, 77.

GOULD, A. 1998 *ApJ* **499**, 728.

GOULD, A. 1999 *ApJ* **525**, 734.

GRAFF, D. S., GOULD, A. P., SUNTZEFF, N. B., SCHOMMER, R. A., & HARDY, E. 2000 *ApJ* **540**, 211.

GROENEWEGEN, M. A. T. & SALARIS, M. 2001 *A&A* **366**, 752.

GUHATHAKURTA, P. & REITZEL, D. B. 1998. In *Galactic Halos: A UC Santa Cruz Workshop* (ed. D. Zaritsky). ASP Conference Series 136, p. 22. ASP.

GYUK, G., DALAL, N., & GRIEST, K. 2000 *ApJ* **535**, 90.

HATZIDIMITRIOU, D., CROKE, B. F., MORGAN, D. H., & CANNON, R. D. 1997 *A&AS* **122**, 507.

HELLER, P. & ROHLFS, K. 1994 *A&A* **291**, 743.

IBATA, R. A., LEWIS, G. F., & BEAULIEU, J.-P. 1998 *ApJ* **509**, L29.

IBATA, R., LEWIS, G. F., IRWIN, M., TOTTEN, E., & QUINN, T. 2001 *ApJ* **551**, 294.

IRWIN, M. J. 1991. In *The Magellanic Clouds* (eds. R. Haynes, & D. Milne). Proceedings IAU Symposium 148, p. 453. Kluwer.

JETZER, P., MANCINI, L., & SCARPETTA, G. 2002 *A&A* **393**, 129.

JONES, B. F., KLEMOLA, A. R., & LIN, D. N. C. 1994 *AJ* **107**, 1333.

KIM, S., STAVELEY-SMITH, L., DOPITA, M. A., FREEMAN, K. C., SAULT, R. J., KESTEVEN M. J., & McCONNELL, D. 1998 *ApJ* **503**, 674.

KIM, S., STAVELEY-SMITH, L., DOPITA, M. A., SAULT, R. J., FREEMAN, K. C., LEE, Y., & CHU, Y.-H. 2003 *ApJS* **148**, 473.

KOCHANEK, C. S. 1996 *ApJ* **457**, 228.

KONTIZAS, E., DAPERGOLAS, A., MORGAN, D. H., & KONTIZAS, M. 2001 *A&A* **369**, 932.

KONTIZAS, M., MORGAN, D. H., HATZIDIMITRIOU, D., & KONTIZAS, E. 1990 *A&AS* **84**, 527.

KORNREICH, D. A., HAYNES, M. P., & LOVELACE, R. V. E. 1998 *AJ* **116**, 2154.

KROUPA, P. & BASTIAN, U. 1997 *New Astronomy* **2**, 77.

KROUPA, P., RÖSER, S., & BASTIAN, U. 1994 *MNRAS* **266**, 412.

KUNKEL, W. E., IRWIN, M. J., & DEMERS, S. 1997a *A&AS* **122**, 463.

KUNKEL, W. E., DEMERS, S., & IRWIN, M. J. 2000 *AJ* **119**, 2789.

KUNKEL, W. E., DEMERS, S., IRWIN, M. J., & ALBERT, L. 1997b *ApJ* **488**, L129.

LAMBAS, D. G., MADDOX, S. J., & LOVEDAY, J. 1992 *MNRAS* **258**, 404.

LANEY, C. D. & STOBIE, R. S. 1986 *MNRAS* **222**, 449.

LIN, D. C. N., JONES, B. F., & KLEMOLA, A. R. 1995 *ApJ* **439**, 652.

LIN, D. N. C. & LYNDEN-BELL, D. 1982 *MNRAS* **198**, 707.

LIU, Y.-Z. 1992 *A&A* **257**, 505.

LU, L., SARGENT, W. L. W., SAVAGE, B. D., WAKKER, B. P., SEMBACH, K. R., & OOSTER-LOO, T. A. 1998 *AJ* **115**, 162.

LUKS, TH. & ROHLFS, K. 1992 *A&A* **263**, 41.

LYNGA, G. & WESTERLUND, B. E. 1963 *MNRAS* **127**, 31.

MAJEWSKI, S. R., SKRUTSKIE, M. F., WEINBERG, M. D., & OSTHEIMER, J. C. 2003 *ApJ* **599**, 1082.

MARAGOUDAKI, F., KONTIZAS, M., MORGAN, D. H., KONTIZAS, E., DAPERGOLAS, A., & LIVANOU, E. 2001 *A&A* **379**, 864.

MARIGO, P., GIRARDI, L., & CHIOSI, C. 2003 *A&A* **403**, 225.

MASTROPIETRO, C., MOORE, B., MAYER, L., & STADEL, J. 2004. In *Satellites and Tidal Streams* (eds. F. Prada, D. Martinez-Delgado, & T. Mahoney). ASP Conf. Series, in press; astro-ph/0309244.

McGEE, R. X. & MILTON, J. A. 1966 *Austr. Journal of Physics* **19**, 343.

MEATHERINGHAM, S. J., DOPITA, M. A., FORD, H. C., & WEBSTER, B. L. 1988 *ApJ* **327**, 651.

MINNITI, D., BORISSOVA, J., REJKUBA, M., ALVES, D. R., COOK, K. H., FREEMAN, K. C. 2003 *Science* **301**, 1508.

MOORE, B. & DAVIS, M. 1994 *MNRAS* **270**, 209.

MURAI, T. & FUJIMOTO, M. 1980 *PASJ* **32**, 581.

NIKOLAEV, S., DRAKE, A. J., KELLER, S. C., COOK, K. H., DALAL, N., GRIEST, K., WELCH, D. L., & KANBUR, S. M. 2004 *ApJ* **601**, 260.

NIKOLAEV, S. & WEINBERG, M. D. 2000 *ApJ* **542**, 804.

OLSEN, K. A. G. & SALYK, C. 2002 *AJ* **124**, 2045.

PEDREROS, M. H., ANGUITA, C., & MAZA, J. 2002 *AJ* **123**, 1971.

PUTMAN, M. E., ET AL. 1998 *Nature* **394**, 752.

PUTMAN, M. E., STAVELEY-SMITH, L., FREEMAN, K. C., GIBSON, B. K., & BARNES, D. G. 2003 *ApJ* **586**, 170.

RIBAS, I., FITZPATRICK, E. L., MALONEY, F. P., GUINAN, E. F., & UDALSKI, A. 2002 *ApJ* **574**, 771.

RIX, H.-W. & ZARITSKY, D. 1995 *ApJ* **447**, 82.

ROHLFS, K., KREITSCHMANN, J., SIEGMAN, B. C., & FEITZINGER, J. V. 1984 *A&A* **137**, 343.

SAHU, K. C. 2003. In *The Dark Universe: Matter, Energy, and Gravity* (ed. M.Livio). Cambridge University Press.

SALATI, P., TAILLET, R., AUBOURG, E., PALANQUE-DELABROUILLE, N., & SPIRO, M. 1999 *A&A* **350**, L57.

SCHMIDT-KALER, T. & GOCHERMANN, J. 1992. In *Variable Stars and Galaxies* (ed. B. Warner). ASP Conf. Series Vol. 30, p. 203.

SCHOENMAKERS, R. H. M., FRANX, M., & DE ZEEUW, P. T. 1997 *MNRAS* **292**, 349.

SCHOMMER, R. A., SUNTZEFF, N. B., OLSZEWSKI, E. W., & HARRIS, H. C. 1992 *AJ* **103**, 447.

SCHWERING, P. B. W. 1989 *A&AS* **99**, 105.

SHUTER, W. L. H. 1992 *ApJ* **386**, 101.

SKRUTSKIE, M. 1998. In *The Impact of Near-Infrared Sky Surveys on Galactic and Extragalactic Astronomy, Proc. of the 3rd Euroconference on Near-Infrared Surveys* (ed. N. Epchtein) Astrophysics and Space Science Library, Vol. 230, p. 11. Kluwer.

STANIMIROVIC, S., STAVELEY-SMITH, L., DICKEY, J. M., SAULT, R. J., & SNOWDEN, S. L. 1999 *MNRAS* **302**, 417.

STANIMIROVIC, S., STAVELEY-SMITH, L., & JONES, P. 2004 *ApJ* **604**, 176.

STAVELEY-SMITH, L., KIM, S., CALABRETTA, M. R., HAYNES, R. F., & KESTEVEN, M. J. 2003 *MNRAS* **339**, 87.

SUBRAMANIAM, A. 2003 *ApJ* **598**, L19.

SUBRAMANIAM, A. 2004 *ApJ* **604**, L41.

TEUBEN, P. 1987 *MNRAS* **227**, 815.

UDALSKI, A., SZYMANSKI, M., KUBIAK, M., PIETRZYNSKI, G., SOSZYNSKI, I., WOZNIAK, P., & ZEBRUN, K. 2000 *Acta Astronomica* **50**, 307.

UDALSKI, A., SZYMANSKI, M., KUBIAK, M., PIETRZYNSKI, G., WOZNIAK, P., & ZEBRUN, K. 1998 *Acta Astronomica* **48**, 147.

VAN DEN BERGH, S. 2000. *The Galaxies of the Local Group*. Cambridge University Press.

VAN DER MAREL, R. P. 2001 *AJ* **122**, 1827.

VAN DER MAREL, R. P., ALVES, D. R., HARDY, E., & SUNTZEFF, N. B. 2002 *AJ* **124**, 2639.

VAN DER MAREL, R. P. & CIONI, M.-R. 2001 *AJ* **122**, 1807.

WEINBERG, M. D. 2000 *ApJ* **532**, 922.

WEINBERG, M. D. & NIKOLAEV, S. 2001 *ApJ* **548**, 712.

WELCH, D. L., McLAREN, R. A., MADORE, B. F., & McALARY, C. W. 1987 *ApJ* **321**, 162.

WESTERLUND, B. E. 1997. *The Magellanic Clouds*. Cambridge University Press.

WILKINSON, M. I. & EVANS, N. W. 1999 *MNRAS* **310**, 645.

YOSHIZAWA, A. M. & NOGUCHI, M. 2003 *MNRAS* **339**, 1135.

ZARITSKY, D. 1999 *AJ* **118**, 2824.

ZARITSKY, D., HARRIS, J., GREBEL, E. K., & THOMPSON, I. B. 2000 *ApJ* **534**, L53.

ZARITSKY, D., HARRIS, J., & THOMPSON, I. 1997 *AJ* **114**, 1002.

ZARITSKY, D., HARRIS, J., THOMPSON, I. B., GREBEL, E. K., & MASSEY, P. 2002 *AJ* **123**, 855.

ZARITSKY, D. & LIN, D. N. C. 1997 *AJ* **114**, 2545.

ZARITSKY, D., SHECTMAN, S. A., THOMPSON, I., HARRIS, J., & LIN, D. N. C. 1999 *AJ* **117**, 2268.

ZHAO, H. S. 1998 *MNRAS* **294**, 139.

ZHAO, H. S. 1999 *ApJ* **527**, 167.

ZHAO, H. S. 2000 *ApJ* **530**, 299.

ZHAO, H. S. & EVANS, N. W. 2000 *ApJ* **545**, L35.

ZHAO, H. S., GRAFF, D. S., & GUHATHAKURTA, P. 2000 *ApJ* **532**, L37.

ZHAO, H., IBATA, R. A., LEWIS, G. F., & IRWIN, M. J. 2003 *MNRAS* **339**, 701.

The Local Group as an astrophysical laboratory for massive star feedback

By M. S. OEY

Lowell Observatory, 1400 W. Mars Hill Road, Flagstaff, AZ 86001, USA

The feedback effects of massive stars on their galactic and intergalactic environments can dominate evolutionary processes in galaxies and affect cosmic structure in the Universe. Only the Local Group offers the spatial resolution to quantitatively study feedback processes on a variety of scales. Lyman continuum radiation from hot, luminous stars ionizes H II regions and is believed to dominate production of the warm component of the interstellar medium (ISM). Some of this radiation apparently escapes from galaxies into the intergalactic environment. Supernovae and strong stellar winds generate shell structures such as supernova remnants, stellar wind bubbles, and superbubbles around OB associations. Hot (10^6 K) gas is generated within these shells, and is believed to be the origin of the hot component of the ISM. Superbubble activity thus is likely to dominate the ISM structure, kinematics, and phase balance in star-forming galaxies. Galactic superwinds in starburst galaxies enable the escape of mass, ionizing radiation, and heavy elements. Although many important issues remain to be resolved, there is little doubt that feedback processes plays a fundamental role in energy cycles on scales ranging from individual stars to cosmic structure. This contribution reviews studies of radiative and mechanical feedback in the Local Group.

1. Introduction

The Local Group is especially suited as a laboratory for studying the effects of the massive star population on the galactic environment. There are three types of massive star feedback:

(a) Radiative feedback, i.e., ionizing emission, which results in photoionized nebulae and diffuse, warm (10^4 K) ionized gas;

(b) Mechanical feedback, predominantly from supernovae (SNe), resulting in supernova remnants (SNRs), superbubbles, and galactic superwinds; and

(c) Chemical feedback from nucleosynthesis in SNe and massive star evolution, which drives galactic chemical evolution.

Since the last will be reviewed by Don Garnett and Monica Tosi in this volume, I will address here only the radiative and mechanical feedback processes.

Only in the Local Group can we spatially resolve the various physical parameters that determine, and result from, the interaction of massive stars with their immediate environment. Radiative and mechanical feedback return large quantities of energy to the galactic environment, both with luminosities $\log L$ of order 36–41 erg s^{-1} for typical OB associations. These processes may dominate the balance between different temperature phases of the interstellar medium (ISM), profoundly affecting the structure and kinematics of the ISM, star formation, and other evolutionary processes in star-forming galaxies.

2. Radiative feedback

Photoionization results in H II regions and ionized gas, which are especially familiar and photogenic in the Milky Way and Magellanic Clouds.

2.1. *Nebular emission-line diagnostics*

The nebular emission-line spectra offer vital diagnostics of conditions in these regions, and are especially powerful if modeled with tailored photoionization models. However, a class of "semi-empirical" line diagnostics are widely used to probe nebular parameters. For example, the parameter (Vílchez & Pagel 1988)

$$\eta' \equiv \frac{[\text{O \textsc{ii}}]\lambda3727/[\text{O \textsc{iii}}]\lambda\lambda4959, 5007}{[\text{S \textsc{ii}}]\lambda6724/[\text{S \textsc{iii}}]\lambda\lambda9069, 9532} \tag{2.1}$$

and [Ne \textsc{iii}]/Hβ (Oey et al. 2000) probe the ionizing stellar effective temperature, while the parameters (Pagel et al. 1979)

$$R23 \equiv \left([\text{O \textsc{ii}}]\lambda3727 + [\text{O \textsc{iii}}]\lambda\lambda4959, 5007\right)/\text{H}\beta \tag{2.2}$$

and (Vílchez & Esteban 1996; Christensen et al. 1997)

$$S23 \equiv \left([\text{S \textsc{ii}}]\lambda6724 + [\text{S \textsc{iii}}]\lambda\lambda9069, 9532\right)/\text{H}\beta \tag{2.3}$$

estimate the oxygen and sulfur abundances. Using observations of LMC H \textsc{ii} regions, Oey & Shields (2000) question the reliability of $S23$ and show that

$$S234 \equiv \left([\text{S \textsc{ii}}]\lambda6724 + [\text{S \textsc{iii}}]\lambda\lambda9069, 9532 + [\text{S \textsc{iv}}]10.5\mu\right)/\text{H}\beta \quad , \tag{2.4}$$

a superior diagnostic of sulfur abundance, can be easily estimated from optical line strengths. All of these semi-empirical diagnostics must first be similarly tested and calibrated using objects with independently constrained parameters. The Local Group offers by far the best nebular samples in which to simultaneously: obtain spectral classifications of the ionizing stars; evaluate the gas morphology relative to the ionizing stars, thereby constraining the nebular ionization parameter; and estimate the abundances. These are the three primary parameters that determine the nebular emission. Thus, having empirical constraints on photoionization models, the behavior of the emission-line diagnostics can be calibrated. The nebular emission can also be used to constrain the hot stellar atmosphere models themselves, since these NLTE, expanding atmospheres are complicated and difficult to model. Such studies have been carried out for O and WR stars in the Magellanic Clouds and the Galaxy (Oey et al. 2000; Kennicutt et al. 2000; Crowther et al. 1999).

Another important test is to evaluate the degree to which H \textsc{ii} regions are indeed radiation-bounded, as is normally assumed; the escape of ionizing radiation is a critical question for the ionization of the diffuse, warm ionized medium (see below), as well as the use of H recombination emission as a star formation tracer. The escape of ionizing radiation from the host galaxies themselves is also vital to understanding the ionization state of the intergalactic medium (IGM) and the reionization of the early Universe. We can test whether H \textsc{ii} regions are radiation-bounded by simply comparing the observed stellar spectral types, and thus inferred ionizing flux, with the observed nebular emission. Currently, it appears that while most nebulae are radiation-bounded, a large subset apparently are density-bounded (Oey & Kennicutt 1997; Hunter & Massey 1990).

With adequate calibrations of nebular diagnostics against Local Group objects, we can infer various physical parameters for distant, unresolved star-forming regions.

2.2. *Statistical properties of H\textsc{ii} regions*

Statistical properties of H \textsc{ii} region populations offer important quantitative characterizations of global star formation in galaxies. The H \textsc{ii} region luminosity function (H \textsc{ii} LF) reveals the relative importance of major star-forming events, hosting super star clusters,

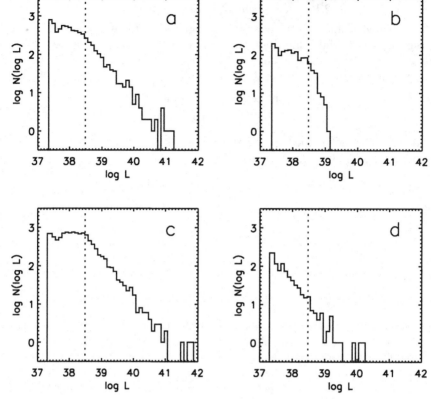

FIGURE 1. Monte Carlo models of the H II LF from Oey & Clarke (1998a): (*a*) constant nebular creation rate for a full power law in N_* given by equation 2.6; (*b*) continuous creation with an upper cutoff in N_* of 10 ionizing stars; (*c*) zero-age instantaneous burst for all objects, with a full power law in N_*; (*d*) the evolved burst in (*c*) after 7 Myr. The maximum stellar ionizing luminosity in the IMF here corresponds to log $\mathcal{L} = 38.5$ (vertical dotted line).

and smaller, ordinary OB associations. The H II LF has been determined for many nearby galaxies, including all of the star-forming galaxies in the Local Group (Milky Way: Smith & Kennicutt 1989, McKee & Williams 1997; Magellanic Clouds, M31, M33: Kennicutt et al. 1989, Walterbos & Braun 1992, Hodge et al. 1999; IC 10, Leo A, Sex A, Sex B, GR8, Peg, WLM: Youngblood & Hunter 1999). There is agreement that the H II LF universally appears to be described by a power law:

$$N(\mathcal{L}) \, d\mathcal{L} \propto \mathcal{L}^{-a} \, d\mathcal{L} \quad , \tag{2.5}$$

with a power-law index $a \sim 2$ for the differential LF. Figure 1 presents Monte Carlo models by Oey & Clarke (1998a) that show the existence of a flatter slope below log $\mathcal{L} \sim 37.5$–38.5, owing to a transition at low luminosity to objects dominated by small number statistics in the ionizing stellar population. We also see that the H II LF can offer some insights on the nature and history of the very most recent global star formation, within the last $\sim$10 Myr.

The observed behavior of the H II LF is consistent with the existence of a universal power law for the number of ionizing stars N_* per cluster (Oey & Clarke 1998a):

$$N(N_*) \, dN_* \propto N_*^{-2} \, dN_* \quad . \tag{2.6}$$

This is consistent with direct observations of the cluster luminosity and mass functions in a variety of regimes (see Chandar, this volume; Elmegreen & Efremov 1997; Meurer et al.

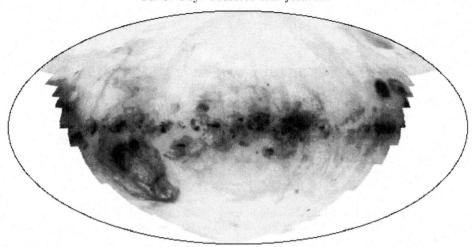

FIGURE 2. WHAM Hα survey of the Milky Way warm ionized medium, centered at $l = 120°$ (Reynolds 1998; http://www.astro.wisc.edu/wham/).

1995; Harris & Pudritz 1994). Such a universal power-law for the cluster mass function is fundamental, similar to the stellar initial mass function (IMF; e.g., Oey & Muñoz-Tuñon 2003).

The nebular size distribution has also been determined in many galaxies, although it has been studied in less detail than the H II LF (Milky Way, Magellanic Clouds, M31, M33, NGC 6822: van den Bergh 1981, Hodge et al. 1999; IC 10, Leo A, Sex A, Sex B, GR8, Peg, WLM: Hodge 1983, Youngblood & Hunter 1999). In his pioneering work, van den Bergh (1981) described the size distribution as an exponential:

$$N_R \propto e^{-R/R_0} \quad . \tag{2.7}$$

However, this relation is difficult to reconcile with the power-law form of the H II LF. To first order, the Hα luminosity $\mathcal{L}$ should scale with the volume emission as R^3, and thus the size distribution should have a similar power-law form to that of the H II LF, but with exponent $b = 2 - 3a$. We can see in Figure 1a and c that the existence of the turnover in the H II LF described above can cause the entire H II LF to mimic an exponential form. Thus we suggest that the intrinsic form of the size distribution is also a power-law relation:

$$N(R) \, dR \propto R^{-b} \, dR \quad . \tag{2.8}$$

With a slope also flattening below a value of $\log R \sim 130$ pc, corresponding to the transition in the H II LF above, this form of the size distribution is also a good description of the available data (Oey et al. 2003). Our initial investigation shows good agreement between observations and the predicted value for $b = 4$, implied by the H II LF slope $a = 2$.

2.3. *Warm ionized medium*

Another global effect of radiative feedback is the diffuse, warm ionized medium (WIM). This 10^4 K component of the ISM contributes $\sim$40% of the total Hα luminosity in star-forming galaxies for a wide variety of Hubble types (e.g., Walterbos 1998). In the Galaxy, the WIM has a scale height of $\sim$1 kpc, temperature of $\sim$8000 K, and mean density $\sim$0.025 cm^{-3} (Minter & Balser 1997). While it has long been thought that massive stars dominate its ionization (e.g., Frail et al. 1991; Reynolds & Tufte 1995), contributions

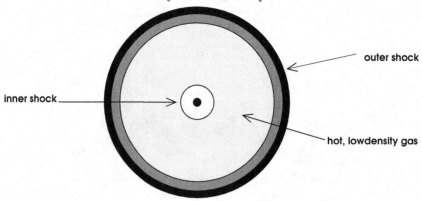

FIGURE 3. The adiabatic wind-driven bubble model (see text). The shell at the outer shock consists of swept-up ISM on the outside and shocked wind material on the inside, separated by a contact discontinuity.

from other processes also appear to be necessary. Dissipation of turbulence (Minter & Spangler 1997; Minter & Balser 1997) and photoelectric heating (Reynolds & Cox 1992) are among the suggested heating candidates in our Galaxy.

The WIM is most often studied through optical nebular emission. For the Galaxy, the largest optical survey is from the Wisconsin Hα Mapper (WHAM) project (Reynolds et al. 1998; Figure 2). In addition to Hα, the WHAM Fabry-Perot data also include observations of [S II]λ6717, [N II]$\lambda\lambda$6583, 5755, [O III]λ5007, He Iλ5876, and other nebular emission lines. The other disk galaxies in the Local Group have also been studied optically: the LMC (Kennicutt et al. 1995), M31 (Galarza et al. 2000; Greenawalt et al. 1997; Walterbos & Braun 1992, 1994), M33 (Hoopes & Walterbos 2000), and NGC 55 (Otte & Dettmar 1999; Ferguson et al. 1996).

Other techniques, notably at radio wavelengths, are available for studying the WIM in the Milky Way. These offer additional probes of the WIM distribution and filling factor. Heiles et al. (1998) observed radio recombination lines in the Galaxy, and Frail et al. (1991) examined lines of sight through the WIM via pulsar dispersion measures. Faraday rotation obtained through radio polarimetry has been exploited by e.g., Uyaniker et al. (2003), Gray et al. (1999), and Minter & Spangler (1996); this technique is also used by Berkhuijsen et al. (2003) for M31.

3. Mechanical feedback

Supernova (SN) explosions of massive stars are a dominant source of kinetic energy in the interstellar medium. Strong, supersonic stellar winds are also an important source in the case of the extreme most massive stars ($\gtrsim 40\ M_\odot$). Mechanical feedback structures the ISM, with immediate consequences corresponding to supernova remnants (SNRs), stellar wind-driven bubbles, and superbubbles resulting from combined SNe and winds from multiple stars.

While the SNRs evolve, in the simplest description, according to the Sedov (1959) model, the wind-driven bubbles and superbubbles are thought to evolve according to a similar, Sedov-like adiabatic model for constant energy input (Pikel'ner 1968; Castor et al. 1977): the central supersonic wind drives a shock into the ambient ISM, piling up a radiatively cooled, dense shell; and a reverse shock near the source thermalizes the wind's kinetic energy, thereby generating a hot (10^6–10^7 K), low-density ($n \sim 10^{-2}$–10^{-3} cm^{-3}) medium that dominates the bubble volume (Figure 3). This heating process is believed to be the origin of the diffuse hot, ionized medium (HIM) in the interstellar medium. Assuming that the hot bubble interior remains adiabatic, the self-similar shell evolution follows the simple analytic relations,

$$R \propto (L/n)^{1/5} \, t^{3/5} \quad ,$$
$$v \propto (L/n)^{1/5} \, t^{-2/5} \quad , \qquad (3.1)$$

where R and v are the shell radius and expansion velocity, L is the input mechanical power, and t is the age. Once SNe begin to explode, they quickly dominate L, and the standard treatment is to consider the discrete SNe as a constant energy input (e.g., Mac Low & McCray 1988). Hence, we may write L in terms of the SN parameters:

$$L = N_* E_{51}/t_e \quad , \qquad (3.2)$$

where N_* is the number of SNe, E_{51} is the SN energy, and t_e is the total time during which the SNe occur.

There are several approaches to testing the standard, adiabatic shell evolution, and by extension, our understanding of mechanical feedback. In the first instance, we can examine the properties and kinematics of individual shell systems and carry out rigorous comparisons with the model predictions. Secondly, we can also examine statistical properties of entire shell populations in galaxies, and compare with model predictions. And thirdly, we can carry out spatial correlations of shells with regions of recent star formation, to confirm the existence of putative stellar progenitors. All three of these methods require high spatial resolution, and thus the Local Group offers by far the best, and often the only feasible, laboratory.

3.1. *Individual shell systems*

A number of individual superbubbles exhibit multi-phase ISM, as is qualitatively predicted by the adiabatic shell model. DEM L152 (N44) in the Large Magellanic Cloud (LMC) is a beautiful example where the nebular (10^4 K) gas in the shell clearly confines the hot, X-ray–emitting (10^6 K) gas within (Magnier et al. 1996; Figure 4). Chu et al. (1994) also confirmed the existence of C IV and Si IV absorption in the lines of sight toward all stars within LMC superbubbles. These tracers of intermediate temperature (10^5 K) gas are expected in interface regions between hot and cold gas. Quantitatively, however, the detected X-ray emission from LMC superbubbles has been an order of magnitude higher than predicted by the adiabatic model. It is therefore thought that the anomalous emission results from impacts to the shell wall by internal SNRs (Chu & Mac Low 1990; Wang & Helfand 1991), a scenario which is supported by other signatures such as enhanced [S II]/Hα and anomalous kinematics (Oey 1996; see below). Many other superbubbles have not been detected in X-rays, although the upper limits tend to be high, and remain within the model predictions. It is hoped that the capabilities of *Chandra* and *XMM* will be applied to these objects.

Since most early-type stars are found in OB associations, wind-driven bubbles of individual massive stars are rare, and consequently few quantitative studies of these objects exist. In principle, O stars offer the most straightforward test of the standard shell evolution, since their wind histories are simple and relatively well-understood. One of the few such studies was carried out by Oey & Massey (1994) on two nebular bubbles around individual late-type O stars in M33. They found crude consistency with the model predictions, but the constraints are limited by lack of kinematic information. Cappa and collaborators (e.g., Cappa & Benaglia 1999; Benaglia & Cappa 1999; Cappa & Herbstmeier 2000) have studied a number of H I shells around Of stars, which are presumably evolved O stars. They generally find a significant growth-rate discrepancy such that the

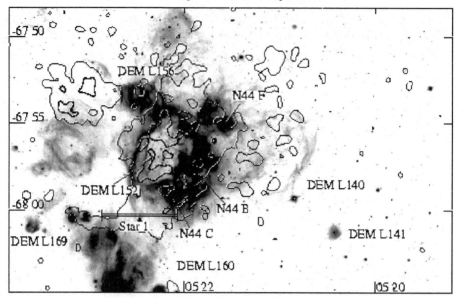

FIGURE 4. Hα image of superbubble DEM L152 (N44) in the LMC, with *ROSAT* X-ray
contours overlaid (from Magnier et al. 1996).

shells appear to be too small for the assumed stellar wind power. Wolf-Rayet (W-R) stars are well-known to have the most powerful stellar winds, and a number of W-R ring nebulae have also been examined kinematically. Optical studies include those by Treffers & Chu (1982), García-Segura & Mac Low (1995), and Drissen et al. (1995); while, e.g., Arnal et al. (1999) and Cappa et al. (2002) examine the neutral and radio continuum properties. The W-R nebulae are also apparently too small, but owing to the complicated stellar wind history and associated environment, it is more difficult to interpret the W-R shell dynamics.

Superbubbles around OB associations are much larger, brighter, and easier to identify than single star wind-driven bubbles, and consequently have been much more actively studied. They are especially prominent in the LMC, where the proximity and high galactic latitude offer a clear, detailed view of the objects. The most comprehensive study of superbubble dynamics was carried out by Oey and collaborators (Oey 1996; Oey & Smedley 1998; Oey & Massey 1995) on a total of eight LMC objects. Saken et al. (1992) and Brown et al. (1995) also examined two Galactic objects. All of these are young, nebular superbubbles having ages ≲ 5 Myr. These studies again consistently reveal a growth-rate discrepancy equivalent to an overestimate in the inferred L/n by up to an order of magnitude. About half of the objects also show anomalously high expansion velocities, implying a strong, rapid shell acceleration from the standard evolution.

A number of factors could be individually, or collectively, responsible for these dynamical discrepancies. The first possibility is a systematic overestimate in L/n: stellar wind parameters remain uncertain within factors of 2–3. The ambient density distribution is also critical to the shell evolution; as shown by Oey & Smedley (1998), a sudden drop in density can cause a "mini-blowout" of the shell, whose kinematics can reproduce those of the high-velocity LMC shells. In the case of those objects, however, the presence of anomalously high X-ray emission favors the SNR impact hypothesis discussed above. In any case, the critical role of the ambient environment motivated us to map the H I distribution around three superbubbles in the LMC sample (Oey et al. 2002). The results were surprisingly inhomogeneous, with one object essentially in a void, another with significant H I in close proximity, and a third with no correspondence at all between the nebular and neutral gas. Thus it is virtually impossible to infer the ambient H I properties for any given object without direct observations. This heterogeneity suggests that a systematic underestimate of n is not responsible for the universal growth-rate discrepancy. However, a related environmental parameter is the ambient pressure. If the ISM pressure has been systematically underestimated, then the superbubble growth would become pressure-confined at an earlier, smaller stage. We are currently exploring this possibility (Oey & García-Segura 2003, in preparation).

Finally, if the superbubble interiors are somehow cooling, then the shells will no longer grow adiabatically. While this possibility has been explored theoretically from several angles, there is as yet no empirical evidence, in particular, radiation, that the objects are cooling. Meanwhile, mass-loading has long been a candidate cooling mechanism (e.g., Cowie et al. 1981; Hartquist et al. 1986), either by evaporating material from the shell wall, or by ablating clumps that are overrun by the shell. The enhanced density would then increase the cooling rate of the hot interior. More recently, Silich et al. (2001) and Silich & Oey (2002) suggested that the metallicity increase expected from the parent SN explosions can significantly enhance the cooling and X-ray emission, especially for extremely low metallicity systems.

3.2. *Global mechanical feedback*

Another approach to understanding superbubble evolution is in examining the statistical properties of superbubble populations in galaxies. It is possible to derive the distributions in, for example, size and expansion velocity from equations 3.1 and assumptions for the mechanical luminosity function (MLF), object creation rate, and ambient parameters. Oey & Clarke (1997) analytically derived the superbubble size distribution for simple combinations of creation rate and MLF. They found that the size distribution is dominated by pressure-confined objects that are no longer growing, following a differential distribution in radius R:

$$N(R) \ dR \propto R^{1-2\beta} \ dR \quad , \tag{3.3}$$

where β is the power-law slope of the MLF $\phi(L)$ for the form, $\phi(L) \propto L^{-\beta}$. Since the mechanical power L is determined by the SN progenitors (equation 3.2), the MLF to first order has the universal power-law slope of -2 given by equation 2.6. Equation 3.3 shows that the form of the MLF turns out to be a vital parameter in determining the size distribution. For the MLF slope $\beta = 2$, the size distribution $N(R) \ dR \propto R^{-3} \ dR$. There is a peak in $N(R)$, corresponding to the stall radius of the lowest-L objects, which would be individual SNRs in this analysis (Figure 5).

This prediction can be compared to the observed size distribution of H I shells that have been catalogued in the largest Local Group galaxies. By far the most complete catalog is that for the Small Magellanic Cloud (SMC; Staveley-Smith et al. 1997). The companion survey for the LMC (Kim et al. 1999) surprisingly shows about four times *fewer* shells than the SMC, in spite of the former's much larger size. The number of H I shells in the LMC is also much smaller than expected from the number of H II regions, whereas their relative numbers are fully consistent with their respective life expectancies in the SMC (Oey & Clarke 1997). Thus it appears that some process may be destroying the LMC shells prematurely, perhaps merging, shearing in the disk, or other ISM dynamical processes related to the high LMC star-formation rate and/or disk morphology. In contrast, the SMC has a more 3-dimensional, solid-body kinematic structure, and thus offers a better ISM for comparison with the crude size distribution predictions.

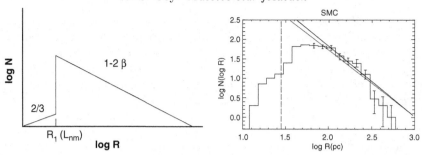

FIGURE 5. *Left:* Schematic representation of the predicted superbubble size distribution for a power-law MLF of slope $-\beta$ and constant object creation rate. *Right:* Comparison of slope (dotted line) fitted to the observed size distribution of H I shells in the SMC and predicted slope (solid line; from Oey & Clarke 1997).

For the SMC, the observed H I shell size distribution has a slope of -2.7 ± 0.6, in remarkable agreement with the predicted slope of -2.8 ± 0.4 derived from the actual H II LF (Oey & Clarke 1997; Figure 5). This suggests that the bubbly structure in the neutral ISM of this galaxy can be entirely attributed to mechanical feedback. It is also worth noting that a size distribution of fractal holes can be derived from the same dataset (Stanimirović et al. 1999). The power-law slope of -3.5 for the holes is similar, but different, from that for the shells, so it will be extremely interesting to make further such comparisons in other galaxies.

H I shell catalogs also have been compiled for M31 (Brinks & Bajaja 1986) and M33 (Deul & den Hartog 1990). These older catalogs lack sensitivity and resolution, but preliminary comparisons of the shell size distributions are broadly in agreement with the prediction (Oey & Clarke 1997). A modern H I survey of M33 by Thilker et al. (2000) is also eagerly anticipated. For the Milky Way, no complete samples of H I shells exist, but the International Galactic Plane Surveys (IGPS; e.g., McClure-Griffiths et al. 2002) may eventually yield data useful for a statistically significant sample.

Similarly, it is possible to derive the distribution in expansion velocities for the shells. Oey & Clarke (1998b) find,

$$N(v)\, dv \propto v^{-7/2}\, dv \quad , \quad \beta > 3/2 \quad . \tag{3.4}$$

Again, comparison with the SMC catalog shows consistency with the prediction, although with much larger uncertainty in the observed slope of -2.9 ± 1.4. Although taken from the same dataset, this comparison examines a different subset of objects, since the shells with non-negligible expansion velocities clearly sample the growing objects, whereas the size distribution is dominated by pressure-confined, stalled objects.

The Local Group galaxies, especially the Magellanic Clouds, also provide an opportunity to examine the detailed morphology of the ISM. An especially compelling discussion is presented by Elmegreen et al. (2001), who compare the structural morphology of the neutral ISM in the LMC with a fractal model. They show that the LMC's ISM is clearly more filamentary than accounted for in their simple fractal model (Figure 6). Such filamentary structure can be attributed at least in part to the superbubble structuring caused by mechanical feedback. Oey (2002) reviews H I structure in the ISM and argues that the origin of filamentary structure is key to understanding the dynamical and evolutionary processes in the ISM, such as phase balance and star formation.

FIGURE 6. H I distribution in a central region of the LMC (white is positive), showing a filamentary ISM morphology that is inconsistent with simple fractal models (from Elmegreen et al. 2001).

3.3. *Correspondence with star formation*

Modern studies that compare the spatial distribution of H I shells and star-forming regions show perhaps surprisingly ambiguous results. The galaxy that has been most actively studied in this respect recently is Holmberg II. At a distance of 3 Mpc, Ho II is not a member of the Local Group, and the conflicting results in the literature may be symptomatic of the difficulty with spatial resolution at that distance. While Rhode et al. (1999) failed to identify H I shell progenitor populations from BVR aperture photometry, Stewart et al. (2000) did find a positive spatial correlation with star-forming regions using FUV data from the *Ultraviolet Imaging Telescope* (*UIT*) and Hα observations. Tongue & Westpfahl (1995) also found that the SN rate implied by the radio continuum emission is consistent with the total, integrated superbubble energy in Ho II.

The Magellanic Clouds, and other nearby Local Group galaxies, are clearly superior candidates for investigating quantitative spatial correlations. Kim et al. (1999) have carried out a preliminary study that compares the H I shell properties with those of the OB associations and nebular emission. They are able to identify an evolutionary sequence such that the shells with associated Hα emission show higher expansion velocities than those showing only the presence of OB stars, and the latter in turn show higher v than the remainder of the shells. The Hα emission also shows smaller radial extent, remaining within that of the H I shells. These trends are consistent with an age sequence and feedback origin for the neutral shells. We are currently carrying out a more detailed follow-up of this study using *UIT* and new optical data (Oey et al. 2003, in preparation).

3.4. *Interstellar porosity*

A conventional parameterization for global mechanical feedback in galaxies is the interstellar porosity, or volume filling factor of superbubbles. Having derived analytic expressions for the superbubble size distribution, it is straightforward to derive expressions for the porosity (Oey & Clarke 1997). This can be written in terms of the star formation

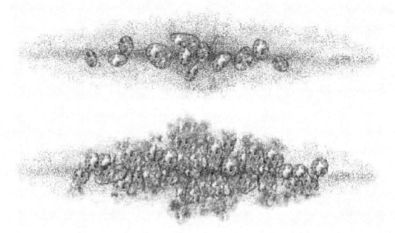

FIGURE 7. Low and high interstellar porosities are shown schematically in the upper and lower panels, respectively. The lower panel shows how $Q = 1$ defines a threshold porosity and star-formation rate for galactic outflows and escape of ionizing radiation.

rate (SFR), galactic scale height h, and galactic star-forming radius R_g (Oey et al. 2001):

$$Q \simeq 16 \frac{\text{SFR}(M_\odot \text{ yr}^{-1})}{hR_g^2(\text{kpc}^3)} \quad , \tag{3.5}$$

for Milky Way ISM parameters and Salpeter (1955) IMF. The porosity also can be written in terms of galactic parameters, i.e., ISM mass and velocity dispersion (Clarke & Oey 2002). Assuming a feedback origin for the HIM, the interstellar porosity quantifies the relative phase balance between the HIM and cooler ISM phases. Q greater than a critical value of unity therefore indicates an outflow condition for the hot gas, as might be encountered in a starburst situation (Figure 7).

    For the Local Group star-forming galaxies, Oey et al. (2001) find $Q \ll 1$ in almost all cases, suggesting that the HIM generally does not dominate the ISM volume. The LMC, however, does show $Q \sim 1$. The Milky Way situation is ambiguous, with different estimates of the SFR yielding Q in the range 0.2 to 1. The truly glaring exception is the starburst galaxy IC 10, for which $Q \gtrsim 20$, clearly fulfilling the outflow or superwind criterion. Clarke & Oey (2002) also posit the critical $Q = 1$ condition as a threshold for radiative feedback to the IGM. Once mechanical outflow is established, the merging and blowout of superbubbles also opens free pathways for ionizing photons to escape from the parent galaxy. They apply this simple model to a variety of phenomena, ranging from giant molecular clouds to Lyman break galaxies.

4. Conclusion

    It is clear that both radiative and mechanical feedback energies can dominate evolutionary processes in star-forming galaxies, although we still lack understanding of important aspects in the feedback processes. The extreme range in scale over which feedback has influence is especially remarkable. In addition, it is also possible to investigate and parameterize chemical feedback with analogous methods (e.g., Oey 2000, 2003).

    Radiative feedback is responsible for the nebular emission-line diagnostics and tracers of star formation, as well as the WIM component of the ISM. On a detailed level, the line diagnostics still need improved calibrations and photoionization modeling to increase their utility and parameter space. The H II LF is emerging as a quantitative diagnostic of global star formation in galaxies, and reveals a universal law in the stellar membership function for massive stars (equation 2.6). The ionization and energy budget of the WIM, while apparently tied to massive stars, remains to be better understood. Ultimately, the escape of ionizing radiation beyond the parent galaxies bears upon the intergalactic environment and reionization of the early Universe.

Mechanical feedback indisputably drives shell structures that are ubiquitous in the ISM: SNRs, stellar wind-driven bubbles, and superbubbles. The standard, adiabatic model for the evolution of the bubbles and superbubbles is broadly consistent with a variety of empirical evidence; yet, quantitatively, a number of outstanding problems remain. The late evolution of the shells is especially enigmatic and critical for understanding the role of feedback in generating the HIM, as well as the global properties of the ISM. Spatial correlations of interstellar shells with star-forming regions still need to firmly establish the physical processes and interactions related to feedback. Preliminary studies of the size and velocity distributions of H I shells appear to confirm a dominant role for mechanical feedback in structuring the ISM. This analytic analysis can be extended to parameterize the interstellar porosity, implying a critical threshold for the outflow of superwinds, heavy elements, and ionizing radiation.

I am grateful to the conference organizers and ST ScI for supporting in part this contribution to the Symposium.

REFERENCES

ARNAL, E. M., CAPPA, C. E., RIZZO, J. R., & CICHOWOLSKI, S. 1999 *AJ* **118**, 1798.

BERKHUIJSEN, E. M., BECK, R., & HOERNES, P. 2003 *A&A* **398**, 937.

BENAGLIA, P. & CAPPA, C. E. 1999 *A&A* **346**, 979.

BRINKS, E. & BAJAJA, E. 1986 *A&A* **169**, 14.

BROWN, A. G. A., HARTMANN, D., & BURTON, W. B. 1995 *A&A* **300**, 903.

CAPPA, C. E. & BENAGLIA, P. 1998 *AJ* **116**, 1906.

CAPPA, C. E., GOSS, W. M., & PINEAULT, S. 2002 *AJ* **123**, 3348.

CAPPA, C. E. & HERBSTMEIER, U. 2000 *AJ* **120**, 1963.

CASTOR, J., MCCRAY, R., & WEAVER, R. 1975 *ApJ* **200**, L107.

CHRISTENSEN, T., PETERSEN, L., & GAMMELGAARD, P. 1997 *AA* **322**, 41.

CHU, Y.-H. & MAC LOW, M-M. 1990 *ApJ* **365**, 510.

CHU, Y.-H., WAKKER, B., MAC LOW, M.-M., & GARCÍA-SEGURA, G. 1994 *AJ* **108**, 1696.

CLARKE, C. J. & OEY, M. S. 2002 *MNRAS* **337**, 1299.

COWIE, L. L., MCKEE, C. F., & OSTRIKER, J. P. 1981 *ApJ* **247**, 908.

CROWTHER, P. A., PASQUALI, A., DE MARCO, O., SCHMUTZ, W., HILLIER, D. J., & DE KOTER, A. 1999 *A&A* **350**, 1007.

DEUL, E. R. & DEN HARTOG, R. H. 1990 *A&A* **229**, 362.

DRISSEN, L., MOFFAT, A. F. J., WALBORN, N. R., & SHARA, M. R. 1995 *AJ* **110**, 2235.

ELMEGREEN, B. G. & EFREMOV, Y. N. 1997 *ApJ* **480**, 235.

ELMEGREEN, B. G., KIM, S., & STAVELEY-SMITH, L. 2001 *ApJ* **548**, 749.

FERGUSON, A. M. N., WYSE, R. F. G., & GALLAGHER, J. S. 1996 *AJ* **112**, 2567.

FRAIL, D. A., CORDES, D. M., HANKINS, T. H., & WEISBERG, J. M. 1991 *ApJ* **382**, 168.

GALARZA, V. C., WALTERBOS, R. A. M., & BRAUN, R. 1999 *AJ* **118**, 2775.

GARCÍA-SEGURA, G. & MAC LOW, M.-M. 1995 *ApJ* **455**, 145.

GRAY, A. D., LANDECKER, T. L., DEWDNEY, P. E., TAYLOR, A. R., WILLIS, A. G., & NORMANDEAU, M. 1999 *ApJ* **514**, 221.

GREENAWALT, B., WALTERBOS, R. A. M., & BRAUN, R. 1997 *ApJ* **483**, 666.

HARRIS, W. E. & PUDRITZ, R. E. 1994 *ApJ* **429**, 177.

HARTQUIST, T. W., DYSON, J. E., PETTINI, M., & SMITH, L. J. 1986 *MNRAS* **221**, 715.

HEILES, C., KOO, B.-C., LEVENSON, N. A., & REACH, W. T. 1996 *ApJ* **462**, 326.

HODGE, P. W., BALSLEY, J., WYDER, T. K., & SKELTON, B. P. 1999 *PASP* **111**, 685.

HOOPES, C. G. & WALTERBOS, R. A. M. 2000 *ApJ* **541**, 597.

HUNTER, D. A. & MASSEY, P. 1990 *AJ* **99**, 846.

KENNICUTT, R. C., BRESOLIN, F., BOMANS, D. J., BOTHUN, G. D., & THOMPSON, I. B. 1995 *AJ* 109, 594.

KENNICUTT, R. C., BRESOLIN, F., FRENCH, H., & MARTIN, P. 2000 *ApJ* **537**, 589.

KENNICUTT, R. C., EDGAR, B. K., & HODGE, P. W. 1989 *ApJ* **337**, 761.

KIM, S., DOPITA, M. A., STAVELEY-SMITH, L., & BESSELL, M. S. 1999 *AJ* **118**, 2797.

MAC LOW, M.-M. & MCCRAY, R. 1988 *ApJ* **324**, 776.

MAGNIER, E. A., CHU, Y.-H., POINTS, S. D., HWANG, U., & SMITH, R. C. 1996 *ApJ* **464**, 829.

MCCLURE-GRIFFITHS, N. M., DICKEY, J. M., GAENSLER, B. M., & GREEN, A. J. 2002 *ApJ* **578**, 176.

MCKEE, C. F. & WILLIAMS J. P. 1997 *ApJ* **476**, 144.

MEURER, G. R., HECKMAN, T. M., LEITHERER, C., KINNEY, A., ROBERT, C., & GARNETT, D. R. 1995 *AJ* **110**, 2665.

MINTER, A. H. & BALSER, D. S. 1997 *ApJ* **484**, L133.

MINTER, A. H. & SPANGLER, S. R. 1996 *ApJ* **458**, 194.

MINTER, A. H. & SPANGLER, S. R. 1997 *ApJ* **485** 182.

OEY, M. S. 1996 *ApJ* **467**, 666.

OEY, M. S. 2000 *ApJ* **542**, L25.

OEY, M. S. 2002. In *Seeing Through the Dust: The Detection of H*I *and the Exploration of the ISM in Galaxies* (eds. R. Taylor, T. Landecker, & A. Willis). p. 195. ASP.

OEY, M. S. 2003 *MNRAS* **339**, 849.

OEY, M. S. & CLARKE, C. J. 1997 *MNRAS* **289**, 570.

OEY, M. S. & CLARKE, C. J. 1998a *AJ*, 115, 1543.

OEY, M. S. & CLARKE, C. J. 1998b. In *Interstellar Turbulence*, (eds. J. Franco & A. Carramiñana). p. 112. Cambridge Univ. Press.

OEY, M. S., CLARKE, C. J., & MASSEY, P. 2001. In *Dwarf Galaxies and their Environment*, (eds. K. de Boer, R.-J. Dettmar, & U. Klein). p. 181. Shaker Verlag.

OEY, M. S., DOPITA, M. A., SHIELDS, J. C., & SMITH, R. C. 2000 *ApJS* **128**, 511.

OEY, M. S., GROVES, B., STAVELEY-SMITH, L., & SMITH, R. C. 2002 *AJ* **123**, 255.

OEY, M. S. & KENNICUTT, R. C. 1997 *MNRAS* **291**, 827.

OEY, M. S. & MASSEY, P. 1994 *ApJ* **425**, 635.

OEY, M. S. & MASSEY, P. 1995 *ApJ* **452**, 210.

OEY, M. S. & MUÑOZ-TUÑÓN, C. 2003. In *Star Formation Through Time*, (eds. E. Pérez, R. González-Delgado, & G. Tenorio-Tagle). (ASP), in press.

OEY, M. S., PARKER, J. S., MIKLES, V. J., & ZHANG, X. 2003 *AJ*, submitted.

OEY, M. S. & SHIELDS, J. C. 2000 *ApJ* **539**, 687.

OEY, M. S. & SMEDLEY, S. A. 1998 *AJ* **116**, 1263.

OTTE, B. & DETTMAR, R.-J. 1999 *A&A* **343**, 705.

PAGEL, B. E. J., EDMUNDS, M. G., BLACKWELL, D. E., CHUN, M. S., & SMITH, G. 1979 *MNRAS* **189**, 95.

PIKEL'NER, S. B. 1968 *ApJ* **2**, L97.

REYNOLDS, R. J. & COX, D. P. 1992 *ApJ* **400**, L33.

REYNOLDS, R. J. & TUFTE, S. L. 1995 *ApJ* **439**, L17.

REYNOLDS, R. J., TUFTE, S. L., HAFFNER, L. M., JAEHNIG, K., & PERCIVAL, J. W. 1998 *PASA* **15**, 14.

RHODE, K. L., SALZER, J. J., WESTPFAHL, D. J., & RADICE, L. A. 1999 *AJ* **118**, 323.

SAKEN, J. M., SHULL, J. M., GARMANY, C. D., NICHOLS-BOHLIN, J., & FESEN, R. A. 1992 *ApJ*, **397**, 537.

SALPETER, E. E. 1955 *ApJ* **121**, 161.

SEDOV, L. I. 1959, *Similarity and Dimensional Methods in Mechanics*. Academic.

SILICH, S. A. & OEY, M. S. 2002. In *Extragalactic Star Clusters* (eds. D. Geisler, E. K. Grebel, & D. Minitti). IAU Symposium 207, p. 459. ASP.

SILICH, S. A., TENORIO-TAGLE, G., TERLEVICH, R., TERLEVICH, E., & NETZER, H. 2001 *MNRAS* **324**, 191.

SMITH, T. R. & KENNICUTT, R. C. 1989 *PASP* **101**, 649.

STANIMIROVIĆ, S., STAVELEY-SMITH, L., DICKEY, J. M., SAULT, R. J., & SNOWDEN, S. L. 1999 *MNRAS* **302**, 417.

STAVELEY-SMITH, L., SAULT, R. J., HATZIDIMITRIOU, D., KESTEVEN, M. J., & MCCONNELL, D. 1997 *MNRAS* **289**, 225.

STEWART, S. G., ET AL. 2000 *ApJ* **529**, 201.

THILKER, D. A., BRAUN, R., & WALTERBOS, R. A. M. 2000 *BAAS* **197**, #38.08.

TONGUE, T. D. & WESTPFAHL, D. J. 1995 *AJ* **109**, 2462.

TREFFERS, R. R. & CHU, Y.-H. 1982 *ApJ* **254**, 569.

UYANIKER, B., LANDECKER, T. L., GRAY, A. D., & KOTHES, R. 2003 *ApJ* **585**, 785.

VAN DEN BERGH, S. 1981 *AJ* **86**, 1464.

VÍLCHEZ, J. M. & ESTEBAN, C. 1996 *MNRAS* **290**, 265.

VÍLCHEZ, J. M. & PAGEL, B. E. J. 1988 *MNRAS* **231**, 257.

WALTERBOS, R. A. M. 1998 *PASA* **15**, 99.

WALTERBOS, R. A. M. & BRAUN, R. 1992 *A&AS* **92**, 625.

WALTERBOS, R. A. M. & BRAUN, R. 1994 *ApJ* **431**, 156.

WANG, Q. & HELFAND, D. J. 1991 *ApJ* **373**, 497.

YOUNGBLOOD, A. J. & HUNTER, D. A. 1999 *ApJ* **519**, 55.

Hot gas in the Local Group and low-redshift intergalactic medium

By KENNETH R. SEMBACH

The Space Telescope Science Institute, 3700 San Martin Drive, Baltimore, MD 21218, USA

There is increasing observational evidence that hot, highly ionized interstellar and intergalactic gas plays a significant role in the evolution of galaxies in the local universe. The primary spectral diagnostics of the warm-hot interstellar/intergalactic medium are ultraviolet and X-ray absorption lines of O VI and O VII. In this paper, I summarize some of the recent highlights of spectroscopic studies of hot gas in the Local Group and low-redshift universe. These highlights include investigations of the baryonic content of low-z O VI absorbers, evidence for a hot Galactic corona or Local Group medium, and the discovery of a highly ionized high velocity cloud system around the Milky Way.

1. Introduction

We live in a wonderful age of discovery and exploration of the universe. As we peer farther and farther back in time, it is becoming ever more important to make sure that we observe the local universe as well as possible. Observations of galactic systems and the intergalactic medium (IGM) in the low-redshift universe are required to study the universe as it has evolved over the last $\sim$5 billion years. They are essential for the interpretation of higher redshift systems, and they form a framework for studies of such key topics as galactic evolution, "missing mass," and the distribution of dark matter. Studies of hot gas and its relationship to galaxies are shedding new light on these and other astronomical topics of interest today. In this review, I summarize some basic information about the elemental species and types of observations that can be used to study hot gas. I also provide overviews of recent spectroscopic observations of hot gas in the low-redshift universe (§2), the Local Group (§3), and the high velocity cloud (HVC) system that surrounds the Milky Way (§4). Some concluding remarks on future observations can be found in §5.

Table 1 contains a summary of some of the most important diagnostics of hot gas in the local universe. The species listed are detectable in either the ultraviolet or X-ray bandpasses accessible with spectrographs aboard the *Hubble Space Telescope* (*HST*), the *Far Ultraviolet Spectroscopic Explorer* (*FUSE*), the *Chandra X-ray Observatory*, and *XMM-Newton*. Most of the diagnostics listed can be observed at their rest wavelength, which means that they can be used to study hot gas in the Local Group. Others must be redshifted into one of the observable bandpasses; the observed wavelength at $z = 0.5$ is listed for for comparison in the table.

The O VI $\lambda\lambda 1031.926, 1037.617$ resonance doublet lines are the best lines to use for kinematical investigations of hot ($T \sim 10^5 - 10^6$ K) gas in the low-redshift universe. O VI has a higher ionization potential than other species observable by *FUSE* and *HST*, and oxygen has the highest cosmic abundance of all elements other than hydrogen and helium. X-ray spectroscopy of the interstellar or intergalactic gas in higher ionization lines (e.g., O VII, O VIII) is possible with *XMM-Newton* and the *Chandra X-ray Observatory* for a small number of sight lines toward AGNs and QSOs, but the spectral resolution (R $\equiv \lambda/\Delta\lambda \lesssim 400$) is modest compared to that afforded by *FUSE* (R $\sim 15,000$) or *HST*/STIS (R $\sim 45,000$). While the X-ray lines provide extremely useful information about the amount of gas at temperatures greater than 10^6 K, the interpretation of where that gas

Ion, λ_{rest}	f	$\log f\lambda$	$\lambda_{z=0.5}$ (Å)	T_{CIE}† (K)	b_{th}‡ (km s^{-1})
FUSE, HST					
H I Ly-series	...	...	1369–1824	...	13–129¶
C IV 1548.195	0.1908	2.470	2322.292	1.0×10^5	11.8
C IV 1550.770	0.0952	2.169	2326.155	1.0×10^5	11.8
N V 1238.821	0.1570	2.289	1858.232	1.8×10^5	14.6
N V 1242.804	0.0782	1.988	1864.206	1.8×10^5	14.6
O IV 787.711	0.111	1.942	1181.567	1.6×10^5	12.9
O V 629.730	0.515	2.511	944.595	2.5×10^5	16.1
O VI 1031.926	0.1329	2.137	1547.889	2.8×10^5	17.1
O VI 1037.617	0.0661	1.836	1556.425	2.8×10^5	17.1
Ne VIII 770.409	0.103	1.900	1155.614	5.6×10^5	21.5
Ne VIII 780.324	0.0505	1.596	1170.486	5.6×10^5	21.5
Chandra, XMM-Newton					
O VII 21.602	0.696	1.177	32.403	8.0×10^5	28.8
O VIII 18.967	0.277	0.720	28.450	2.2×10^6	47.8
O VIII 18.972	0.139	0.421	28.459	2.2×10^6	47.8
Ne IX 13.447	0.0724	0.988	20.170	1.5×10^6	35.2

† Temperature of maximum ionization fraction in collisional ionization equilibrium (Sutherland & Dopita 1993).
‡ Thermal line width, b $= (2kT/m)^{1/2}$, at $T = T_{CIE}$ unless indicated otherwise.
¶ Value of b for $T = 10^4 - 10^6$ K.

TABLE 1. Diagnostics of hot gas at low redshift ($z < 0.5$)
[f-values and wavelengths (in Å) are from Morton (1991), Verner, Barthel, & Tytler (1994), and Verner, Verner, & Ferland (1996).]

is located, or how it is related to the $10^5 - 10^6$ K gas traced by O VI, is hampered at low redshift by the kinematical complexity of the hot interstellar medium (ISM) and IGM along the sight lines observed, as discussed below. Nevertheless, the X-ray diagnostics are of fundamental importance for determining the ionization state of the nearby hot gas, and they provide information about hotter gas that is not traced by species in the ultraviolet wavelength region of the electromagnetic spectrum.

Table 2 contains a high-level summary of key considerations for absorption and emission-line spectroscopy of hot interstellar and intergalactic plasmas. Both types of observations have their strengths, and the combination of ultraviolet and X-ray information holds great promise for studies of the ISM and IGM.

2. Low-redshift O VI absorption systems

One of the recent successes of observational cosmology is the excellent agreement in estimates of the amount of matter contained in baryons derived from measures of the temperature fluctuations in the cosmic microwave background and spectroscopic measures of the primordial abundance of deuterium relative to hydrogen in the high-redshift intergalactic medium (Spergel et al. 2003). At high redshift ($z \gtrsim 3$), most of the baryons in the universe are contained in intergalactic absorbers, which are detected through spectroscopic observations of their H I and He II Lyα absorption against the background light

	UV Absorption	*X-ray Absorption*	*X-ray Emission*
Observatory	*FUSE, HST*	*Chandra* *XMM − Newton*	*Chandra, ROSAT* *XMM − Newton*
$T_{\rm CIE}\,(K)$ *Range*	10^5–10^6	10^6–10^8	10^6–10^8
Density Dependence	$N_{\rm ion} \propto n_e$	$N_{\rm ion} \propto n_e$	$I_{\rm xray} \propto n_e^2$
Limiting Sensitivity†	$\log N(O^{+5}) \sim 13$ $\Rightarrow \log N(H^{+}) \sim 18$	$\log N(O^{+6}) \sim 16$ $\Rightarrow \log N(H^{+}) \sim 20.3$	$\cdots$
Spatial Info.	Point source	Point source	Extended source
Spectral Resolution	R > 15,000 Detailed kinematics	R < 1000 General kinematics	Low (broadband) R < 1000

† Assuming a metallicity of ~ 0.1 solar and peak ionization fractions (see text).

TABLE 2. Spectroscopy of hot interstellar/intergalactic gas

of distant quasars. At lower redshifts, the Lyα forest thins out and galaxies become more prevalent (Penton, Shull, & Stocke 2000). Censuses of these low-redshift Lyα clouds and galaxies reveal a baryon deficit compared to the amount of matter observed in the high-redshift universe (see Fukugita, Hogan, & Peebles 1998). This "missing baryon" problem has led to the suggestion that much of the baryonic material at low redshift is found in the form of hot, highly ionized intergalactic gas that is difficult to detect with existing instrumentation.

Hydrodynamical simulations of the evolution of the IGM in the presence of cold dark matter predict that the intergalactic clouds collapse into coherent sheets and filaments arranged in a web-like pattern. As lower density gas streams into the deeper potential wells at the intersections of these sheets and filaments, clusters of galaxies form. Shocks heat the collapsing structures to temperatures of 10^5–10^7 K, resulting in a pervasive network of hot gas (Cen & Ostriker 1999; Davé et al. 2001). Understanding the physical processes involved in galaxy-formation is the key to determining whether this description of the intergalactic medium is correct. In particular, the simulations do not yet have sufficient observational constraints to accurately model how the cosmic web responds to the formation of galaxies. Feedback between galaxies and the cosmic web affects the kinematics, distribution, metal content, and temperature of the IGM. These processes are probably self-regulating. Detailed studies are most promising at low redshift, where it is possible to sample the IGM absorption on finer scales, identify faint galaxies, and study galactic properties in greater detail than is possible at high redshift. Many fundamental questions remain to be answered:

(*a*) Is the hot IGM a significant repository of baryons at low redshift? How much mass is contained in the cosmic web of hot gas?

(*b*) How does feedback during galaxy formation affect the properties of the IGM and the efficiency of galaxy formation? What are the primary feedback mechanisms?

(*c*) What is the morphology of the cosmic web and the dark matter it traces?

(*d*) Are there fundamental relationships between the hot IGM and the hot gas found in clusters and groups of galaxies?

Observational evidence for a hot IGM at low redshifts remains limited mainly to detections of highly ionized oxygen (primarily O VI) along a small number of sight lines.

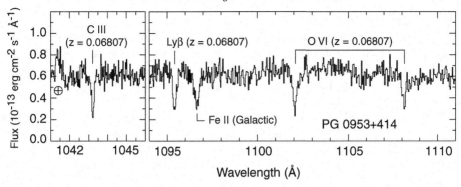

FIGURE 1. A *FUSE* spectrum of PG 0953+415 showing redshifted O VI, C III, and H I Lyβ at $z = 0.06807$ (Savage et al. 2002). This is one of two O VI systems along this sight line. The crossed circle in the left panel marks the location of a terrestrial O I airglow line.

The low-redshift O VI systems that have been observed have a variety of strengths and O VI/H I ratios. Many occur at the redshifts of groups or clusters of galaxies along the same sight lines. Several sight lines have been examined in detail for O VI absorbers: H 1821+643 (Tripp, Savage, & Jenkins 2000; Oegerle et al. 2000), PG 0953+415 (Tripp & Savage 2000; Savage et al. 2002), 3C 273 (Sembach et al. 2001), PKS 2155−304 (Shull, Tumlinson, & Giroux 2003), PG 1259+593 (Richter et al. 2004), PG 1116+215 (Sembach et al. 2004), and PG 1211+143 (Tumlinson et al. 2004). An important piece of information that is lacking for many of the well-observed O VI absorbers is the amount of C IV associated with the O VI. This information can, and should, be obtained with the *HST*.

Figure 1 illustrates a few of the absorption lines observed by *FUSE* in one of the two O VI systems toward PG 0953+415 (Savage et al. 2002). The lines are easily detected and are nearly resolved at the *FUSE* resolution of ∼20 km s^{-1} (FWHM). This absorption system has ionization properties consistent with photoionization by dilute ultraviolet background radiation but may also be collisionally ionized. A good example of a system that is probably collisionally ionized is given by Tripp et al. (2001). Other investigators have suggested that many of the O VI systems are collisionally ionized based on the observed relationship between the line width and O VI column density found for a wide range of environments containing O VI (Heckman et al. 2002).

The baryonic contribution of the O VI absorbers to the closure density of the universe is $\Omega_b(\text{O VI}) = 0.002\ h_{75}^{-1}$ if their typical metallicity is ∼1/10 solar (Savage et al. 2002). Here, h_{75}^{-1} is the Hubble constant in units of 75 km s^{-1} Mpc^{-1}. This baryonic contribution is of the same order of magnitude as the contribution from stars and gas inside galaxies. More information about O VI in the low-redshift IGM can be found in the aforementioned sight line references. See Tripp (2002) for a recent review of the subject.

Information on hot gas at low redshift is also provided by X-ray measurements of O VII and O VIII, although the number of measurements is limited at this time. These results are sometimes conflicting, as in the case of the $z \sim 0.057$ absorbers toward PKS 2155−304 (e.g., Fang et al. 2002; Shull et al. 2003), or are of modest statistical significance (e.g., Cagnoni 2002; Mathur, Weinberg, & Chen 2003; McKernan et al. 2003b). None of the claimed detections to date at $z > 0$ have been particularly convincing. Still, the fact that it is possible to conduct searches for X-ray absorption lines in the hot IGM is promising. To form a large enough database to make firmer statements about the baryonic contribution of gas hotter than that contained in the O VI absorption systems will likely require spectrographs on X-ray telescopes with very large effective areas, such as Constellation-X.

3. Hot Local Group Gas

An interesting development in studies of the gaseous content of Local Group galaxies is the mounting evidence for a hot, extended corona of low density gas around the Milky Way. This corona may extend to great distances from the Galaxy, and may even pervade much of the Local Group. In recent years, several results have strengthened the case for such a medium. These include:

(i) *The detection of* O VI *high velocity clouds in the vicinity of the Galaxy.* The properties of these clouds suggest that the O VI is created at the boundaries of cooler HVCs as they move through a pervasive coronal medium (Sembach et al. 2003; Collins, Shull, & Giroux 2004). The properties of the O VI HVCs and their possible origins are discussed below in §4.

(ii) *Chandra and XMM-Newton detections of* O VII *absorption near zero redshift.* O VII peaks in abundance at $T \sim 10^6$ K in collisional ionization equilibrium. A summary of these O VII detections is given in Table 3, where the observed equivalent width of the O VII $\lambda 21.60$ line is listed. Other high ionization species, such as O VIII $\lambda 18.97$ and Ne IX $\lambda 13.45$ may also be detected along the same sight lines, albeit at somewhat lower significance.

(iii) *Ram-pressure stripping of gas in Local Group dwarf galaxies.* Moore & Davis (1994) postulated a hot, low-density corona to provide ram pressure stripping of some of the Magellanic Cloud gas and to explain the absence of gas in globular clusters and nearby dwarf spheroidal companions to the Milky Way (see also Blitz & Robishaw 2000). The shape and confinement of some Magellanic Stream concentrations (Stanimirovic et al. 2002) and the shapes of supergiant shells along the outer edge of the LMC (de Boer et al. 1998) are also more easily explained if an external medium is present.

(iv) *Drag deflection of the Magellanic Stream on its orbit around the Milky Way.* N-body simulations of the tidal evolution and structure of the leading arm of the Magellanic Stream require a low-density medium ($n_H < 10^{-4}$ cm^{-3}) to deflect some of the Stream gas into its observed configuration (Gardiner 1999).

3.1. An estimate of X-ray absorption produced by the hot gas at $z \approx 0$

Perhaps the most direct measure of the amount of hot gas is given by the X-ray absorption measurements. A tenuous hot Galactic corona or Local Group gas should be revealed through X-ray absorption-line observations of O VII. The medium should have a temperature $T \gtrsim 10^6$ K to avoid direct detection in lower ionization species such as O VI, a density $n_H \lesssim 10^{-4}$ cm^{-2} to prevent orbital decay of the Magellanic Stream, and an extent $L \gtrsim 70$ kpc to explain the detections of O VI in the Magellanic Stream (see Sembach et al. 2003). The column density of O VII in the hot gas is given by

$$\mathrm{N(O\,VII)} = (\mathrm{O/H})_\odot \, Z/Z_\odot \, \mathrm{f_{O\,VII}} \, n_H \, L \, ,$$

where $Z/Z_\odot$ is the metallicity of the gas in solar units, $\mathrm{f_{O\,VII}}$ is the ionization fraction of O VII, L is the path length, and $(\mathrm{O/H})_\odot = 4.90 \times 10^{-4}$ (Allende Prieto, Lambert, & Asplund 2001). At $T \sim 10^6$ K, $\mathrm{f_{O\,VII}} \approx 1$ (Sutherland & Dopita 1993). For $n_H = 10^{-4}$ cm^{-3}, $N(\mathrm{O\,VII}) \sim 1.5 \times 10^{16}$ $Z/Z_\odot$ $(L/100$ kpc) cm^{-2}. This column density is of the order of magnitude derived for the equivalent widths listed in Table 3. Conversion of the observed O VII equivalent widths into an O VII column densities requires assumptions about the Doppler parameters for the lines since the lines are likely unresolved, with some authors preferring to constrain the possible b-values (e.g., Fang et al. 2003; Nicastro et al. 2002), and others preferring to list only lower limits to N(O VII) from the assumption of a linear curve of growth (e.g., Rasmussen et al. 2003).

It is extremely important to realize that there may be sources for the O VII and other high ionization X-ray lines observed that do not involve a hot Galactic corona or Local

Sight Line	l($^\circ$)	b($^\circ$)	W_λ(mÅ)†	Instrument	References‡
PKS 2155–304	17.73	−52.25	$11.6\pm^{7.5}_{6.0}$	*Chandra* LETG	N02,F03
			$15.6\pm^{8.6}_{4.9}$	*Chandra* LETG	F02
			16.3 ± 3.3	*XMM* RGS	R03
Mrk 421	179.83	+65.03	15.4 ± 1.7	*XMM* RGS	R03
3C 273	289.95	+64.36	$28.4\pm^{12.5}_{6.2}$	*Chandra* LETG	F03
			26.3 ± 4.5	XMM RGS	R03
NGC 4593	297.48	+57.40	$18.0\pm^{9.4}_{15.8}$	Chandra HETG	M03

† Equivalent width in mÅ. For those studies listing equivalent widths in energy units rather than wavelength units, the following conversion was used: $W_\lambda(\text{mÅ}) = (\lambda^2/hc)W_E = 37.5W_E(\text{eV})$. Errors for all values are quoted are at 90% confidence, except the C02 value for Mrk 421, which is a 1σ estimate.

‡ References: C02 = Cagnoni (2002); F02 = Fang et al. (2002); F03 = Fang, Sembach, & Canizares (2003); M03 = McKernan et al. (2003a); N02 = Nicastro et al. (2002); R03 = Rasmussen, Kahn, & Paerels (2003).

TABLE 3. Reported detections of O VII λ21.60 absorption near $z \sim 0$

Group medium. Alternative locations for some of the hot gas along the sight lines observed include the thick disk and low halo of the Galaxy (see Savage et al. 2003) and large Galactic structures that are filled with hot gas, such as Loop I (see Snowden et al. 1997). The possible contributions of these different regions to the observed O VII absorption along the 3C 273 sight line are discussed by Fang et al. (2003); a portion of their X-ray spectrum is reproduced in Figure 2. Isolating the distant Galactic corona or Local Group contributions to the X-ray absorption from nearby regions of hot gas is difficult to do kinematically because the spectral resolution of the X-ray data is insufficient to resolve the velocities of most objects in the Local Group from the velocity of the Galaxy.

A hot extended Galactic corona or Local Group medium has several testable predictions. O VII absorption should be observed in essentially any direction where an X-ray bright background continuum source can be observed with sufficient signal-to-noise to detect the λ21.60 line with a strength $W_\lambda \gtrsim 15$ mÅ. The amount of absorption observed may vary based on direction and the types of foreground structures probed. In general, observations of nearby X-ray sources should yield interstellar O VII column densities less than those observed toward extragalactic X-ray sources. Futamoto et al. (2003) have recently reported the detection of strong O VII absorption toward the low mass X-ray binary 4U1820–303 in NGC 6624, which is at odds with this prediction since 4U1820–303 is only ~ 1 kpc from the Galactic plane. However, Futamoto et al. (2003) did not detect O VII toward Cyg-X2, which is at a similar Galactic altitude. Additional observations of this type would be valuable for determining the distribution of hot gas near the plane of the Galaxy. Investigators working in this field should report significant non-detections whenever possible.

3.2. *Problems with a single-phase ultra-low density medium model*

Given the probability that several regions contribute to the observed absorption, it is dangerous to assume that the X-ray bearing gas is uniform, at a single temperature,

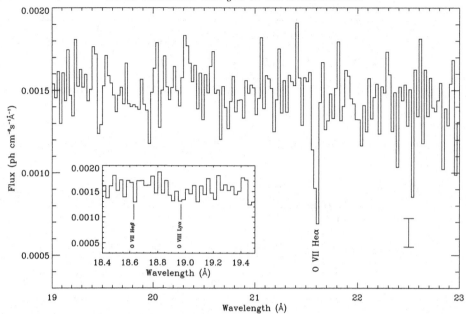

FIGURE 2. A portion of the *Chandra* LETG-ACIS spectrum of 3C 273 showing a well-detected zero-redshift O VII Heα line at 21.6 Å, with $W_\lambda = 28.4\pm^{12.5}_{6.3}$ mÅ at 90% confidence (Fang et al. 2003). The O VII may arise in hot gas within the Milky Way or within a hot Local Group medium. The error bar in the lower right corner indicates the typical 1σ photon-counting error on each data point.

or confined to one type of region. The same holds true for the lower temperature O VI gas, which is known to have a patchy spatial distribution and line profiles that contain multiple components (see Savage et al. 2003; Sembach et al. 2003; Howk et al. 2002). An example of the pitfalls that can be encountered in analyses of the X-ray absorption is provided by the PKS 2155–304 sight line. Nicastro et al. (2002) associated the X-ray absorption near $z \sim 0$ along the PKS 2155–304 sight line with the high-velocity O VI absorption observed by Sembach et al. (2000). They found that a relatively uniform, ultra-low density, single-phase plasma with a density $n_e \sim 6 \times 10^{-6}$ cm^{-3} and a size of roughly 3 Mpc was able to explain the O VI–VIII absorption, modulo some difficulties in reproducing the observed Ne/O ratio in the hot gas. From this, they concluded that the X-ray bearing gas must be distributed throughout the Local Group. However, this conclusion rested strongly on the supposition that the high velocity O VI absorption and the X-ray absorption in this direction are uniquely related, with no other significant sources of O VI or O VII–VIII inside or outside the Galaxy. This conclusion fails to account for the complex distribution of gas along the sight line and the ample evidence that the O VI HVCs in this direction have kinematical signatures similar to lower ionization species whose presence is not consistent with gas at $T = (0.5-1.0) \times 10^6$ K and $n_e \sim 10^{-6}$ cm^{-3} [see Sembach et al. (1999) and Collins et al. (2004) for information on the lower ionization species such as C II, C IV, Si II, Si III, Si IV]. The ionization parameter in the proposed medium is far too high to explain the strengths of the low ionization lines observed at essentially the same velocities as the O VI HVCs. Other types of inhomogeneities in the hot gas distribution may also be present, as they are needed to reconcile the observed Ne/O ratio with a solar ratio of Ne/O (Nicastro et al. 2002). (The possibility of explaining the factor of two discrepancy between the observed Ne/O ratio and the solar ratio by preferential incorporation of oxygen into dust grains seems dubious given the much higher cosmic abundance of oxygen). If one invokes inhomogeneities in the gas distribution to explain the lower ionization gas, then such regions must necessarily account for some, or perhaps even all, of the O VI. As a result, the ionization constraints used to deduce the existence of an ultra-low density medium break down, and the adoption of a uniform single temperature medium is cast in doubt.

The low ionization high velocity clouds toward PKS 2155–304 should have densities typical of those of clouds located near the Galaxy (i.e., $n \sim 10^{-3}$–10^{-1} cm^{-3}). An example of one such region is high velocity cloud Complex C, whose ionization conditions suggest that the cloud is interacting with gas in an extended Galactic corona (Sembach et al. 2003; Fox et al. 2004; Collins et al. 2004). Interactions of this type are consistent with the wider distribution of high velocity O VI seen on the sky (see §4) and provide a dynamic, non-equilibrium view of the gas ionization that is far more amenable to creating a range of ionization conditions than is possible in a static, ultra-low density medium having enormous cooling times.

Thus, while some of the highly ionized oxygen may be located in the Local Group, consideration of all of the available information indicates that an ultra-low density ($n \sim 10^{-6}$ cm^{-3}, $\delta \sim 60$) Local Group medium is not a viable explanation for all of the X-ray and high velocity O VI absorption toward PKS 2155–304. A low density (but not too low density) medium with $n \sim 10^{-4}$–10^{-5} cm^{-3} is still a possible repository for much of the highly ionized gas, but is in all likelihood only part of a complex distribution of hot gas along this sight line. Additional discussion of the observational and theoretical implications of a Local Group medium and the complications that can arise in interpreting the existing data can be found in Maloney (2003).

4. High velocity clouds in the vicinity of the milky way

We have conducted an extensive study of the highly ionized high velocity gas in the vicinity of the Milky Way using data from the *FUSE* satellite. Below, I summarize the results for the sight lines toward 100 AGNs/QSOs and two distant halo stars (see Sembach et al. 2003; Wakker et al. 2003). The *FUSE* observations represent the culmination of ~ 4 megaseconds of actual exposure time and many years of effort by the *FUSE* science team. For the purposes of this study, gas with $|v_{LSR}| \gtrsim 100$ km s^{-1} is typically identified as "high velocity," while lower velocity gas is attributed to the Milky Way disk and halo. Sample spectra from the survey are shown in Figure 3. Complementary talks on this subject were presented at the Sydney IAU in Symposium 217 by Blair Savage (a description of the *FUSE* O VI survey) and Bart Wakker (an overview of high velocity clouds) (see Savage et al. 2004; Wakker 2004).

4.1. *Detections of high velocity O VI*

Sembach et al. (2003) identified 84 individual high velocity O VI features along 102 sight lines observed by *FUSE*. A critical part of this identification process involved detailed consideration of the absorption produced by O VI and other species (primarily H$_2$) in the thick disk and halo of the Galaxy, as well as the absorption produced by low-redshift intergalactic absorption lines of H I and ionized metal species along the lines of sight studied. Our methodology for identifying the high velocity features and the possible complications involved in these identifications are described in detail by Wakker et al. (2003). We searched for absorption in an LSR velocity range of ± 1200 km s^{-1} centered on the O VI $\lambda 1031.926$ line. With few exceptions, the high velocity O VI absorption is confined to $100 \leqslant |v_{LSR}| \leqslant 400$ km s^{-1}, indicating that the identified O VI features are either associated with the Milky Way or are nearby clouds within the Local Group.

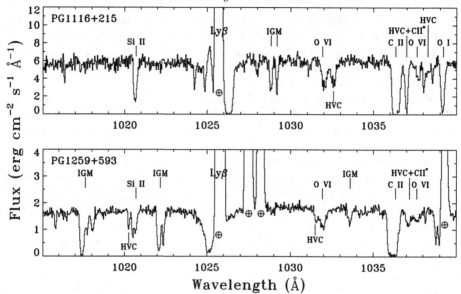

FIGURE 3. Observations of two objects in the *FUSE* high-velocity O VI survey. The data have a velocity resolution of $\sim$20 km s^{-1} (FWHM) and are binned to $\sim$10 km s^{-1} ($\sim$0.033 Å) samples. Prominent interstellar lines, including the two lines of the O VI doublet at 1031.926 Å and 1037.617 Å, are identified above each spectrum at their rest wavelengths. High velocity O VI absorption is present along both sight lines. H I and metal lines from intervening intergalactic clouds are marked in both panels. Unmarked absorption features are interstellar H$_2$ lines. Crossed circles mark the locations of terrestrial airglow lines of H I and O I. From Sembach et al. (2003).

Information about the lower velocity ($|v_{\mathrm{LSR}}| \leqslant 100$ km s^{-1}) gas inside the thick disk and low halo of the Milky Way can be found in Savage et al. (2003).

We detect high velocity O VI λ1031.926 absorption with total equivalent widths $W_\lambda >$ 30 mÅ at $\geqslant 3\sigma$ confidence along 59 of the 102 sight lines surveyed. For the highest quality sub-sample of the dataset, the high velocity detection frequency increases to 22 of 26 sight lines. Forty of the 59 sight lines have high velocity O VI λ1031.926 absorption with $W_\lambda > 100$ mÅ, and 27 have total equivalent widths $W_\lambda > 150$ mÅ. Converting these O VI equivalent width detection frequencies into estimates of $N(\mathrm{H}^+)$ in the hot gas indicates that $\sim$60% of the sky (and perhaps as much as $\sim$85%) is covered by hot ionized hydrogen at a level of $N(\mathrm{H}^+) \gtrsim 10^{18}$ cm^{-2}, assuming an ionization fraction $f_{\mathrm{O\,VI}} < 0.2$ and a gas metallicity similar to that of the Magellanic Stream ($Z/Z_\odot \sim 0.2$–0.3). This detection frequency of hot H$^+$ associated with the high velocity O VI is larger than the value of $\sim$37% found for warm, high velocity neutral gas with $N(\mathrm{H\,I}) \sim 10^{18}$ cm^{-2} traced through 21 cm emission (Lockman et al. 2002).

4.2. *Velocities*

The high velocity O VI features have velocity centroids ranging from $-372 < v_{\mathrm{LSR}} < -90$ km s^{-1} to $+93 < v_{\mathrm{LSR}} < +385$ km s^{-1}. There are an additional six confirmed or very likely (> 90% confidence) detections and two tentative detections of O VI between $v_{\mathrm{LSR}} = +500$ and $+1200$ km s^{-1}; these very high velocity features probably trace intergalactic gas beyond the Local Group. Most of the high velocity O VI features have velocities incompatible with those of Galactic rotation (by definition).

We plot the locations of the high velocity O VI features in Figure 4, with squares indicating positive velocity features and circles indicating negative velocity features. The

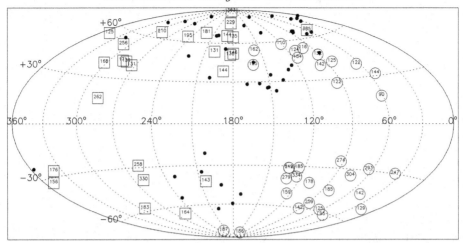

FIGURE 4. All-sky Hammer-Aitoff projection of the high velocity O VI features along the sight lines in the *FUSE* O VI survey. Squares indicate positive velocity features. Circles indicate negative velocity features. The magnitude of the velocity is given inside each circle or square for the highest column density O VI HVC component along each sight line. Solid dots indicate sight lines with no high velocity O VI detections.

magnitude of the LSR velocity (without sign) is given inside each square or circle for the dominant high velocity O VI feature along each sight line. For sight lines with multiple high velocity features, we have plotted the information for the feature with the greatest O VI column density. Solid dots indicate directions where no high velocity O VI was detected.

There is sometimes good correspondence between high velocity H I 21 cm emission and high velocity O VI absorption. When high velocity H I 21 cm emission is detectable in a particular direction, high velocity O VI absorption is usually detected if a suitable extragalactic continuum source is bright enough for *FUSE* to observe. For example, O VI is present in the Magellanic Stream, which passes through the south Galactic pole and extends up to $b \sim -30°$, with positive velocities for $l \gtrsim 180°$ and negative velocities for $l \lesssim 180°$. O VI is present in high velocity cloud Complex C, which covers a large portion of the northern Galactic sky between $l = 30°$ and $l = 150°$ and has velocities of roughly -100 to -170 km s^{-1}. The H 1821+643 sight line ($l = 94.0°, b = 27.4°$) contains O VI absorption at the velocities of the Outer Arm ($v \sim -90$ km s^{-1}) as well as at more negative velocities.

In some directions, high velocity O VI is observed with no corresponding high velocity H I 21 cm emission. For example, at $l \sim 180°, b > 0°$ there are many O VI HVCs with $\bar{v} \sim +150$ km s^{-1}. Some of these features are broad absorption wings extending from the lower velocity absorption produced by the Galactic thick disk/halo. High velocity O VI features toward Mrk 478 ($l = 59.2°, b = +65.0°, \bar{v} \approx +385$ km s^{-1}), NGC 4670 ($l = 212.7°, b = +88.6°, \bar{v} \approx +363$ km s^{-1}), and Ton S180 ($l = 139.0°, b = -85.1°, \bar{v} \approx +251$ km s^{-1}) stand out as having particularly unusual velocities compared to those of other O VI features in similar regions of the sky. They too lack counterparts in H I 21 cm emission. These features may be located outside the Local Group (i.e., in the IGM). We are currently investigating the H I content of some of these clouds through their H I Lyman-series absorption in the *FUSE* data.

The segregation of positive and negative velocities in Figure 4 is striking, indicating that the clouds and the underlying (rotating) disk of the Galaxy have very different

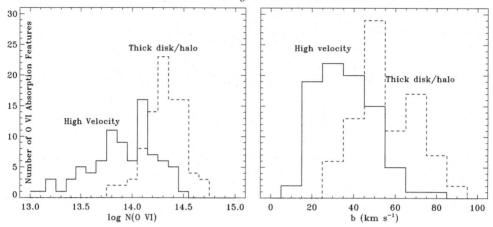

FIGURE 5. Histograms of the high velocity O VI column densities and line widths (solid lines). The bin sizes are 0.10 dex and 10 km s^{-1}, respectively. For comparison, the distributions for the O VI absorption arising in the thick disk and halo of the Galaxy are also shown (dashed lines) (from Sembach et al. 2003).

kinematics. A similar velocity pattern is seen for high velocity H I 21 cm emission and has sometimes been used to argue for a Local Group location for the high velocity clouds (see Blitz et al. 1999). The kinematics of the high velocity O VI clouds are consistent with a distant location, but do not necessarily require an extended Local Group distribution as proposed by Nicastro et al. (2003). The dispersion about the mean of the high velocity O VI centroids decreases when the velocities are converted from the Local Standard of Rest (LSR) into the Galactic Standard of Rest (GSR) and the Local Group Standard of Rest (LGSR) reference frames. While this reduction is expected if the O VI is associated with gas in a highly extended Galactic corona or in the Local Group, it *does not* provide sufficient proof by itself of an extended extragalactic distribution for the high velocity gas because the correction to the LGSR reference frame requires proper knowledge of the total space velocities of the clouds. Only one component of motion—the velocity toward the Sun—is observed. Additional information, such as the gas metallicity, ionization state, or parallax resulting from transverse motion across the line of sight, is needed to constrain whether the clouds are located near the Galaxy or are farther away in the Local Group.

4.3. *Column densities and line widths*

The high velocity O VI features have logarithmic column densities (cm^{-2}) of 13.06 to 14.59, with an average of $\langle \log N \rangle = 13.95 \pm 0.34$ and a median of 13.97 (see Figure 5, left panel). The average high velocity O VI column density is a factor of 2.7 times lower than the typical low velocity O VI column density found for the same sight lines through the thick disk/halo of the Galaxy (see Savage et al. 2003).

The line widths of the high velocity O VI features range from ~16 km s^{-1} to ~81 km s^{-1}, with an average of $\langle b \rangle = 40 \pm 14$ km s^{-1} (see Figure 5, right panel). The lowest values of b are close to the thermal width of 17.1 km s^{-1} expected for O VI at its peak ionization fraction temperature of $T = 2.8 \times 10^5$ K in collisional ionization equilibrium (Sutherland & Dopita 1993). The higher values of b require additional non-thermal broadening mechanisms or gas temperatures significantly larger than 2.8×10^5 K.

4.4. *Origin of the high velocity* O VI

One possible explanation for some of the high velocity O VI is that transition temperature gas arises at the boundaries between cool/warm clouds of gas and a very hot ($T > 10^6$ K) Galactic corona or Local Group medium. Sources of the high velocity material might include infalling or tidally disturbed galaxies. A hot, highly extended ($R > 70$ kpc) corona or Local Group medium might be left over from the formation of the Milky Way or Local Group, or may be the result of continuous accretion of smaller galaxies over time. Evidence for a hot Galactic corona or Local Group medium is given in §3. Hydrodynamical simulations of clouds moving through a hot, low-density medium show that weak bow shocks develop on the leading edges of the clouds as the gas is compressed and heated (Quilis & Moore 2001). Even if the clouds are not moving at supersonic speeds relative to the ambient medium, some viscous or turbulent stripping of the cooler gas likely occurs. An alternative explanation for the O VI observed at high velocities may be that the clouds and any associated H I fragments are simply condensations within large gas structures falling onto the Galaxy. Cosmological structure formation models predict large numbers of cooling fragments embedded in dark matter, and some of these structures should be observable in O VI absorption as the gas passes through the $T = 10^5 - 10^6$ K temperature regime (Davé et al. 2001).

4.5. *Are the O VI HVCs extragalactic clouds?*

Some of the high velocity O VI clouds may be extragalactic clouds, based on what we currently know about their ionization properties. However, claims that essentially *all* of the O VI HVCs are extragalactic entities associated with an extended Local Group filament based on kinematical arguments alone appear to be untenable. Such arguments fail to consider the selection biases inherent in the O VI sample, the presence of neutral (H I) and lower ionization (Si IV, C IV) gas associated with some of the O VI HVCs, and the known "nearby" locations for at least two of the primary high velocity complexes in the sample—the Magellanic Stream is circumgalactic tidal debris, and Complex C is interacting with the Galactic corona (see Fox et al. 2004). Furthermore, the O VII X-ray absorption measures used to support an extragalactic location have not yet been convincingly tied to either the O VI HVCs or to a Local Group location. The O VII absorption may well have a significant Galactic component in some directions (see Fang et al. 2003). The Local Group filament interpretation (Nicastro et al. 2003) may be suitable for some of the observed high velocity O VI features, but it clearly fails in other particular cases (e.g., the Magellanic Stream or the PKS 2155–304 HVCs—see §3). It also does not reproduce some of the properties of the ensemble of high velocity O VI features in our sample. For example, the "Local Supercluster Filament" model (Kravtsov, Klypin, & Hoffman 2002) predicts average O VI velocity centroids higher than those observed ($\langle \bar{v} \rangle \sim 1000$ km s^{-1} vs. $\langle \bar{v} \rangle < 400$ km s^{-1}) and average O VI line widths higher than those observed (FWHM $\sim 100-400$ km s^{-1} vs. FWHM $\sim 30-120$ km s^{-1}). Additional absorption and emission-line observations of other ions at ultraviolet wavelengths, particularly C IV and Si IV, would provide valuable information about the physical conditions, ionization, and locations of the O VI clouds.

5. Summary and future prospects

The detection of hot gas locally and at low redshift has altered our perspective on the presence of highly ionized gas outside of galaxies and has led to refined estimates of the baryonic content of the present-day universe. Clearly, there is much work yet to be done in determining the spatial distribution of the hot gas, its physical conditions, and its association with galaxies and the larger-scale gaseous structures from which galaxies form.

A deeper understanding of the spatial distribution of the hot gas would be possible with ultra-sensitive maps of the diffuse O VI emission associated with the cosmic web. Emission maps would enable estimates of the filling factor of the hot gas, and would set the context for ongoing absorption-line studies of the hot gas. Experiments to produce O VI emission maps over large regions of the sky are well-suited to small and medium Explorer-class missions and have been proposed to NASA.

The Hubble Space Telescope Cosmic Origins Spectrograph, which is to be installed as part of the next *HST* servicing mission, will greatly enhance studies of O VI systems at redshifts $0.14 < z < 0.5$ by increasing the number of accessible background sources to use for the absorption-line measurements. Larger redshift paths will be probed, and more sight lines will be examined. In addition, the high sensitivity of the spectrograph will permit the acquisition of very high signal-to-noise spectra for bright extragalactic sources. In the longer term, larger (6–20 m) ultraviolet/optical telescopes in space will make it possible to select many closely spaced lines of sight for detailed spectroscopic and imaging investigations of galactic and intergalactic structures associated with the hot gas. A key scientific motivation for such investigations will be to determine feedback mechanisms operating on the hot gas and the degree to which these processes affect the formation and evolution of galaxies.

I thank Blair Savage, Bart Wakker, Philipp Richter, and Marilyn Meade for their efforts in making the *FUSE* O VI survey a reality. I acknowledge lively discussions about O VI absorption with Todd Tripp and Mike Shull.

REFERENCES

ALLENDE PRIETO, C., LAMBERT, D. L., ASPLUND, M. 2001 *ApJ* **556,** L63.

BLITZ, L. & ROBISHAW, T. 2000 *ApJ* **541,** 675.

BLITZ, L., SPERGEL, D. N., TEUBEN, P., HARTMANN, D., & BURTON, W. B. 1999 *ApJ* **514,** 818.

CAGNONI, I. 2002; astro-ph/0212070.

CEN, R. & OSTRIKER, J. P. 1999 *ApJ* **514,** 1.

COLLINS, J. A., SHULL, J. M., & GIROUX, M. L. 2004 *ApJ*, submitted.

DAVÉ, R., ET AL. 2001 *ApJ* **552,** 473.

DE BOER, K. S., BRAUN, J. M., VALLENARI, A., & MEBOLD, U. 1998 *A&A* **329,** L49.

FANG, T. T., MARSHALL, H. L., LEE, J. C., DAVIS, D. S., & CANIZARES, C. R. 2002 *ApJ* **572,** L127.

FANG, T., SEMBACH, K. R., & CANIZARES, C. R. 2003 *ApJ*, **586,** L49.

FUTAMOTO, K., MITSUDA, K., TAKEI, Y., FUJIMOTO, R., & YAMASAKI, N. 2003 *ApJ* **605,** 793.

FOX, A., SAVAGE, B. D., WAKKER, B. P., RICHTER, P., TRIPP, T. M., & SEMBACH, K. R. 2004 *ApJ* **602,** 738.

FUKUGITA, M., HOGAN, C., & PEEBLES, J. 1998 *ApJ* **503,** 518.

GARDINER, L. T., 1999, in ASP Conf. Ser. 166, *The Stromlo Workshop on High Velocity Clouds* (eds. B. K. Gibson & M. E. Putman), p. 292. ASP.

HECKMAN, T. M., NORMAN, C. A., STRICKLAND, D. K., & SEMBACH, K. R. 2002 *ApJ* **577,** 691.

HOWK, J. C., SAVAGE, B. D., SEMBACH, K. R., & HOOPES, C. G. 2002 *ApJ* **572,** 264.

KRAVTSOV, A. V., KLYPIN, A., & HOFFMAN, Y. 2002 *ApJ* **571,** 563.

LOCKMAN, F. J., MURPHY, E. M., PETTY-POWELL, S., & URICK, V. 2002 *ApJ* **140,** 331.

MATHUR, S., WEINBERG, D. H., & CHEN, X. 2003 *ApJ* **582,** 82.

MALONEY, P. R. 2003, in *The IGM/Galaxy Connection* (eds. J. L. Rosenberg & M. E. Putman), p. 299. Kluwer.

MCKERNAN, B., YAQOOB, T., GEORGE, I. M., & TURNER, T. J. 2003a, *ApJ* **593**, 142.

MCKERNAN, B., YAQOOB, T., MUSHOTZKY, R., GEORGE, I. M., & TURNER, T. J. 2003b *ApJ* **598**, L83.

MOORE, B. & DAVIS, M. 1994, *MNRAS* **270**, 209.

MORTON, D. C. 1991, *ApJS* **77**, 119.

NICASTRO, F., ET AL. 2002 *ApJ* **573**, 157.

NICASTRO, F., ET AL. 2003 *Nature* **421**, 719.

OEGERLE, W. R., ET AL. 2000 *ApJ* **538**, L23.

PENTON, S. V., SHULL, J. M., & STOCKE, J. T. 2000 *ApJ* **544**, 150.

QUILIS, V., & MOORE, B. 2001 *ApJ* **555**, L95.

RASMUSSEN, A., KAHN, S. M., & PAERELS, F. 2003, in *The IGM/Galaxy Connection* (eds. J. L. Rosenberg & M. E. Putman), p. 109. Kluwer.

RICHTER, P., SAVAGE, B. D., TRIPP, T. M., & SEMBACH, K. R. 2004 *ApJ*, submitted.

SAVAGE, B. D., SEMBACH, K. R., TRIPP, T. M., & RICHTER, P. 2002 *ApJ* **564**, 631.

SAVAGE, B. D., ET AL. 2003 *ApJS* **146**, 125.

SAVAGE, B. D, WAKKER, B. P., SEMBACH, K. R., RICHTER, P., & MEADE, M. 2004, in IAU Symp. 217, *Recycling Intergalactic and Interstellar Matter* (eds. P. Duc, J. Brain, & E. Brinks), in press.

SEMBACH, K. R., SAVAGE, B. D., LU, L., & MURPHY, E. M. 1999 *ApJ* **515**, 108.

SEMBACH, K. R., ET AL. 2000 *ApJ* **538**, L31.

SEMBACH, K. R., HOWK, J. C., SAVAGE, B. D., SHULL, J. M., & OEGERLE, W. R. 2001 *ApJ* **561**, 573.

SEMBACH, K. R., ET AL. 2003 *ApJS* **146**, 165.

SEMBACH, K. R., TRIPP T. M., & SAVAGE, B. D. 2004 *ApJ*, in prep.

SHULL, J. M., TUMLINSON, J., & GIROUX, M. L. 2003 *ApJ* **594**, L107.

SNOWDEN, S. L., ET AL. 1997 *ApJ* **485**, 125.

SPERGEL, D., ET AL. 2003 *ApJS* **148**, 175.

STANIMIROVIC, S., DICKEY, J. M., KRCO, M., & BROOKS, A. M. 2002 *ApJ* **576**, 773.

SUTHERLAND, R. S., & DOPITA, M. A. 1993 *ApJS* **88**, 253.

TRIPP, T. M. 2002, in ASP Conf. Ser. 254, *Extragalactic Gas at Low Redshift* (eds. J. S. Mulchaey & J. T. Stocke), p. 323. ASP.

TRIPP, T. M., GIROUX, M. L., STOCKE, J. T., TUMLINSON, J., & OEGERLE, W. R. 2001 *ApJ* **563**, 724.

TRIPP, T. M., & SAVAGE, B. D. 2000 *ApJ* **542**, 42.

TRIPP, T. M., SAVAGE, B. D., & JENKINS, E. B. 2000 *ApJ* **534**, L1.

TUMLINSON, J., SHULL, J. M., GIROUX, M. L., & STOCKE, J. T. 2004, in prep.

VERNER, D. A, BARTHEL, P. & TYTLER, D. 1994 *ApJ* **430**, 186.

VERNER, D. A., VERNER, E. M., & FERLAND, G. J. 1996, *Atomic Data Nucl. Data Tables* **64**, 1.

WAKKER, B. P. 2001 *ApJS* **136**, 463.

WAKKER, B. P. 2004, in IAU Symp. 217, *Recycling Intergalactic and Interstellar Matter* (eds. P. Duc, J. Brain), in press.

WAKKER, B. P., ET AL. 2003 *ApJS* **146**, 1.

Stages of satellite accretion

By M. E. PUTMAN

Center for Astrophysics and Space Astronomy, University of Colorado, Boulder, CO
80309-0389, USA; Hubble Fellow; mputman@casa.colorado.edu

The Galaxy's extended halo contains numerous satellites which are in the process of being disrupted. This paper discusses the stages of satellite accretion onto the Galaxy with a focus on the Magellanic Clouds and Sagittarius dwarf galaxy. In particular, a possible gaseous component to the stellar stream of the Sgr dwarf is presented that has a total neutral hydrogen mass between 4–10×10^6 $M_\odot$ at the distance to the stellar debris in this direction (36 kpc). This gaseous stream was most likely stripped from the main body of the dwarf 0.2–0.3 Gyr ago during its current orbit after a passage through a diffuse edge of the Galactic disk with a density $> 10^{-4}$ cm^{-3}. This gas represents the dwarf's last source of star formation fuel and explains how the galaxy was forming stars 0.5–2 Gyr ago. This is consistent with the star formation history and H I content of the other Local Group dwarf galaxies.

---

1. Introduction

Our Galaxy has built itself up by accreting satellite galaxies. This process if evident today through the satellites currently found in the extended Galactic halo. There are nine satellite galaxies within 150 kpc interacting with our Galaxy at various levels. These are in order of distance (Grebel, Gallagher, & Harbeck 2003): the Fornax dSph (138 kpc), the Carina dSph (94 kpc), the Sculptor dSph (88 kpc), the Sextans dSph (86 kpc), the Draco dSph (79 kpc), the Ursa Minor dSph (69 kpc), the Small Magellanic Cloud (63 kpc), the Large Magellanic Cloud (50 kpc), and the closest example of a recognizable accreting satellite is the Sagittarius Dwarf (28 kpc; hereafter Sgr dwarf).† Many of these satellites show evidence of the Galaxy's tidal forces through elongation and/or the presence of extra-tidal stars (e.g., Carina [Majewski et al. 2002] and Ursa Minor [Palma et al. 2003]). The Galaxy's closest satellite, the Sgr dwarf, demonstrates accumulating evidence for a continuous stream of stellar tidal debris as it spirals into the Milky Way (e.g., Ibata et al. 1994; Newberg et al. 2002; Majewski et al. 2003; hereafter M03). The stars in the Sgr dwarf stream span a wide range of ages, with the youngest population between 0.5–3 Gyr old (Layden & Sarajedini 2000; Dolphin 2002; M03). This indicates that within the past Gyr this Galactic satellite was forming stars and had a source of star formation fuel.

Neutral hydrogen is a principal source of star formation fuel for a galaxy. Galaxies which contain H I are commonly currently forming stars (e.g., Lee et al. 2002; Meurer et al. 2003) and those without detectable H I tend to have a primarily older stellar population and thus appear to have exhausted their star formation fuel (e.g., Gavazzi et al. 2002). This is evident in the dwarf galaxies of the Local Group. The stars in the Local Group dwarf galaxies vary from being almost entirely ancient (>10 Gyr; Ursa Minor) to a number of systems which are actively forming stars (e.g., WLM, Phoenix, LMC). The H I content of the dwarfs is summarized by Mateo (1998) and Grebel et al. (2003), but it may be possible to make several additions that cause their star formation histories and H I content to more closely agree. The Sculptor dSph galaxy contains predominantly older stars, and though there have been claims of associated H I (Carignan et al. 1998;

† Some may argue the closest recognizable example of an accreting satellite is the recently discovered ring-like stellar structure in the direction of the Galactic anti-center, often called the Monoceros structure (e.g., Newberg et al. 2002; Crane et al. 2003).

Bouchard et al. 2003a), it is also possible that the clouds are part of a large complex of high-velocity clouds found in this region (Putman et al. 2003; hereafter P03). Recent results also indicate the Fornax and Carina galaxies may have H I gas associated with them (Bouchard et al. 2003b), which agrees with their recent star formation activity (<2 Gyr ago). With these additions, all of the Local Group dwarf galaxies that have detectable H I have formed stars within the past two Gyr and those without H I show no evidence for recent star formation (e.g., Mateo 1998; Dolphin 2002).

The H I gas associated with Galactic satellites can be an important tracer of their interaction with the Galaxy. This has been previously evident with the Magellanic System. The Magellanic Clouds are trailed by a 100° continuous H I stream, known as the Magellanic Stream (Mathewson et al. 1974; P03), and are led by a 25° −50° Leading Arm (Putman et al. 1998). Pointed H I observations on the central region of the Sgr dwarf (α, δ = 19^h 00^m, −30° 25′ (J2000); l, b = 6°, −15°) indicate that our closest satellite galaxy does not currently contain a significant amount of star formation fuel (M_{HI} < 1.5×10^4 $M_\odot$ (3σ); Koribalski, Johnston, & Otrupcek 1994). The search for H I associated with the Sgr dwarf was continued along the leading arm of the Sgr stream by Burton & Lockman (1999); but they also found no associated gas over 18 deg² between b = −13° to −18.5° with limits of M_{HI} < 7×10^3 $M_\odot$ (3σ). These results are surprising considering the Sgr dwarf was forming stars within the last Gyr. Since the orbit of the Sgr dwarf is approximately 0.7 Gyr (Ibata & Lewis 1998), one might expect to find the fuel stripped along the dwarfs orbit, possibly in a similar location to the stellar trail. The trailing stellar tidal tail of the Sgr dwarf has recently been found to extend for over 150° across the South Galactic Hemisphere with a mean distance between 20 to 40 kpc from the Sun (M03). This proceedings presents H I data from HIPASS (HI Parkes All-Sky Survey†) along the entire Sgr dwarf galaxy orbit to investigate the possibility that a gaseous Sgr trail is also present (see also Putman et al. 2004). The Magellanic Stream and the stripping mechanisms responsible for forming gaseous interactive streams are also discussed. Understanding how gas accretes from Galactic satellites onto our Galaxy will aid our comprehension of its current structure and metallicity.

2. Observations

The neutral hydrogen data presented here are from the H I Parkes All-Sky Survey (HIPASS) reduced with the MINMED5 method (P03). HIPASS is a survey for H I in the Southern sky, extending from the South celestial pole to Decl. = +25°, over velocities from −1280 to +12700 km s⁻¹ (see Barnes et al. 2001 for a full description). The survey utilized the 64-m Parkes radio telescope, with a focal–plane array of 13 beams arranged in a hexagonal grid, to scan the sky in 8° zones of Decl. with Nyquist sampling. The spectrometer has 1024 channels for each polarization and beam, with a velocity spacing of 13.2 km s⁻¹ between channels and a spectral resolution, after Hanning smoothing, of 26.4 km s⁻¹. The survey was completed with a repetitive scanning procedure which provides source confirmation, mitigates diurnal influences and aids interference excision.

The MINMED5 reduction method was designed to recover extended emission in the HIPASS data for the purpose of studying HVCs. This procedure greatly increases the sensitivity of the data to large-scale structure without substantial loss of flux density, except when the emission fills the entire 8° scan (i.e., the Galactic Plane). H I emission in the LSR velocity range −700 to +1000 km s⁻¹ was reduced in this manner and the

† The Parkes Telescope is part of the Australia Telescope which is funded by the Commonwealth of Australia for operation as a National Facility managed by CSIRO.

calibrated scans were gridded with the median method described by Barnes et al. (2001), without the weighting which overcorrects the fluxes for extended sources. The spatial size of the gridded cubes is $24° \times 24°$ with a few degrees of overlap between each cube. For extended sources, the RMS noise is 10 mJy beam^{-1} (beam area 243 arcmin^2), corresponding to a brightness temperature sensitivity of 8 mK. The northern extension of the survey from $+2°$ to $+25°$ was only recently completed and the noise in these cubes is elevated compared to the southern data (11 mK vs. 8 mK). This may be due a combination of low zenith angles during the observations and an inability to avoid solar interference as effectively. The details involved in making the map of the Magellanic Stream shown in Figure 1 are described in P03. To make Figure 3, integrated intensity maps of the high negative velocity gas (generally $v_{lsr} < -80$ km s^{-1}, as long as the gas was clearly separate from Galactic emission) were made for the cubes which lie along the Sgr orbit. The positive velocity cubes had very little emission in them, so we concentrated on the negative velocity cubes. The noise at the edges of the negative velocity maps were blanked within AIPS and the images were then read into IDL to create the map of the entire orbit shown in Figure 3. At 20–40 kpc the MINMED5 reduced HIPASS data has a sensitivity to clouds of gas with $M_{HI} > 80 - 320\ M_\odot$ ($\Delta v = 25$ km s^{-1}; 3σ).

3. H I associated with the Magellanic System

The Magellanic Clouds depict their interaction with the Galaxy through a $100°$ trailing stream containing $2 \times 10^8\ M_\odot$ of H I (known as the Magellanic Stream) and a Leading Arm containing a tenth of that mass (see Figure 1). These two counter streams show there are tidal forces affecting the Clouds (Putman et al. 1998). The Magellanic Stream is complex, but has a smooth velocity gradient from head to tail which is 400 km s^{-1} greater than that expected from Galactic rotation alone. It also shows a decrease in column density from head to tail that may represent the lower column density of the first gas stripped from the outer regions of the Clouds, or the Stream being overwhelmed by a halo medium. The 2MASS data presented by van der Marel (2001) indicates the stellar component of the LMC is also being tidally stretched; however, the stellar disruption of the Clouds is limited compared to the H I. The Magellanic Stream does not appear to have a stellar counterpart (e.g., Guhathakurta & Reitzel 1998). The Magellanic Bridge, an H I link between the LMC and SMC, does contain stars (e.g., Demers & Battinelli 1998), but many of them appear to have formed within the Bridge. This lack of stellar streams is most likely due to the gaseous component initially being more extended than the stellar component, but as shown in Figure 2, the current distribution of stars (seen by 2MASS) and gas in the LMC is equally extended.

Tidal forces on the Clouds are most likely being supplemented by an interaction of the H I gas with a diffuse Galactic halo medium. This statement is made based on a number of features of the Magellanic System, including the column density gradient and lack of stellar stream mentioned in the previous paragraph.

The Leading Arm does not follow the orbit of the Magellanic Clouds, but has a deflection angle of $\sim 30°$ which can be explained by adding a degree of ram pressure stripping to a tidal model (Gardiner 1999). There are quite a number of head-tail HVCs along the Magellanic Stream that are easily explained as the interaction of clouds with a halo medium (Quilis & Moore 2001). There are also [O VI] absorption line detections along the length of the Magellanic Stream that can be produced as the Stream interacts with a hot halo medium (Sembach et al. 2003). Finally, there are bright Hα detections along the Stream that have yet to be explained (Putman et al. 2003b).

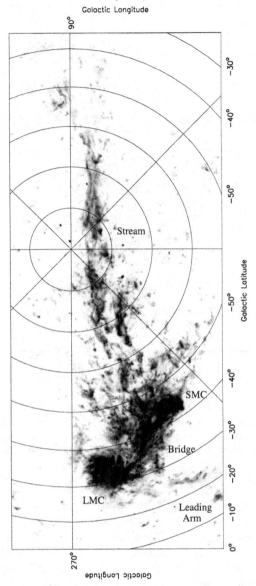

FIGURE 1. The interaction of the Magellanic Clouds with each other and the Galaxy is shown in Galactic coordinates with the main features labeled (P03). $\ell = 0°$ is on the right. This is a column density map, with the lowest levels corresponding to 2×10^{18} cm^{-2}.

A diffuse halo medium with a density on the order of 10^{-4} cm^{-3} at the distance of the Magellanic Clouds is likely. The precise distance of the Magellanic Stream is unknown, but most models place it between 20–80 kpc away. Estimates of the halo density at these distances come from many sources: $\sim 5 \times 10^{-4}$ cm^{-3} is the density needed to produce the deflection angle of the Leading Arm (Gardiner 1999), $> 10^{-4}$ cm^{-3} to create the head-tail HVCs (Quilis & Moore 2001), $< 3 \times 10^{-4}$ cm^{-3} to confine the tail of the Magellanic Stream (Stanimivovic et al. 2002), the [O VI] absorption line results indicate 10^{-4} to 10^{-5} cm^{-3} within 70 kpc, and finally, though assumptions must be made on the gas distribution, the detection of x-ray emission (Wang & McCray 1993) and dispersion

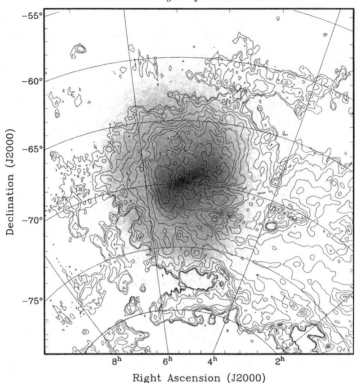

FIGURE 2. The LMC's distribution of H I (P03) and the red giant branch and asymptotic giant branch stars from 2MASS (van der Marel 2001) in Celestial coordinates. Contours are 0.4, 0.8, 1.6, 3.2×10^{19} cm^{-2}, etc. up to the central contour of 2.2×0^{21} cm^{-2}. The Magellanic Bridge is shown to extend to the right in this image, as it extends toward the Small Magellanic Cloud and Stream. The high velocity H I gas at Decl. $\sim -61°$ reaches velocities up to 400 km s^{-1} (LSR).

measures towards LMC pulsars (Taylor, Manchester, & Lyne 1993) suggest a halo density of approximately 10^{-4}. This type of halo medium should have a significant affect on the current distribution of gas in the Magellanic System and may also be responsible for affecting the gas in the other satellites of the Galaxy. This was recently discussed by Blitz & Robishaw (2000), and with the changes in H I content discussed in the introduction, the lack of H I gas present in most dwarf galaxies within 250 kpc of the Milky Way (excluding the Magellanic Clouds) is an interesting relationship. The next two sections discuss the effect this halo medium may have on the closest satellite to the Milky Way.

4. H I associated with the Sgr Dwarf

The large scale H I map which includes all of the high negative velocity gas along the orbit of the Sgr dwarf is shown in Figure 3. This plot is in Celestial coordinates as it depicts the main features found along the orbit better than Galactic coordinates. The Galactic Plane is shown by the solid line with the dotted lines on each side representing $\pm 5°$ from $b = 0$. The orbit of the Sgr dwarf, as calculated by Ibata & Lewis (1998), is shown by the solid line extending across the plot horizontally, and the current position of the Sgr dwarf is represented by the solid circle. The stream of M giants presented by M03 has quite a broad width (commonly several degrees) along this orbit. The orbit of the Sgr dwarf crosses the Galactic Plane at $\ell \sim 185°$ and $\ell \sim 0°$ which corresponds to

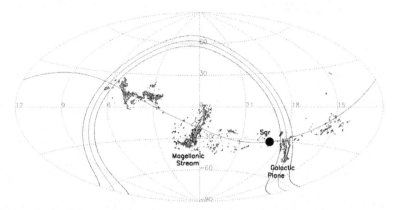

FIGURE 3. The negative high-velocity H I (~ -85 to -400 km s^{-1}; LSR) along the orbit of the Sgr dwarf galaxy in Celestial coordinates. The current position of the Sgr dwarf is shown by the solid point, and the orbit of the Sgr dwarf (Ibata & Lewis 1998) is plotted as the solid line through this point. The negative velocity gas attributed to the Magellanic Stream and Galactic Center is labeled, and the Galactic Plane is indicated by the solid line with the two dotted lines on each side representing $b = +5°$ and $-5°$. The negative velocity carbon stars extend from approximately $\alpha, \delta = 0^h, -20°$ to $12^h, 20°$ along the Sgr orbit (Ibata et al. 2001). Contours represent column density levels of 0.5, 1.0, 5.0, and 10.0×10^{19} cm^{-2}.

approximately $(\alpha, \delta) = 6^h$, $25°$ and 18^h, $-30°$ respectively. At the Galactic Center high negative and positive velocity gas is present, thus we have labeled the negative velocity H I emission within $\pm 5°$ of the Galactic plane at that location. The Sgr orbit crosses the Magellanic Stream (labeled) at the South Galactic Pole or $(\alpha, \delta) = 0^h$, $-15°$. The majority of the Sgr dwarf orbit does not have both stars and H I gas at high negative velocities, with the exception of the H I gas between $\alpha = 3$ to 4.5 h and $\delta = 0$ to $30°$, or $\ell \approx 155$ to $195°$ and $b \approx -5$ to $-50°$. A close up of these H I clouds in Galactic coordinates is shown in Figures 4 and 5. These plots show the column density and velocity distribution of the high-velocity gas, respectively. The M giants in this region of the Sgr stellar stream fill a large percentage of Figs. 4 and 5. This high velocity H I gas was previously identified as part of the Anti Center complexes (ACHV and ACVHV; Wakker & van Woerden 1991) and was discovered almost 40 years ago (e.g., Mathewson et al. 1966), but its relationship to the Sgr dwarf was not previously noted. There are some small negative velocity clouds along the Sgr dwarf orbit, but besides the labeled gas, the only other substantial complex of negative velocity gas along the orbit of the Sgr dwarf is in a region which has predominantly positive velocity stars (Ibata et al. 2001). This is the group of high-velocity clouds known as Complex L at $\alpha \approx 15.25$ h, $\delta \approx -19.5°$. The amount of high positive velocity gas along the orbit of the Sgr dwarf is very limited, and not correlated with the position of the positive velocity stars along the Sgr stellar stream.

The complex of gas shown in Figs. 4 and 5 has a velocity range that extends from to -375 to -75 km s^{-1} (LSR). This region of sky begins to be dominated by Galactic emission at -125 km s^{-1} however, so only velocities from -375 to -125 km s^{-1} are included. The filament closer to the Galactic Plane is restricted to the velocity range of -245 to -75 km s^{-1} and has a velocity gradient across its long axis that is increasingly negative towards the Galactic Plane. The emission extending from -375 km s^{-1} to -180 km s^{-1} at $b < -20°$ is oriented along the orbit of the Sgr dwarf and is completely isolated from gas that merges with lower velocity emission (see the channel maps in

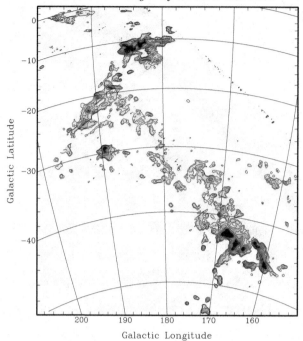

FIGURE 4. Integrated intensity map of the negative velocity H I clouds found along the trailing stellar stream of the Sgr dwarf in Galactic coordinates. The M giants extend across this plot in a $\sim 15°$ band from $\ell, b = 185°, 0°$ to the bottom of the plot at $\ell = 170°$. Contours are 0.6, 1.2, 2.4, 4.8, and 9.6×10^{19} cm^{-2}.

Putman et al. 2004). The carbon stars associated with the Sgr dwarf at this position have velocities between -140 to -160 km s^{-1} (LSR; Dinescu et al. 2002; Totten & Irwin 1998; Green et al. 1994). Thus the velocity range of this gas tends to be more negative than that of the limited number of stars which have velocity determinations. This gas lies approximately at the position where the Sgr dwarf would have passed through the extreme edge of the Galactic disk (assuming the gas is at the same distance of the stellar debris, ~ 45 kpc from the Galactic Center; M03).

All of the gas in the region of the stellar debris shown in Figs. 4 and 5 would amount to $M_{HI} = 1.2 \times 10^{7}\ M_{\odot}$ of neutral hydrogen at 36 kpc, the approximate distance to the stars in the Sgr stream in this direction. If the filament closest to the Galactic Plane is isolated, its mass is $5.4 \times 10^{6}\ M_{\odot}$. The highest negative velocity emission which is aligned with the orbit of the Sgr dwarf has a total H I mass of $4.3 \times 10^{6}\ M_{\odot}$. Both of these complexes have peak column densities on the order of 10^{20} cm^{-2} at the 15.5′ resolution of HIPASS and extend to the column density limits of the data ($5\sigma \sim 3 \times 10^{18}$ cm^{-2}; $\Delta v = 25$ km s^{-1}). The distance of 36 kpc was adopted based on the continuous stellar trail presented in M03 at those distances in this region.

5. Discussion of the Sgr Dwarf H I

The H I gas presented in Figs. 3–5 lies at negative velocities and is at approximately the same position as the stellar Sgr stream, indicating that the gas was stripped from the Sgr dwarf in its current orbit. The present orbit of the Sgr dwarf is estimated to be 0.7 Gyr (Ibata & Lewis 1998). Using this orbit, the core of the Sgr dwarf was at the position of the H I complex approximately 0.2 to 0.3 Gyr ago. It would make sense if

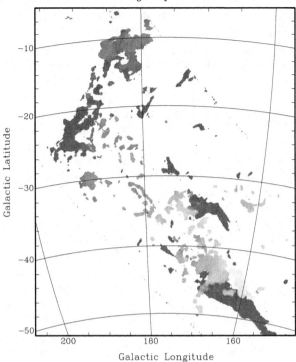

FIGURE 5. The velocity distribution of the H I clouds found along the trailing stellar stream of the Sgr dwarf. Black represents -125 km s^{-1} (LSR) and light grey represents -375 km s^{-1}. Carbon stars in the Sgr stellar stream in this region have velocities which range from -140 to -160 km s^{-1}.

the gas was part of the Sgr dwarf within that timescale considering the age of part of the Sgr dwarf stellar population (0.5–2 Gyr; Layden & Sarajedini 2000; Dolphin 2002; M03). Not surprisingly star formation surveys find that those galaxies which are actively forming stars, also contain substantial amounts of star formation fuel in the form of neutral hydrogen (Lee et al. 2002 [KISS]; Meurer et al. 2003 [SINGG]).

A large percentage of the H I gas is along a sight line close to the plane of our Galaxy, and at this position the Sgr dwarf stellar debris is approximately 40–50 kpc from the Galactic Center (M03). We have adopted 45 kpc as the typical distance from the Galactic Center at the position of the bulk of the gas. This distance is significantly beyond the typical radius quoted for our Galaxy ($\sim$26 kpc), however it is possible that an extended ionized disk exists for our Galaxy (e.g., Savage et al. 2003), as found in other systems (Bland-Hawthorn, Freeman & Quinn 1997). At 36 kpc from the sun the H I gas most likely once associated with the Sgr dwarf has a mass between 4–10×10^6 $M_\odot$. A typical total H I mass for a dwarf galaxy is on the order of 10^7 to 10^8 $M_\odot$ (Grebel et al. 2003), so this amount of gas originally being associated with the dwarf is certainly plausible and may represent anywhere from the majority to 10% of the total amount of gas once associated with the Sgr dwarf. Using a total Sgr dwarf mass of 5×10^8 $M_\odot$ (M03), this gas represents 1–2% of its total mass.

The Sgr dwarf most likely originally had more than 10^7 $M_\odot$ of neutral hydrogen associated with it. The outer gaseous component of the Sgr dwarf would have been the first thing stripped from our closest satellite, as predicted by tidal simulations (e.g., Mihos 2001) and evident in the interaction of the Magellanic Clouds with the Milky Way which exhibits a gaseous stream with no stellar counterpart (see Section 3). This outer gas would have column densities between 10^{18}–10^{19} cm^{-2} and has most likely either already dispersed or been ionized, although remnants of this gas may be present as small HVCs along the Sgr orbit. Since there is currently no H I associated with the core of the Sgr dwarf (Koribalski et al. 1994), and the dwarf has stars which are 0.5–2 Gyr old, the H I gas presented here most likely represents the high column density gas from the core of the Sgr dwarf. This is supported by the relatively high peak column densities of the H I filament ($\sim 2 \times 10^{20}$ cm^{-2}). The passage through an extended disk of our Galaxy, in addition to the tidal forces already obviously at work as evident from the stellar tidal stream, might have been enough to disrupt the H I in the core of the galaxy and cause the dwarf to lose all remaining star formation fuel. The gas will be stripped from a galaxy if $\rho_{IGM} v^2 > \sigma^2 \rho_{gas}/3$ (Mori & Burkert 2000; Gunn & Gott 1972). We can use this equation to estimate the density needed at the edge of the disk to strip the gas from the core of the Sgr dwarf. We use a tangential velocity of 280 km s^{-1} for the Sgr dwarf and a velocity dispersion of 11.4 km s^{-1} (Ibata et al. 2004; Ibata & Lewis 1998). If the column densities and size of the H I distribution in the core of the Sgr dwarf were on the order of 5×10^{20} cm^{-2} (averaged over the core) and 1 kpc, the typical ρ_{gas} is 0.16 cm^{-3}. An extended disk density greater than 3×10^{-4} cm^{-3} is then needed to strip the gas via ram pressure stripping. Based on estimates of the density of the Galactic halo discussed in section 3, this density would be easily achieved in the plane of Galaxy, 20 kpc from the currently observed edge of the disk.

The stripping of gas from a Galactic satellite by passing it through the edge of the Galaxy's disk was proposed as a possible solution for the origin of the Magellanic Stream by Moore & Davis (1994). They discuss the possibility of an extended ionized disk at 65 kpc from the Galactic center with column densities $< 10^{19}$ cm^{-2}. Indeed, rotating, ionized disks extending out to 100 kpc beyond the extent of the spiral galaxies observed in emission are consistent with the results summarized above, as well as absorption line observations (e.g., Steidel et al. 2002) and theoretical arguments to explain the H I column densities at the edges of spirals with the extragalactic UV radiation field (Maloney 1993). As in the case of the Magellanic System, the forces on the Sgr dwarf are most likely a combination of tidal and ram pressure. Though the H I trail of the Sgr dwarf has a very different structure from the Magellanic Stream which contains $\sim 2 \times 10^8$ $M_\odot$ of neutral hydrogen (Fig. 1), it does show similarities to the Leading Arm of the Magellanic System and depending on the actual distance to the Leading Arm, it may have a similar mass (Putman et al. 1998; Wakker, Oosterloo & Putman 2002). For the Sgr H I filament to survive for 0.2–0.3 Gyr (the time since the last passage of the core Sgr dwarf), not to mention the Magellanic Stream which is thought to be ~ 2 Gyr old, the gas should either be confined by an existing halo medium or associated with significant amounts of dark matter.

6. Overview

Our Galaxy's halo is made up of numerous streams of satellite debris. The finding of H I gas associated with the Sagittarius dwarf galaxy, along with the Magellanic Stream, emphasizes the number of dynamical processes occurring within the Galactic halo. Since the gas is responsible for the formation of the stars which form the stellar trail, these two satellite components should be studied in unison. The H I stream found along the orbit of the Sgr dwarf galaxy contains between 4–10×10^6 $M_\odot$ at the distance of the Sgr stellar stream in this direction (~ 36 kpc). This is approximately 1–2% of the total mass of the dwarf, and 20–50 times lower than the H I mass of the Magellanic Stream, and approximately the same H I mass as the Leading Arm of the Magellanic System (P03). The H I gas along the Sgr orbit was most likely the last gas stripped from the core of the Sgr dwarf and was finally stripped ~ 0.2 to 0.3 Gyr ago due to the dwarf's passage through the edge of the Galaxy's disk with densities $> 10^{-4}$ cm^{-3}. This is supported by the matching spatial and kinematic distribution of the gas and stars along the Sgr orbit, the location of the filament along a sight line close to the plane of our Galaxy, the relatively high column densities of the HI gas, and the star formation history of the Sgr dwarf. A density greater than a few $\times 10^{-4}$ cm^{-3} at a distance of 36 kpc in the Galactic Plane fits with estimates of the Galactic halo density at the distance of the Magellanic System. The association of H I gas with the Sgr stellar stream argues that the dwarf spheroidal classification for the Sagittarius galaxy may need to be reconsidered. It also suggests that slightly offset H I and stellar streams may be a common feature of disrupted satellites in the Galactic halo. With the combination of the accretion of this Sgr H I stream, the Magellanic Stream, Complex C (Wakker et al. 1999), and possibly other high-velocity clouds, there is ample fuel for our Galaxy's continuing star formation and understanding the distribution of stellar metallicities (e.g., the G-dwarf problem is not a problem). Determining the metallicity, distance, and ionization properties of the H I filament presented here will aid in confirming if this H I gas was indeed stripped from the Sgr dwarf during its current orbit.

I would like to thank my collaborators on this work, Chris Thom, Brad Gibson, and Lister Staveley-Smith, as well as Phil Maloney and Steve Majewski for useful discussions. Roeland van der Marel is also thanked for providing the 2MASS data of the LMC to create Figure 2. M. E. P. acknowledges support by NASA through Hubble Fellowship grant HST-HF-01132.01 awarded by the Space Telescope Science Institute, which is operated by AURA, Inc. under NASA contract NAS 5-26555.

REFERENCES

BARNES, D. G., ET AL. 2001 *MNRAS* **322**, 486.

BLAND-HAWTHORN, J., FREEMAN, K. C., & QUINN, P. J. 1997 *ApJ* **490**, 143.

BLITZ, L. & ROBISHAW, T. 2000 *ApJ* **541**, 675.

BOUCHARD, A., CARIGNAN, C., & MASHCHENKO, S. 2003a *AJ* **126**, 1295.

BOUCHARD, A., ET AL. 2003b *AJ*, in preparation.

BURTON, W. B. & LOCKMAN, F. J. 1999 *A&A* **349**, 7.

CARIGNAN, C., BEAULIEU, S., COTE, S., DEMERS, S., & MATEO, M. 1998 *AJ* **116**, 1690.

CRANE, J. D., MAJEWSKI, S. R., ROCHA-PINTO, H. J., FRINCHABOY, P. M., SKRUTSKIE, M. R., LAW, D. R. 2003 *ApJL*, **594**, L119.

DEMERS, S. & BATTINELLI, P. 1998 *AJ* 115, 154.

DINESCU, D. I., ET AL. 2002 *ApJ* **575**, 67.

DOLPHIN, A. E. 2002 *MNRAS* **332**, 91.

GARDINER, L. T. 1999. In *Stromlo Workshop on High-Velocity Clouds* (eds. B. K. Gibson & M. E. Putman). ASP Conf. Ser. 166, (ASP). p. 292.

GAVAZZI, G., BONFANTI, C., SANVITO, G., BOSELLI, A., & SCODEGGIO, M. 2002 *ApJ* **576**, 135.

GREBEL, E. K., GALLAGHER, J. S. III, & HARBECK, D. 2003 *AJ* **125**, 1926.

GREEN, P. J., MARGON, B., ANDERSON, S. F., & COOK, K. H. 1994 *ApJ* **434**, 319.

GUHATHAKURTA, P. & REITZEL, D. B. 1998. In *Galactic Halos: A UC Santa Cruz Workshop* (ed. D. Zaritsky), ASP Conf. Ser. 136. ASP. p. 22

GUNN, J. E. & GOTT, J. R. III 1972 *ApJ* **176**, 1.

MEURER, G., ET AL. 2004 *ApJ*, in preparation.

IBATA, R. A., GILMORE, G., & IRWIN, M. J. 1994 *Nature* **370**, 194.

IBATA, R. A., GILMORE, G., IRWIN, M., LEWIS, G., WYSE, R., & SUNTZEFF, N. 2004 *ApJ*, in preparation.

IBATA, R. A. & LEWIS, G. F. 1998 *ApJ* **500**, 575.

IBATA, R. A., LEWIS, G. F., IRWIN, M., TOTTEN, E., & QUINN, T. 2001 *ApJ* **551**, 294.

KORIBALSKI, B., JOHNSTON, S., & OTRUPCEK, R. 1994 *MNRAS* **270**, 43.

LAYDEN, A. C. & SARAJEDINI, A. 2000 *AJ* **119**, 1760.

LEE, J. C., SALZER, J. J., IMPEY, C., THUAN, T. X., & GRONWALL, C. 2002 *AJ* **124**, 3088.

MAJEWSKI, S. R., OSTHEIMER, J. C., PATTERSON, R. J., DUNKEL, W. E., JOHNSTON, K. V., & GEISLER, D. 2000 *AJ* **119**, 760.

MAJEWSKI, S. R., SKRUTSKIE, M. F., WEINBERG, M. D. & OSTHEIMER, J. C. 2003 *ApJ* **599**, 1082 (M03).

MALONEY, P. R. 1993 *ApJ* **414**, 41.

MATEO, M. 1998 *ARA&A* **36**, 435.

MATHEWSON, D. S., CLEARY, J. D., & MURRAY, M. N. 1974 *ApJ* **190**, 291.

MATHEWSON, D. S., MENG, S. Y., BRUNDAGE, W. G., & KRAUS, J. D. 1966 *AJ* **71**, 863.

MIHOS, C. J. 2001 *ApJ* **550**, 94.

MOORE, B. & DAVIS, M. 1994 *MNRAS* **270**, 209.

MORI, M. & BURKERT, A. 2000 *ApJ* **538**, 559.

NEWBERG, H. J., ET AL. 2002 *AJ* **569**, 245.

PALMA, C., MAJEWSKI, S. R., SIEGEL, M. H., PATTERSON, R. J., OSTHEIMER, J. C., & LINK, R. 2003 *AJ* **125**, 1352.

PUTMAN, M. E., BLAND-HAWTHORN, J., VEILLEUX, S., GIBSON, B. K., FREEMAN, K. C., & MALONEY, P. R. 2003b *ApJ* **597**, 948.

PUTMAN, M. E., GIBSON, B. K., STAVELEY-SMITH, L., ET AL. 1998 *Nature* **394**, 752.

PUTMAN, M. E., ET AL. 2002 *AJ* **123**, 873.

PUTMAN, M. E., STAVELEY-SMITH, L., FREEMAN, K. C., GIBSON, B. K., & BARNES, D. G. 2003 *ApJ* **586**, 170 (P03).

PUTMAN, M. E., THOM, C., GIBSON, B. K., & STAVELEY-SMITH, L. 2004 *ApJL* **603**, 77.

QUILIS, V. & MOORE, B. 2001 *ApJ* **555**, L95.

SAVAGE, B. D., SEMBACH, K. R., WAKKER, B. P., RICHTER, P., MEADE, M., JENKINS, E. B., SHULL, J. M., MOOS, H. W., & SONNEBORN, G. 2003 *ApJS* **146**, 125.

SEMBACH, K. R., WAKKER, B. P., SAVAGE, B. D., RICHTER, P., MEADE, M., SHULL, J. M., JENKINS, E. B., SONNEBORN, G., & MOOS, H. W. 2003 *ApJS* **146**, 165.

STANIMIROVIC, S., DICKEY, J. M., KRCO, M., & BROOKS, A. M. 2002 *ApJ* **576**, 773.

STEIDEL, C. C., KOLLMEIER, J. A., SHAPLEY, A. E., CHURCHILL, C. W., DICKINSON, M., & PETTINI, M. 2002 *ApJ* **570**, 526.

TAYLOR, J. H., MANCHESTER, R. N., & LYNE, A. G. 1993 *ApJS* **88**, 529.

TOTTEN, E. J. & IRWIN, M. J. 1998 *MNRAS* **294**, 1.

VAN DER MAREL, R. 2001 *AJ* **122**, 1827.

WAKKER, B. P., ET AL. 1999 *Nature*, **402**, 388.

WAKKER, B. P., OOSTERLOO, T. A., & PUTMAN, M. E. 2002 *AJ* **123**, 1953.

WAKKER, B. P. & VAN WOERDEN, H. 1991 *A&A* **250**, 509.

WANG, Q. D. & MCCRAY, R. 1993 *ApJ* **409**, L37.

The star formation history in the Andromeda halo

By THOMAS M. BROWN

Space Telescope Science Institute, 3700 San Martin Drive, Baltimore, MD 21218, USA

I present the preliminary results of a program to measure the star formation history in the halo of the Andromeda galaxy. Using the Advanced Camera for Surveys (ACS) on the *Hubble Space Telescope*, we obtained the deepest optical images of the sky to date, in a field on the southeast minor axis of Andromeda, 51' (11 kpc) from the nucleus. The resulting color-magnitude diagram (CMD) contains approximately 300,000 stars and extends more than 1.5 mag below the main sequence turnoff, with 50% completeness at $V = 30.7$ mag. We interpret this CMD using comparisons to ACS observations of five Galactic globular clusters through the same filters, and through χ^2-fitting to a finely-spaced grid of calibrated stellar population models. We find evidence for a major ($\sim 30\%$) intermediate-age (6–8 Gyr) metal-rich ([Fe/H] > −0.5) population in the Andromeda halo, along with a significant old metal-poor population akin to that in the Milky Way halo. The large spread in ages suggests that the Andromeda halo formed as a result of a more violent merging history than that in our own Milky Way.

1. Introduction

One of the primary quests of observational astronomy is understanding the formation history of galaxies. An impediment to this research is the relative paucity of galaxies in the Local Group, which contains no giant ellipticals, and only two giant spirals—our own Milky Way and Andromeda. Fortunately, Andromeda (M31, NGC 224) is well situated for studying the formation of giant spiral halos, due to its proximity (770 kpc; Freedman & Madore 1990), small foreground reddening ($E_{B-V} = 0.08$ mag; Schlegel, Finkbeiner, & Davis 1990), and low inclination ($i \approx 12.5°$; de Vaucouleurs 1958). Andromeda is similar to the Milky Way in many respects (Hubble type, absolute magnitude, mass, and size; van den Bergh 1992; Klypin, Zhao, & Somerville 2002), but we have long known that its halo is more metal rich than that of the Milky Way; the metallicity distribution in the Milky Way halo peaks near [Fe/H] ≈ -1.8 (Ryan & Norris 1991), while that in the Andromeda halo peaks near [Fe/H] ≈ -0.5 (Mould & Kristian 1986; Holland, Fahlman, & Richer 1996; Durrell, Harris, & Pritchet 2001). Although the Milky Way halo is dominated by old stars (e.g., VandenBerg 2000), the formation history of the Andromeda halo has been unknown until now.

Physical processes possibly at work in forming spiral halos include rapid dissipative collapse in the early universe (Eggen, Lynden-Bell, & Sandage 1962), slower accretion of separate subclumps (Larson 1969; Searle & Zinn 1978), and dissolution of globular clusters (Aguilar, Hut, & Ostriker 1988). More recent hierarchical models suggest that spheroids (bulges and halos) form in a repetitive process during the mergers of galaxies and protogalaxies, while disks form by slow accretion of gas between merging events (e.g., White & Frenk 1991). The discovery of the Sgr dwarf galaxy (Ibata, Gilmore, & Irwin 1994) sparked renewed interest in halo formation through accretion of dwarf galaxies, and ambitious programs are now underway to map out the spatial distribution, kinematics, and chemical abundance in the halos of the Milky Way (e.g., Morrison et al. 2000; Majewski et al. 2000) and Andromeda (e.g., Ferguson et al. 2002). In the meantime, the realization that hierarchical models based on cold dark matter almost inevitably predict many more dwarf galaxies than are actually seen around the Milky Way (Moore et al.

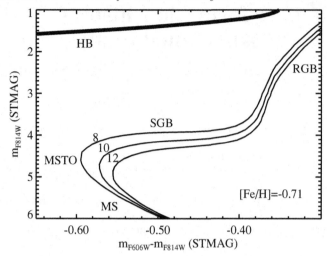

FIGURE 1. The variation in optical photometry of a stellar population with age at fixed metal-licity (*labeled*). Isochrones at 8, 10, and 12 Gyr (*labeled*) are shown in a CMD constructed using the ACS bandpasses F606W and F814W. Note that the subgiant branch (SGB) and main se-quence turnoff (MSTO) are more sensitive to age than the red giant branch (RGB). The primary age indicators in the CMD are the luminosity difference between the MSTO and the horizontal branch (HB), which becomes larger as age increases, and the color difference between the MSTO and the RGB, which becomes smaller as age increases. However, in order to clearly define the MSTO, photometry must extend to the main sequence (MS) stars below the MSTO.

1999) has led to suggestions that most of the dwarf galaxies formed in the early Universe have dissolved into the halo (e.g., Bullock, Kravtsov, & Weinberg 2000). Whether or not such accretion is the dominant source of stars in the stellar halo, it is likely that dwarf galaxies do contribute, and at large galactocentric distances their stars can remain in coherent orbital streams for many Gyr. Indeed, one such stream has been found in the halo of Andromeda (Ibata et al. 2001; McConnachie et al. 2003).

The most direct way to measure ages in a stellar population is to construct a color-magnitude diagram (CMD) that reaches to stars below the main-sequence turnoff. For decades, researchers have used such data to determine the ages of Galactic globular clus-ters and satellite galaxies of the Milky Way, but until now there has been no instrument capable of resolving these stars in a massive galaxy outside of our own. With the instal-lation of the Advanced Camera for Surveys (ACS; Ford et al. 1998) on the *Hubble Space Telescope* (*HST*), this is now possible. We have obtained deep ACS observations of a minor axis field in Andromeda's halo, ≈ 51 arcmin (11 kpc) from the nucleus. In these proceedings, I present the preliminary results of our program (Brown et al. 2003). I start with a brief review of the age diagnostics available in a deep CMD, then describe our ACS observations and their analysis, and finish with the implications for the formation history of Andromeda.

2. Age diagnostics

Stellar population ages are best determined from resolved optical photometry reaching the main sequence. Although the location of the main sequence turnoff in the CMD is the primary age indicator, photometry reaching well below the turnoff is required to define the turnoff accurately. Ages among the Galactic globular clusters are usually determined via one of two diagnostics (Figure 1): the luminosity difference between the

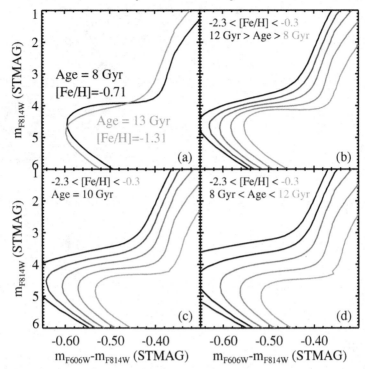

FIGURE 2. *Panel a:* An isochrone of intermediate age and metallicity (*dark curve, labeled*) and an older, more metal-poor isochrone (*light curve, labeled*) intersect, demonstrating that two-color photometry for a single star does not uniquely determine age and metallicity. Both isochrones also have the same color at the main sequence turnoff, and thus unresolved populations with these parameters would appear very similar. However, resolved photometry could distinguish between two such populations. *Panel b:* Isochrones spanning 8–12 Gyr at 1 Gyr intervals, with a range in metallicity (*labeled*) anticorrelated with age, such that the youngest isochrone is the most metal-rich, as might be expected for a simple model of chemical evolution. The subgiant branch and main sequence turnoff both become fainter and redder at increasing age or increasing metallicity, while the red giant branch is more sensitive to metallicity than age. Because the effects of age and metallicity counteract each other at the subgiant branch, it appears very narrow compared to the red giant branch. *Panel c:* The same metallicities as in *b*, but all of the isochrones have the same age. *Panel d:* The same metallicities and ages as in *b*, but now age is correlated with metallicity, such that the older populations are more metal-rich. This is somewhat unphysical, but demonstrates a situation of renewed star formation from the infall of primordial gas. Note the width of the subgiant branch compared to the red giant branch. Panels *b–d* demonstrate that resolved photometry of a mixed population can disentangle age and metallicity, because each evolutionary phase responds differently to these parameters.

turnoff and the horizontal branch (Sandage 1982; Iben & Renzini 1984), and the color difference between the turnoff and the red giant branch (VandenBerg, Bolte, & Stetson 1990). The luminosity of the horizontal branch and the color of the red giant branch are both relatively insensitive to age (at fixed composition), while the main sequence turnoff becomes both fainter and redder at increasing age. Our ACS observations used the F606W (broad V) and F814W (I) filters; for roughly every Gyr beyond 10 Gyr, the main sequence turnoff becomes fainter by $\sim$0.1 mag in m_{F814W} and redder by $\sim$0.01 mag in $m_{F606W} - m_{F814W}$ (see Figure 1).

The age-metallicity degeneracy is a major uncertainty when characterizing individual stars via photometry, or unresolved populations via low-resolution spectra (Figure 2a).

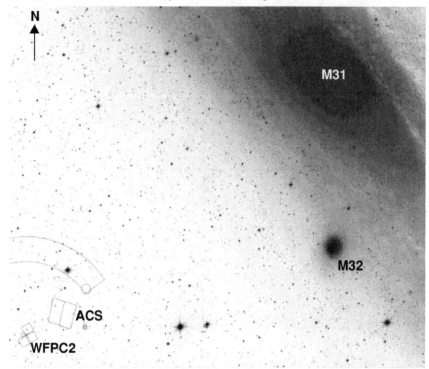

FIGURE 3. A $50' \times 60'$ Digital Sky Survey image of Andromeda, showing the positions of our ACS and WFPC2 images (labeled). The ACS field is 50 arcmin (11 kpc) from the center of M31. The nearly edge-on ($i \approx 12.5°$; de Vaucouleurs 1958) disk should contribute $\lesssim 3\%$ of the stars in the ACS field.

However, the situation is better for a resolved stellar population. Because the red giant branch is much more sensitive to metallicity than age, and the subgiant branch and main sequence turnoff are sensitive to both age and metallicity, the distribution of stars on the red giant branch, subgiant branch, and main sequence turnoff can be simultaneously reproduced only for specific combinations of age and metallicity (Figure 2).

3. Observations

Using the ACS Wide Field Camera (WFC), we obtained deep optical images of a field along the southeast minor axis of the M31 halo, at $\alpha_{2000} = 00^h 46^m 07^s$, $\delta_{2000} = 40°42'34''$ (Figure 3). The field was previously imaged by Holland et al. (1996) with the Wide Field Planetary Camera 2 (WFPC2); the field is not associated with the tidal streams and substructure found by Ferguson et al. (2002), and lies just outside the "flattened inner halo" in their maps. Given the nearly edge-on disk, the contribution of disk stars at this position should be $\lesssim 3\%$ (Walterbos & Kennicutt 1988; Holland et al. 1996 and references therein). We chose this field to optimize the crowding (trading off population statistics versus photometric accuracy) and to place an interesting M31 globular cluster (GC312; Sargent et al. 1977) near the edge of our images. The metallicity of GC312 ([Fe/H] = -0.7; Huchra, Brodie, & Kent 1991) is near the peak in the metallicity distribution for the M31 halo; this should simplify our attempts to derive relative ages for this cluster and the halo (currently underway). Due to scheduling and orientation constraints, one bright

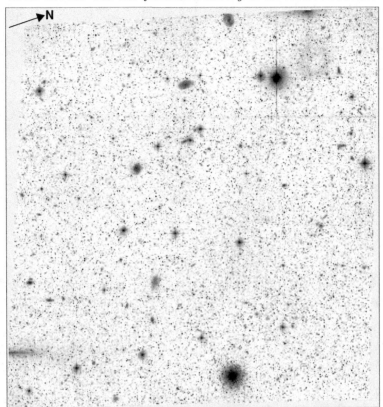

FIGURE 4. The ACS F606W image of the Andromeda halo, shown at a log stretch, and binned for display purposes by a factor of 20, to $0.6''$ pixel^{-1}. The field subtends $211'' \times 221''$ (not much larger than the WFC field of view, because the largest dither was about $6''$). For our preliminary analysis, we excluded the area around GC312 (*bottom*) and around a bright ($V = 14$ mag) foreground star (*upper right*) and its window reflection (to the right of the star). Many background galaxies can be seen through the 300,000 stars detected in Andromeda's halo, but they are a small ($< 5\%$) contamination. The image has been corrected for the strong geometric distortion in the camera, giving the field its unusual shape.

foreground star ($V \sim 14$ mag) was unavoidable; the star and its window reflection affect a few percent of the total image area, which we discard.

From 2 Dec 2002 to 11 Jan 2003, we obtained 39.1 hours of ACS images in the F606W filter (broad V) and 45.4 hours in the F814W filter (I), with every exposure dithered to allow for hot pixel removal, optimal sampling of the point spread function, smoothing of the spatial variations in the detector response, and filling the gap between the two halves of the 4096×4096 pix detector. We co-added the M31 images using the IRAF DRIZZLE package, with masks for the cosmic rays and hot pixels, resulting in geometrically-correct images with a plate scale of $0.03''$ pixel^{-1} and an area of approximately $210'' \times 220''$ (Figure 4). We then performed both aperture and PSF-fitting photometry using the DAOPHOT-II package (Stetson 1987), assuming a variable PSF constructed from the most isolated stars in the images. The aperture photometry on isolated stars was corrected to true apparent magnitudes using TINYTIM models of the *HST* PSF (Krist 1995) and observations of the standard star EGGR 102 (a $V = 12.8$ mag DA white dwarf) in the same filters, with agreement at the 1% level. The PSF-fitting photometry was then compared to the corrected aperture photometry, in order to derive the offset between

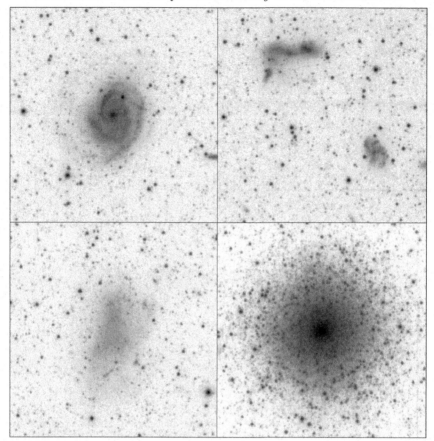

FIGURE 5. Subsections ($15'' \times 15''$) of the ACS image show the globular cluster GC312 in Andromeda's halo (*lower right*), background galaxies, and the level of crowding in the halo population. The images have been rebinned to a scale of $0.06''$ pixel^{-1} for display purposes.

the PSF-fitting photometry and true apparent magnitudes. Our photometry is in the STMAG system: $m = -2.5 \times \log_{10} f_\lambda - 21.1$. For those more familiar with the Johnson V and Cousins I bandpasses, a 5,000 K stellar spectrum has $V - m_{F606W} = -0.05$ mag and $I - m_{F814W} = -1.28$ mag.

Of the $\sim 300{,}000$ stars detected, we discarded those within the GC312 tidal radius ($10''$; Holland, Fahlman, & Richer 1997), within $14.5''$ of a bright foreground star, within $12.6''$ of this star's window reflection, and near the image edges, leaving $\approx 223{,}000$ stars in the final catalog. Using the SExtractor code (Bertin & Arnouts 1996), we estimate $\lesssim 5\%$ of the stars are contaminated by extended sources (Figure 5). Extensive artificial star tests determine the photometric scatter and completeness as a function of color and luminosity. The CMD shows no obvious differences when comparing the population in a 10–$100''$ annulus around GC312 to that beyond $100''$; the cluster does not appear to be associated with an extended underlying system. By integrating our catalog, we estimate that the surface brightness in our field is $\mu_V \approx 26.3$ mag arcsec^{-2}.

We also obtained coordinated parallel WFPC2 observations of a second field along the minor axis of Andromeda (Figure 3) using the same bandpasses as those in the ACS observations. The WFPC2 data are not nearly as deep as the ACS data, but they are about as deep as the Hubble Deep Field (Williams et al. 1996), and thus much deeper

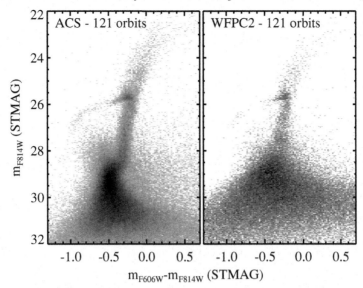

FIGURE 6. *Left panel:* The CMD constructed from the ACS images of the Andromeda halo, using aperture photometry. There are too many stars to plot them individually; instead, this is a Hess diagram at a logarithmic stretch. *Right panel:* The CMD constructed from the parallel WFPC2 images, in a field a few arcmin further out along the minor axis of Andromeda, again using aperture photometry. The filters and exposure times in the WFPC2 data are approximately the same as those in the ACS data. Although the WFPC2 data reach stars at the main sequence turnoff, they are not deep enough to characterize the subgiant branch and the turnoff well.

than any Andromeda observations prior to our program. Although the WFPC2 images resolve stars at the main sequence turnoff, the resulting CMD is not deep enough to characterize the turnoff and subgiant branch well. The striking differences between the WFPC2 and ACS CMDs are a testament to the technical advances achieved by the *HST* servicing missions (Figure 6).

Because the ACS is a new instrument, our program includes ACS observations of five Galactic globular clusters spanning a wide metallicity range (Table 1), using the same filters utilized in our Andromeda halo observations. These cluster images allow the construction of empirical isochrones in the ACS bandpasses, which can be directly compared to the Andromeda CMD and used to calibrate the transformation of the theoretical isochrones to the ACS bandpasses. Figure 7 shows the ACS F606W image of M92. M92, NGC 6752, and 47 Tuc are the most useful calibrators in our program, because their parameters are known very well; NGC 5927 and NGC 6528 are also useful because of their high metallicities, but their parameters are less secure, and they suffer from high, spatially variable foreground reddening (Heitsch & Richtler 1999). To increase the dynamic range, three images were taken in each bandpass for each cluster, with the exposure times varying by an order of magnitude. To minimize the number of orbits required for the cluster data, these observations were not dithered. Thus, we drizzled the data without plate scale changes, in order to remove cosmic rays and to correct for geometric distortion. The cluster images are significantly less crowded (on average) than those of M31, and the 0.05″ pixels undersample the PSF, so we performed aperture photometry but no PSF fitting, and corrected the aperture photometry to true apparent magnitudes (Figure 8).

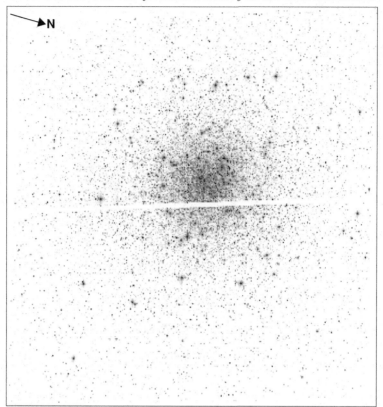

FIGURE 7. The ACS F606W image of M92, shown at a log stretch and binned by 20 to 0.6″ pixel⁻¹ for display purposes. The globular cluster images in our program were not dithered, so a gap appears between the two WFC chips.

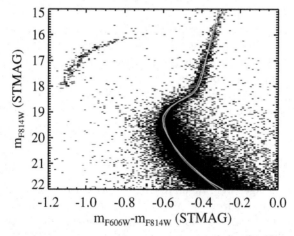

FIGURE 8. The CMD created from our ACS images of M92. The ridge line for this cluster (*dark gray curve*) was created by taking the median $m_{F606W} - m_{F814W}$ color and median m_{F814W} for a series of points along the main sequence and red giant branch. A 14 Gyr isochrone at $[Fe/H] = -2.14$ (*light gray curve*), transformed to the ACS bandpasses, agrees well with the ridge line over the CMD region used for fitting the Andromeda halo population.

Name	$(m - M)_V$ (mag)	E_{B-V} (mag)	[Fe/H]	Exposure time (s) F606W	F814W
M31	24.68[a]	0.08[b]	−0.6[c]	140870	163570
M92	14.60[d]	0.023[b]	−2.14[d]	95.5	106.5
NGC 6752	13.17[e]	0.055[b]	−1.54[f]	44.5	49.5
47 Tuc	13.27[g]	0.032[b]	−0.83[f]	76.5	78
NGC 5927	15.81[h]	0.45[h]	−0.37[h]	532	355.7
NGC 6528	16.15[i]	0.55[i]	−0.2[i]	504	371

[a]Freedman & Madore (1990). [b]Schlegel et al. (1998). [c]Mould & Kristian (1986). [d]VandenBerg & Clem (2003). [e]Renzini et al. (1996). [f]VandenBerg (2000). [g]Zoccali et al. (2001). [h]Harris (1996). [i]Momany et al. (2003).

TABLE 1. M31 and globular cluster parameters

4. Analysis

The ACS CMD reveals, for the first time, the main sequence population in the M31 halo. The horizontal branch extends from a well-populated red clump to a minority blue population ($\sim 10\%$ of the total horizontal branch population). The horizontal branch is not noticeably extended—the hot horizontal branch stars that are seen in clusters like NGC 6752 are missing. The broad red giant branch indicates a wide metallicity distribution extending to near-solar metallicities, long known to be characteristic of the M31 halo (Mould & Kristian 1986). The luminosity function "bump" on the red giant branch is another metallicity indicator, becoming fainter as metallicity increases; it slopes away from the red horizontal branch until the luminosity difference reaches $\approx$0.5 mag—another indication of near-solar metallicities. There is also a prominent blue plume of stars significantly brighter than the main sequence turnoff; this minority population ($\sim 2\%$ the size of the population at the turnoff ±1 mag) may include binaries, blue stragglers, or a residual young stellar population. Although the blue horizontal branch stars are characteristic of very old, metal-poor globular clusters (such as M92), the luminosity difference between the turnoff and horizontal branch is smaller than expected for a purely old stellar population. This is shown in Figure 9, which shows the ridge lines and horizontal branch loci for the five globular clusters we observed with ACS, superimposed upon the CMD of Andromeda. The M31 subgiant branch is nearly horizontal, indicating a high metallicity, while its ridge is appreciably brighter than those of 47 Tuc, NGC 5927, and NGC 6528, implying the presence of a significantly younger population in the M31 halo.

The comparison between the globular clusters and Andromeda in Figure 9 provides a completely empirical indication of the age spread in Andromeda's halo. M92, NGC 6752, and 47 Tuc were shifted in color and magnitude according to the differences in reddening and distance between the clusters and Andromeda. Those shifts naturally aligned the horizontal branch loci of these clusters to the horizontal branch of Andromeda, thus demonstrating the accuracy of the parameters in Table 1, but we could have simply aligned the horizontal branches without any knowledge of the relative distances and reddenings. Indeed, forcing alignment at the horizontal branch was required for NGC 5927 and NGC 6528, because the parameters of those clusters are very uncertain. Once the clusters are all aligned at the horizontal branch, the red giant branches of the clusters span the broad red giant branch of Andromeda, empirically demonstrating the wide

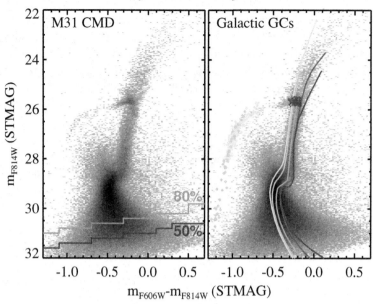

FIGURE 9. *Left panel:* The CMD of Andromeda constructed from the PSF-fitting photometry of the ACS images. Completeness limits determined from artificial star tests are marked. *Right panel:* The CMD of Andromeda, with the ridge lines (*gray curves*) and horizontal branch loci (*gray diamonds*) of five Galactic globular clusters superimposed. The shading of the ridge lines and horizontal branch points range from light gray to dark gray as the cluster metallicity increases (see Table 1). The data for M92, NGC 6752, and 47 Tuc have only been shifted by the differences in distance and reddening between the clusters and M31, yet their horizontal branch loci agree well with the horizontal branch of M31. Those parameters are very uncertain for NGC 6528 and NGC 5927, so the data for the two metal-rich clusters were shifted to align their horizontal branch loci with the M31 horizontal branch: NGC 6528 was shifted 0.16 mag brighter, while NGC 5927 was shifted 0.11 mag brighter and 0.05 mag redder. Moving from the most metal-poor cluster (M92) to the most metal-rich (NGC 6528), the red giant branches of the clusters span the width of the red giant branch in M31, yet the subgiant branches of the clusters become increasingly fainter than that of M31, indicating that the metal-rich stars in the M31 halo are much younger than those in the clusters.

metallicity distribution in Andromeda's halo. For the most metal-poor cluster (M92; *lightest gray curve* in Figure 9), the subgiant branch luminosity agrees well with that in Andromeda, and the turnoff of the cluster agrees well with the blue edge of the turnoff in Andromeda; this indicates that Andromeda contains a significant population of metal-poor stars with ages similar to the oldest Galactic globular clusters. However, as one compares clusters at increasing metallicity to Andromeda, discrepancies at the subgiant branch and turnoff become more apparent. Although the red giant branches of 47 Tuc, NGC 5927, and NGC 6528 straddle a large fraction of the red giant branch population in Andromeda, the cluster subgiant branches are well below the bulk of the subgiant population in Andromeda. This is a strong indication that the metal-rich populations in Andromeda are significantly younger than the cluster, given that the subgiant branch and turnoff become fainter by approximately 0.1 mag for every 1 Gyr increase in age.

Our full analysis of the star formation history is in progress. Here, I will supplement the comparisons shown in Figure 9 with examples from the modeling to date (Figure 10). Our modeling is based upon the isochrones of VandenBerg, Bergbusch, & Dowler (2003, in preparation), which show good agreement with cluster CMDs spanning a wide range of metallicity and age (e.g., Figure 8). To transform these isochrones from the ground-based

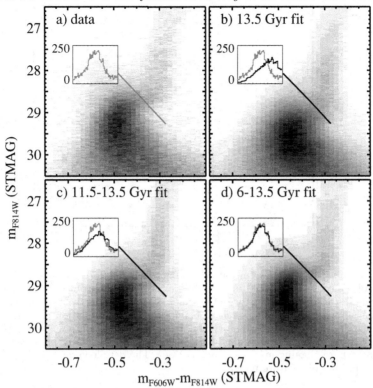

FIGURE 10. *Panel a:* The region of the Andromeda halo CMD that we used for fitting the star formation history. Restricting the fit to this region avoids parts of the CMD that are seriously incomplete, sparsely populated, or poorly constrained by the theoretical models. A histogram (*inset*) of the number of stars along a cut through the subgiant branch (*gray line*) highlights the differences between the data and the models in the subsequent panels. *Panel b:* The fit to the red giant branch using only 13.5 Gyr isochrones spanning a wide range in metallicity. Although such models can reproduce the width of the Andromeda red giant branch, the subgiant branch and turnoff are much fainter than those in the data. The histogram (*inset*) compares the number of stars along the cut (*black line*) through the subgiant branch to that in the data (*gray histogram*); they are poorly matched. *Panel c:* The fit to the entire region shown in panel *a*, using isochrones with a wide metallicity range but ages of 11.5–13.5 Gyr. The histogram (*inset*) again highlights that this model does not reproduce the distribution seen in the data. *Panel d:* The fit to the data shown in panel *a*, using a wide metallicity range and ages of 6–13.5 Gyr. This model reproduces the data well, as highlighted by the histograms (*inset*).

bandpasses to the *HST*-based bandpasses, we used the spectra of Lejeune, Cuisinier, & Buser (1997) to calculate $V - m_{F606W}$ and $I - m_{F814W}$, as a function of effective temperature and gravity along the isochrones, then applied those differences to the ground-based magnitudes of the isochrones, with a small ($\lesssim 0.05$ mag) empirical color correction to force agreement with our globular cluster CMDs. After this correction, the isochrones match the ridge lines within $\lesssim 0.02$ mag over the region of the CMD we are fitting. Our observed CMDs of these clusters are reproduced by a 12.5 Gyr isochrone for 47 Tuc and 14 Gyr isochrones for NGC 6752 and M92. The isochrones do not include core He diffusion, which would reduce their ages by 10–12%, thus avoiding discrepancies with the age of the Universe (VandenBerg et al. 2002).

Using isochrones with a range of ages and metallicities, we fit the region of the M31 CMD shown in Figure 10a using the StarFish code of Harris & Zaritsky (2001). Restrict-

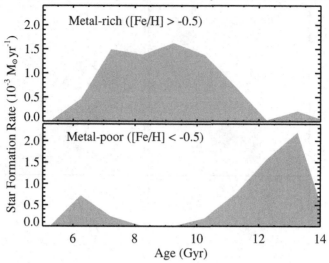

FIGURE 11. *Upper panel:* The star formation history of the metal-rich stars in the best model (Figure 10d) obtained to date. The bulk of the metal-rich population is of intermediate age. *Lower panel:* The star formation history of the metal-poor stars in the best-fit model. Most of the metal-poor stars are old, but this model includes a small population of young metal-poor stars. Note that this star formation history is not unique; we can obtain a fit that is nearly as good as this fit, using two completely distinct population components (a purely old metal-poor population and a purely young metal-rich population). The detailed star formation history will be explored more fully in a future paper.

ing the fit to this region of the CMD focuses on the most sensitive age and metallicity indicators while avoiding regions of the CMD that are seriously incomplete, sparsely populated, or poorly constrained by the models (e.g., the horizontal branch morphology). Note that we do not include the red giant branch luminosity function bump in our fitting; although theory predicts that this bump becomes fainter with increasing metallicity, the theoretical zeropoint is fairly uncertain. The StarFish code fits the observed CMD through a linear combination of input isochrones, using χ^2 minimization, where the isochrones are scattered according to the results of the artificial star tests. We first followed the standard method used when researchers determine the metallicity distribution from the red giant branch (Figure 10b). We fit the red giant branch using a set of old (13.5 Gyr) isochrones spanning a wide metallicity range ($-2.31 < [Fe/H] < 0$). The resulting metallicity distribution was similar to that found by Holland et al. (1996) in our same field, and Durrell et al. (2001) in a field 20 kpc from the nucleus. However, it is very obvious that these purely old isochrones do not match the subgiant branch or the turnoff. Next, as shown in Figure 10c, we tried fitting the entire region of Figure 10a with a wider age range (11.5–13.5 Gyr), but found no acceptable combination. The insets in Figure 10 contain histograms showing the number of stars (data: *gray*; model: *black*) along a cut through the subgiant branch (*thick line*). In Figure 10c, the residual subgiant stars that are not reproduced by this old model (*inset*) suggest that at least 20% of the stars in the halo must be younger than 11.5 Gyr. We conclude that a purely old stellar population cannot explain the CMD of the M31 halo.

Next, we expanded the age range to 6–13.5 Gyr and repeated the fit (see Figure 10d). The width of the red giant branch is now matched without a mismatch at the subgiant branch (*inset*). The best-fit model can be broadly characterized by a combination of two dominant populations (Figure 11): 56% of the stars (*red*) are metal-rich ([Fe/H] >

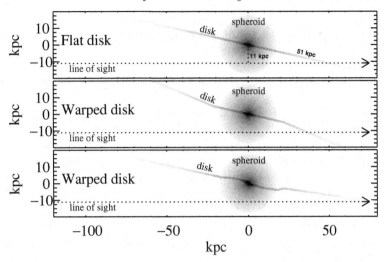

FIGURE 12. *Top panel:* A simple schematic showing the intersection of our sight-line with the Andromeda disk. The plane of the disk is inclined 12.5° with respect to our line of sight (de Vaucouleurs 1958), so that we are viewing the halo population at 11 kpc from the nucleus but the disk population at 51 kpc from the nucleus. The disk has an exponential scale length of 5.3 kpc (Walterbos & Kennicutt 1988). Note that the southeastern half of the disk is the side furthest from the observer (Iye & Richter 1985). *Middle panel:* The isophotes for Andromeda show evidence of a warp near 22 kpc on the major axis (Walterbos & Kennicutt 1987). Here, I insert a hypothetical 10° warp at 22 kpc on the minor axis, such that more of the disk population is moved into our sight-line. Our sight-line has effectively moved inward by about two scalelengths, increasing the disk population by a factor of eight. However, even with this large warp into our sight-line, the increase in the disk population would not explain the large population of intermediate-age stars we find in our field. *Bottom panel:* A more realistic depiction of the warp in the Andromeda disk, by interpolating the measurements of Braun (1991) along the minor axis. Early models of the warp (e.g., Henderson 1979) imply our sight-line would not intersect the disk at all, because of the increasing inclination at large distances from the nucleus. Later models (e.g., Braun 1991) indicate that the inclination varies from 25° close to the nucleus to 10–15° further out, as depicted here. The result is that our effective position within the disk is significantly beyond the 51 kpc shown in the top panel.

−0.5) and of intermediate age (6–11 Gyr), while 30% of the stars (*blue*) are metal-poor ([Fe/H] < −0.5) and old (11–13.5 Gyr). About half of this metal-rich population (i.e., 28% of the total population) is 6–8 Gyr old. Note that these models only illustrate, in a broad sense, the dominant populations present in the M31 halo. Other combinations of young metal-rich and old metal-poor stars are possible. For example, we produced a similar fit by combining two very distinct isochrone groups: one at 6–8 Gyr with [Fe/H] > −1, and one at 11.5–13.5 Gyr with [Fe/H] < −1. We are currently investigating detailed constraints on the star formation history.

5. Disk contamination

Note that our halo field should be relatively free of stars moving in the stellar disk. Several estimates for the disk contribution at this distance range from 1–3%, even with a significant thick disk component (Holland et al. 1996 and references therein). On average, the plane of the disk is inclined from our sight-line by 12.5° (de Vaucouleurs 1958; Figure 12, *top panel*), so we are observing halo stars 11 kpc from the nucleus but disk stars 51 kpc from the nucleus. A warp in the outer disk, if it angled the disk into our sight-line, would move the sight-line intersection with the disk inward, but even a large warp could not explain the number of young stars seen in our field. For example, Walterbos & Kennicutt (1987) found a possible warping of the stellar disk at 100′ (22 kpc) on the southwest major axis. If we assume that the stellar disk along the southeast minor axis is warped such that the disk beyond 22 kpc is inclined by an additional 10° into our sight-line, this would move our effective disk position inward by about two scalelengths (Figure 12, *middle panel*), increasing the disk population by a factor of 8 (i.e., the disk would contribute ≈10–20% of the total population in our field). However, we find that 56% of the stars in our field are metal-rich and of intermediate age, so this hypothetical increase in the disk contamination would not account for our surprisingly young population. In reality, studies of the neutral gas kinematics (e.g., Braun 1991; Henderson 1979) find that the outer disk is actually viewed significantly closer to edge-on than the inner disk (Figure 12, *bottom panel*), such that the warp in the disk would move our sight-line further out, beyond the intersection given by a flat disk. Finally, the fact that the metallicity distribution in our field is very similar to that twice as far out (Holland et al. 1996; Durrell, Harris, & Pritchet 2001) provides another strong indication that we are viewing a halo-dominated population.

6. Summary and discussion

The CMD of the M31 halo is evidently inconsistent with a population composed solely of old (globular cluster age) stars; instead, it is dominated by a population of metal-rich intermediate-age stars. Although the high metallicity in the M31 halo is well-documented, the large age spread required to simultaneously reproduce the red giant branch, subgiant branch, and main sequence came to us as a surprise. Earlier studies of the red giant branch were insensitive to this age spread. For example, Durrell et al. (2001) were able to explain the metallicity distribution 20 kpc from the nucleus, with a simple chemical evolution model forming most of the stars at very early times (see also Côté et al. 2000). Although our field is relatively small in sky coverage, it appears representative; the metallicity in our field (Holland et al. 1996) agrees well with that much further out (Durrell et al. 2001), and there are no indications of substructure or tidal streams in the region we surveyed (Ferguson et al. 2002).

It seems unlikely that star formation in the halo proceeded for ∼6 Gyr from gas *in situ*; instead, the broad age dispersion in the halo is likely due to contamination from the disruption of satellites or of disk material into the halo during mergers. Indeed, our current analysis of the data cannot rule out the possibility that M31 and a nearly-equal-mass companion galaxy experienced a violent merger when the Universe was half its present age. If the 6–8 Gyr population in the halo represents the remnants of a disrupted satellite, the relatively high metallicity suggests that it must have been fairly massive. On the other hand, the stars could represent disruption of the M31 disk, either by a major collision when M31 was ∼ 6 Gyr old, or by repeated encounters with smaller satellites. The resulting halo would be a mix of the old metal-poor stars formed earliest in M31's halo, disk stars that formed prior to the merger(s) that were subsequently dispersed into the halo, stars formed during the merger(s), and the remnant populations of the disrupted satellite(s).

This work was done in collaboration with H. C. Ferguson, E. Smith (STScI), R. A. Kimble, A. V. Sweigart (NASA/GSFC), R. M. Rich, D. Reitzel (UCLA), A. Renzini (ESO), and D. A. VandenBerg (U. of Victoria). I am grateful to J. Harris (STScI) and P. Stetson (DAO) for respectively providing the StarFish and DAOPHOT-II codes and assistance in their use. Thanks to the members of the scheduling and operations teams at STScI (especially P. Royle, D. Taylor, and D. Soderblom) for their efforts in executing a large program during a busy *HST* cycle. Support for program 9453 was provided by NASA through a grant from the Space Telescope Science Institute, which is operated by the Association of Universities for Research in Astronomy, Inc., under NASA contract NAS 5-26555. The Digitized Sky Survey was produced at the Space Telescope Science Institute under U.S. Government grant NAG W-2166. The images of these surveys are based on photographic data obtained using the Oschin Schmidt Telescope on Palomar Mountain and the UK Schmidt Telescope. The plates were processed into the present compressed digital form with the permission of these institutions. This research has made use of the SIMBAD database, operated at CDS, Strasbourg, France.

REFERENCES

AGUILAR, L., HUT, P., & OSTRIKER, J. P. 1988 *ApJ* **335**, 720.

BERTIN, E., & ARNOUTS, S. 1996 *A&AS* **117**, 393.

BRAUN, R. 1991 *ApJ* **372**, 54.

BROWN, T. M., FERGUSON, H. C., SMITH, E., KIMBLE, R. A., SWEIGART, A. V., RENZINI, A., RICH, R. M., & VANDENBERG, D. A. 2003 *ApJ* **592**, L17.

BULLOCK, J. S., KRAVTSOV, A. V., & WEINBERG, D. H. 2000 *ApJ* **539**, 517.

CÔTÉ, P., MARZKE, R. O., WEST, M. J., & MINNITI, D. 2000 *ApJ* **533**, 869.

DE VAUCOULEURS, G. 1958 *ApJ* **128**, 465.

DURRELL, P. R., HARRIS, W. E., & PRITCHET, C. J. 2001 *AJ* **121**, 2557.

EGGEN, O. J., LYNDEN-BELL, D., & SANDAGE, A. R. 1962 *ApJ* **136**, 748.

FERGUSON, A. M. N., IRWIN, M. J., IBATA, R. A., LEWIS, G. F., & TANVIR, N. R. 2002 *AJ* **124**, 1452.

FORD, H. C., ET AL. 1998 *SPIE* **3356**, 234.

FREEDMAN, W. L. & MADORE, B. F. 1990 *ApJ* **365**, 186.

HARRIS, W. E. 1996 *AJ* **112**, 1487.

HARRIS, J. & ZARITSKY, D. 2001 *ApJS* **136**, 25.

HEITSCH, E. & RICHTLER, T. 1999 *A&A* **347**, 455.

HENDERSON, A. P. 1979 *A&A* **75**, 311.

HOLLAND, S., FAHLMAN, G. G., & RICHER, H. B. 1996 *AJ* **112**, 1035.

HOLLAND, S., FAHLMAN, G. G., & RICHER, H. B. 1996 *AJ* **114**, 1488.

HUCHRA, J. P., BRODIE, J. P., & KENT, S. M. 1991 *ApJ* **370**, 495.

IBATA, R. A., GILMORE, G., & IRWIN, M. J. 1994 *Nature* **370**, 194.

IBATA, R. A., IRWIN, M., LEWIS, G., FERGUSON, A. M. N., & TANVIR, N. 2001 *Nature* **412**, 49.

IBEN, I. & RENZINI, A. 1984 *PhR* **105**, 329.

IYE, M. & RICHTER, O.-G. 1985 *A&A* **144**, 471.

KLYPIN, A., ZHAO, H. S., & SOMERVILLE, R. S. 2002 *ApJ* **573**, 597.

KRIST, J. 1995 In *Astronomical Data Analysis Software and Systems IV* (eds. R. A. Shaw, H. E. Payne, & J. J. E. Hayes). ASP Conference Series, vol. 77, p. 349. ASP.

LARSON, R. B. 1969 *MNRAS* **145**, 405.

LEJEUNE, T., CUISINIER, F., & BUSER, R. 1997 *A&AS* **125**, 229.

MAJEWSKI, S. R., OSTHEIMER, J. C., KUNKEL, W. E., & PATTERSON, R. J. 2000 *AJ* **120**, 2550.

MCCONNACHIE, A. W., IRWIN, M. J., IBATA, R. A., FERGUSON, A. M. N., LEWIS, G. F., & TANVIR, N. 2003 *MNRAS*, **343**, 1335.

MOMANY, Y., ET AL. 2003 *A&A* **402**, 607.

MOORE, B., GHIGNA, S., GOVERNATO, F., LAKE, G., QUINN, T., STADEL, J., & TOZZI, P. *ApJ* **524**, L19.

MORRISON, H. L., MATEO, M., OLSZEWSKI, E. W., HARDING, P., DOHM-PALMER, R. C., FREEMAN, K. C., NORRIS, J. E., & MORITA, M. 2000 *AJ* **119**, 2254.

MOULD, J. & KRISTIAN, J. 1986 *ApJ* **305**, 591.

RENZINI, A., ET AL. 1996 *ApJ* **465**, L23.

RYAN, S. G. & NORRIS, J. E. 1991 *AJ* **101**, 1865.

SANDAGE, A. 1982 *ApJ* **242**, 553.

SARGENT, W. L. W., KOWAL, C. T., HARTWICK, F. D. A., VAN DEN BERGH, S. 1977 *AJ* **82**, 947.

SCHLEGEL, D. J., FINKBEINER, D. P., & DAVIS, M. 1998 *ApJ* **500**, 525.

SEARLE, L. & ZINN, R. 1978 *ApJ* **225**, 357.

STETSON, P. 1987 *PASP* **99**, 191.

VAN DEN BERGH, S. 1992 *A&A* **264**, 75.

VANDENBERG, D. A. 2000 *ApJS* **129**, 315.

VANDENBERG, D. A., BOLTE, M., & STETSON, P. B. 1990 *AJ* **100**, 445.

VANDENBERG, D. A. & CLEM, J. L. 2003 *AJ* **126**, 778.

VANDENBERG, D. A., RICHARD, O., MICHAUD, G., & RICHER, J. 2002 *ApJ* **571**, 487.

WALTERBOS, R. A. M. & KENNICUTT, R. C., JR. 1987 *A&AS* **69**, 311.

WALTERBOS, R. A. M. & KENNICUTT, R. C., JR. 1988 *A&A* **198**, 61.

WHITE, S. D. M. & FRENK, C. S. 1991 *ApJ* **379**, 52.

WILLIAMS, R. E., ET AL. 1996 The Hubble Deep Field: Observations, Data Reduction, and Galaxy Photometry. *AJ* **112**, 1335.

ZOCCALI, M., ET AL. *ApJ* **553**, 733.

Bulge populations in the Local Group

By R. MICHAEL RICH

Division of Astronomy and Astrophysics, Department of Physics and Astronomy,
Math-Sciences 8979, UCLA, Los Angeles, CA 90095-1562, USA

The spatial resolution and multiwavelength capability of the *Hubble Space Telescope* has advanced greatly the study of spheroidal populations of galaxies. The main sequence turnoff point of the Galactic bulge has been clearly measured and the proper motions of stars used to separate the foreground disk from the bulge. The study of bulge globular clusters by *HST* shows that the bulge field population is coeval with the halo. In the obscured Galactic Center, infrared imaging with NICMOS reveals a continuous star formation history. NICMOS imaging of the nucleus of M31 has settled a long standing debate about whether luminous AGB stars (possibly indicative of an intermediate age stellar population) are present. Measurements of the composition from the ground, and of the age from *HST*, are consistent with a scenario in which spheroidal populations formed very early. *HST* spectroscopy has also revealed the presence of central black holes in many bulges and spheroids. The formation of the stellar populations and black holes in spheroids was probably one of the earliest events in galaxy formation.

1. Introduction

The bulge populations of the Local Group are a window in the nearby Universe to some of the most important aspects of galaxy formation: The age (formation epoch) of bulge populations, their metal content, and their coevolution and connection with nuclear black holes. Spheroidal populations (including bulges) account for more than half of the stellar mass at the present epoch (Fukugita, Hogan, & Peebles 1998) and also even at redshift 0.7 (Bell et al. 2004). The connection between nearby and distant bulge populations is not a new revelation. Whitford (1978) demonstrated (using newly commissioned photon-counting detectors at CTIO) that the bulge of the Milky Way has an integrated spectrum identical to that of distant ellipticals and bulges.

While much can be learned from integrated spectroscopy, one cannot consider to have actually studied a stellar population until it has been resolved into stars. Their high surface brightness makes bulge and spheroid populations very difficult if not impossible to resolve with ground-based observations (Renzini 1998; see Stephens et al. 2003). The high central surface brightness of these populations is evident in the study of Kent (1987) which plots the surface brightness profiles of Local Group galaxies. As Section 3 discusses, ground-based infrared studies were unable to measure with indisputable precision the luminosity of the AGB tip in the bulge of M31 and M32. Using NICMOS on board the *Hubble Space Telescope* (*HST*), the bulge of M31 is easily resolved and the issue settled. The capability to image the Local Group bulge populations with *HST* has been indispensable. While distant enough that their integrated properties may be studied and compared with more distant galaxy samples, the local group bulges are also near enough to permit the resolution of, and detailed study of, their individual stars. In the case of the Milky Way bulge, we have photometric access to the faintest hydrogen-burning stars. The red giant branch, to the level of the horizontal branch, can be accessed in M31, M32, and M33. We can measure the luminosity function of the brightest stars, giving an age constraint. From the colors, we can constrain the mean abundance and its dispersion, giving information about the enrichment history and formation timescale.

Imaging using *HST* has provided much of the data by which we now define and understand the bulge populations in the Local Group. It is our good fortune that the Local

Group includes the central bulge of M31, a good representative example of more distant spheroid populations. Resolution of the M31 bulge to the main sequence turnoff point is the long-term goal, but that will require apertures larger than 8–10m in space.

The review is organized as follows. The next section discusses studies of the Galactic bulge including the bulge globular clusters and abundance issues. Section 3 treats the case of M31, while the following, Section 4, discusses M32 and M33. We conclude in Section 5 with implications for the study of galaxy formation and evolution.

2. The Galactic Bulge

Stellar populations may be described by the parameters of age, abundance, kinematics, and structure. Hodge (1989) presents the concept of a population box, which is a useful concept in dealing with stellar populations with a common spatial location, but which have evolved through a range of age and metallicity (such as the Magellanic Clouds). Ground-based spectroscopy shows that the bulge is metal rich, with enhanced alpha abundances, while *HST* studies show that the bulge is old, with little age range. Both of these properties lead us to conclude that the bulge formed in an early, brief starburst (Section 5).

The Galactic bulge extends out from the nuclear region to roughly one kpc, where it merges with the inner halo. The *Cosmic Background Explorer* produced a striking image of the bulge in the infrared (Dwek et al. 1995), analysis of which finds that the structure is bar-shaped (Binney, Gerhard, & Spergel 1997). 500 pc South of the nucleus is Baade's Window, a region that has been the focus of considerable study, because Baade believed that the presence of the metal poor globular cluster NGC 6522 would give a more precise estimate of reddening toward the bulge. There are, in fact, wide areas of low extinction in the bulge, but at latitudes below 3° the optical extinction becomes severe. Blanco, McCarthy, & Blanco (1984) found hundreds of M giants in Baade's Window (and in other bulge fields). These M giants are evolved from stars more metal rich than those in globular clusters and indicate the presence of the classical bulge population. Most *HST* imaging examines the bulge population within one kpc from the nucleus, with most studies focusing on the minor axis. The mass of the bulge is constrained from models of the integrated light and by using the statistics of microlensing to constrain the extent and mass of the bar (roughly $2 \times 10^{10} \, M_\odot$; Zhao, Rich, & Spergel 1996).

The bulge of the Galaxy has frequently been imaged by *HST*. The well-known Baade's Window and Sgr I fields were reserved by the instrument teams for both WFPC1 and WFPC2 (Holtzman et al. 1993; 1998), These studies easily resolved the main sequence population, eventually leading to age constraints not possible using ground-based imaging. The author is collaborating in a new major program (GO-9750) led by K. Sahu to image the Galactic bulge with the Advanced Camera for Surveys (ACS) for 120 *HST* orbits. Taking advantage of the very large number of relatively bright main sequence stars in Baade's Window, we expect to discover a number of planetary transit events. Parallel infrared imaging using NICMOS will yield the deepest luminosity function ever obtained in the bulge.

2.1. Age Constraints

The most powerful age constraints for the bulge can be achieved only because *HST* has done such a fine job of imaging the globular clusters in the bulge. As was first realized by Shapley, most of the Milky Way's globular clusters concentrate toward the center and are both distant and superposed against a very substantial population of field stars. *HST* resolution is a gain for three reasons. First, the cores of these clusters are so crowded

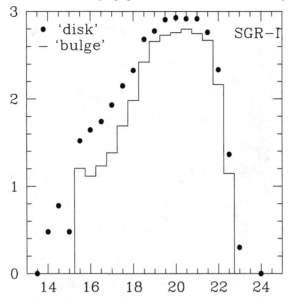

FIGURE 1. Proper motions toward the Galactic bulge are used to separate out the oldest stellar population by kinematics. The solid line shows the luminosity function at the turnoff point for stars which are kinematic members of the bulge (Kuijken & Rich 2002). The filled points show the luminosity function for stars separated by evidence of rotation to be foreground disk stars. Notice that the turnoff rise is far steeper for the old population. The Sgr I bulge field is roughly 300 pc S of the nucleus.

that they cannot be studied from the ground; the images are blended. Second, *HST* not only resolves these cores but permits work in a location where the cluster dominates over the field. Finally, by working in relatively small regions on the sky, one can control (if not eliminate) differential reddening, which tends to broaden the sequences in the CMD. Among the notable papers are Ortolani et al. (1995), Zoccali et al. (2001) and Feltzing & Johnson (2002), which use proper motion studies to produce cleaned color-magnitude diagrams of NGC 6553 and 6528. In the infrared, Ortolani et al. (2001) show that a number of the highly reddened bulge globular clusters are halo age. Cohn et al. (2002) does a comprehensive analysis of Ter 5 using NICMOS. The best age constraints still come from the high quality V, I turnoff photometry in the astrometric studies, which give a very narrow, high quality definition of the main sequence turnoff point. The Ortolani et al. (1995) analysis continues to be confirmed by subsequent studies (bulge and halo globular clusters coeval to within precision of data and models).

In principle, these studies might be accomplished using adaptive optics. In practice, AO has two problems. First, the technique does not work well shortward of 1 μm. Optical bandpasses are more sensitive to age and metallicity, the key parameters of importance in measuring a stellar population. The second problem is the areal stability of the point spread function. In experiments performed using Keck with a very bright (mag 8) guide star, we find that limiting magnitude varies strongly with radial distance from the guide star, even in good conditions. Space imaging will remain the gold standard for precision, high-angular resolution photometry for the foreseeable future.

The metal-rich population dominates the bulge within 500 pc of the nucleus, but the crowding is so severe that the main-sequence turnoff cannot be securely resolved from the ground. Combining the just-mentioned age measurements of globular clusters with improved ground and *HST* observations of the bulge field, *HST* studies clearly show that

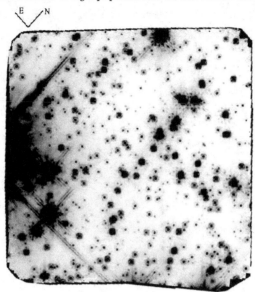

FIGURE 2. A 14-orbit *HST*/NICMOS integration in the Galactic bulge near NGC 6588 ($b = -6°$). The faintest stars are $H \sim 23$. The image is the sum of F110W (J) and F160W (H) images from the NIC2 camera, and it covers approximately $22.5'' \times 22.5''$ (Zoccali et al. 2000).

the bulge population is dominated by a population of globular cluster age. Ortolani et al. (1995) tie the horizontal branch of the field luminosity function to that of the well-studied metal-rich globular cluster NGC 6553, and they constrain the age dispersion in Baade's Window (500 pc S of the nucleus) to <10%.

However, foreground disk stars, and a blue population brighter than the turnoff, pose a problem in every bulge field that has been studied to date. Feltzing & Gilmore (2000) dealt with the problem by arguing that the blue stars do not follow the density of the bulge and must lie in the foreground. Kuijken & Rich (2002) separate the foreground disk from the background bulge using proper motions, showing that the blueward extension of the main sequence population belongs to the foreground disk (figure 1). The approach of Zoccali et al. (2003) statistically subtracts disk fields from the bulge to remove the foreground contribution. These studies uphold the determination that the bulge is old.

Having constrained the age of the bulge, it is now possible to use the brightest evolved stars (Frogel & Whitford 1987) as a standard comparison sample by which one may constrain the age of more distant stellar populations.

2.2. *Mass function*

One area of recent progress has been the measurement of the deep luminosity function of the bulge. As the low luminosity stars are faint and red, the best approach to this problem is via the infrared. The high microlensing optical depth toward the Galactic bulge spurred this problem on, as a huge population of brown dwarfs in the bulge offered one possible explanation (a fine historical discussion is given in Han & Gould 2003). The WFPC2 study of Holtzman et al. (1998) finds an apparent break in the mass function in the 0.5–0.7 $M_\odot$ range. The problem appears solved with the deep luminosity function of Zoccali et al. (2000), which has a power law slope of -1.3 (versus Salpeter of -2.35) and reaches 0.15 $M_\odot$. Figure 2 shows the NICMOS image of their field. If the mass function slope is confirmed with deeper ground- and space-based observations, it will decisively

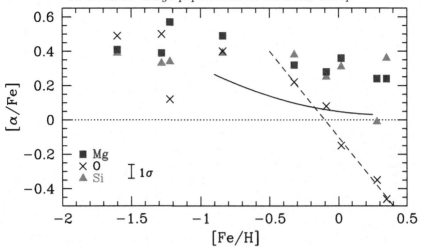

FIGURE 3. Keck spectroscopy of giants in the bulge field population (McWilliam & Rich (2003) gives trends of alpha elements [Mg/Fe] (filled squares), [Si/Fe] (filled triangles), and [O/Fe] (crosses), versus [Fe/H]. The trend for disk stars (solid curve) is from Edvardsson et al. (1993). The dashed line indicates the locus of constant [O/H] above [Fe/H] = −0.5 dex. The drop in oxygen abundance at high metallicity is puzzling, as Mg remains high. Both Mg and O are thought to be produced in the hydrostatic burning zones of massive stars. However, the trends are broadly consistent with an early and rapid formation history, as is inferred from *HST* color-magnitude diagrams.

rule out earlier models for the microlensing results that required a >50% population of brown dwarfs.

2.3. *Abundances*

As the Milky Way bulge is 100 times closer than the nearest major bulge population (Andromeda), at the present time only the Milky Way offers the possibility of obtaining abundances based on high-resolution spectroscopy of its field and globular cluster population.

Following the initial survey of Rich (1988), the first detailed abundances for bulge stars were obtained by McWilliam & Rich (1994; MR94) based on echelle spectra from the 4m Blanco telescope at CTIO. The distinguishing characteristic of the bulge composition is that while the population is old, the most metal-rich stars exceed the Solar iron abundance. Further, even at Solar metallicity, bulge stars (and clusters) are seen to be enhanced in alpha elements. MR94 calibrate Rich's (1988) low resolution study and find the mean [Fe/H] = −0.15.

The defining characteristic of bulge stars is their enhancement of alpha elements. MR94 found that Mg and Ti are enhanced in bulge stars. Two recent studies based on Keck spectra (Rich & McWilliam 2000, McWilliam & Rich 2003) confirm this picture, but find that oxygen has its own peculiar behavior. Although one would expect oxygen to follow Mg, it appears rather than [O/Fe] tends sharply toward the Solar composition near [Fe/H] ∼ 0.0. Figure 3 shows the peculiar behavior of Mg and O. The decline of [O/Fe] is most consistent with a scenario in which the bulge stopped producing oxygen. It is difficult to imagine any nucleosynthetic process in massive stars, or supernova model, which produces abundance Mg without producing O.

One long-standing problem in the bulge is the nature of the large numbers of M giants seen toward Baade's Window. The initial catalog is given in Blanco, McCarthy, &

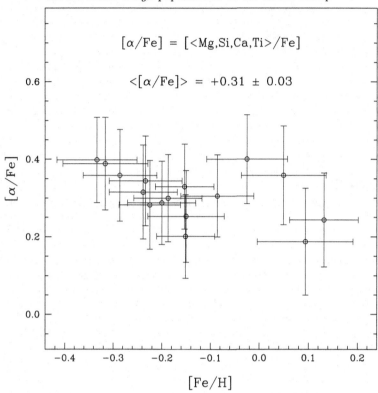

FIGURE 4. Infrared spectroscopy (using Nirspec at the Keck II telescope) of late M giants in the Galactic bulge gives the trend of alpha/Fe vs. [Fe/H] (Rich & Origlia 2004). This is the first instance of abundance determinations for these members of the bulge population. The trend follows that expected from an early formation scenario (Figure 3). Additional study will be needed to determine whether there are composition differences between the bulge K and M giants.

Blanco (1984). Frogel & Whitford (1987) studied the population in the infrared. Terndrup, Frogel, & Whitford (1990) argued that the M giants must be metal rich, based on a comparison of their TiO bands with stars of similar temperature in the Solar vicinity. Figure 4 shows that the composition of 11 M giants, obtained from lines in the 1.6 μm region (Rich & Origlia 2004). In contrast to what had been expected, the M giants are not iron rich. However, they do show alpha enhancement across the board, and even at the Solar abundance. The M giants are not the most metal-rich stars, but they might well represent the evolved progeny of the most alpha-enhanced stars in the bulge.

2.4. *Star formation history from NICMOS studies of the Central 100 pc*

In contrast to M31, the central 100 pc of the Milky Way is distinct from the rest of the bulge. Only infrared studies are possible, due to the $\sim$30 mag of visual extinction toward the Galactic Center. In a seminal paper, Catchpole, Whitelock, & Glass (1990) mapped the central two degrees of the Milky Way and found that the most luminous AGB stars are distinctly concentrated toward the nucleus, relative the rest of the light. Morris & Serabyn (1996) propose that this region is the r^{-2} cluster, while Launhardt, Zylka, & Mezger (2002) find a flattened central stellar disk with radius 230 pc and scale height of 45 pc. In the central 100 pc, massive star formation is evidenced in 10^6 $L_\odot$ LBV stars (Figer et al. 1999a) and two-Myr-old star clusters with $\sim$$10^3$ O stars (Figer et al. 1999b).

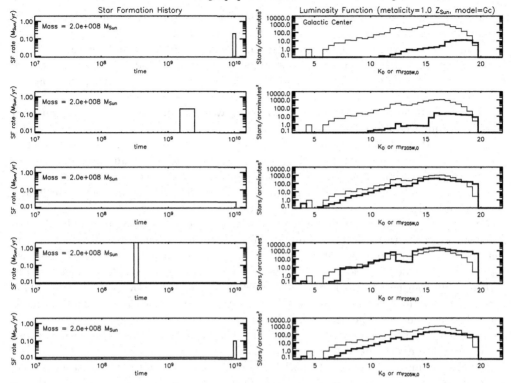

FIGURE 5. Model luminosity functions from the Padova models are compared with the K band luminosity function for stars within 50 pc of the Galactic Center (Figer et al. 2004). Alternative star formation histories are given in the left panel. Notice that a single burst old (10^{10}) stellar population fits very poorly. The best fits are from continuous star formation histories. In contrast to the Galactic bulge, which is old and metal rich, the central region appears to have a more continuous star formation history.

Has the star formation been ongoing, or is most of the mass contained in an old stellar population?

The nuclear population poses a special consideration. The correlations between black hole mass and bulge properties (velocity dispersion, luminosity) would suggest a connection between the central black hole and the earliest epoch of bulge formation. Given the overwhelming evidence for the early and dramatic emergence of supermassive black holes, one would expect the bulk of the mass in the bulge to have been in place early on. By contrast, most of the other spheroidal populations in which black holes have been discovered resemble more closely the bulge of M31, which is "old, red, and dead."

Once again, *HST* observations have been crucial. The central region is so crowded as to require either adaptive optics or *HST* imaging. However, the regular point-spread function delivered by NICMOS is a great advantage in crowded stellar fields. We undertook a systematic NICMOS imaging campaign to constrain the star formation history of the nuclear region. The primary aim of the campaign was to image the youthful Arches and Quintuplet star clusters (Figer et al. 1999b). Additionally, a series of field locations near the nucleus were imaged with the goal of studying the star formation history of the central 100 pc.

The field studies gave initially puzzling results (Figer et al. 2004). In the first instance, our color-magnitude diagrams appeared consistent—a globular cluster-like, old stellar population. Figure 5 shows our early results: The red clump is visible in every field, although occasionally the high differential extinction disperses the red clump locus along the reddening vector. The magnitude difference between the clump and the turnoff point is ~ 4 mag, roughly that expected for a stellar population of halo globular cluster age (12–13 Gyr). This observation led us to conclude initially that the nucleus must be old. The most reasonable interpretation favored a dominant old stellar population with a "frosting" of massive star formation, the latter having a top heavy mass function (and possibly forming no low mass stars).

As we investigated further, we discovered that the observed clump is one mag brighter than that seen in old populations. The presence of both a bright clump and a > 3.5 mag gap between the clump and the turnoff point posed a serious contradiction and it appeared as if our interpretation of the population had reached an impasse. This was finally resolved by undertaking a full model of the star formation history.

We use both Padova and Geneva models, with the emphasis being on the luminosity function. We constrain the total number of stars from the enclosed mass within the central 50 pc required by the best dynamical models of the bulge/nucleus. When the mass constraint is included in modeling the stellar population, we find acceptable matches only with roughly continuous star formation history (Figure 5). In summary, any old stellar population that matches the observed luminosity requires $\sim$ order of magnitude more stars than is observed. The continuous star formation model solves the problem of the anomalously luminous red clump, although it is important to confirm this with a secure detection of the main sequence turnoff point.

Perhaps the most vexing problem with the continuous star formation history is that evidently only a small fraction of the total mass of the nucleus was in place early on when the Galaxy's supermassive black hole formed. This would be in contrast with every other documented case of a dynamically detected SMBH.

3. M31 and M32

In the bulge of M31, the high spatial resolution of *HST* has helped to settle a number of long-standing problems. The distinct properties of the bulge have been known since Morgan's early spectroscopy (Baade 1963). The properties of the old stars in the M31 bulge helped to build the stellar populations paradigm. Sandage, Becklin, & Neugebauer (1969) obtained the first modern broadband spectrophotometry of the M31 nucleus (U through K) and showed that the nucleus is similar in color to the reddest ellipticals in Virgo.

The brightest members of an old, metal-rich stellar population are the asymptotic giant branch stars, the late M giants. These stars reach $M_K < -8$ and have very short lifetimes ($\sim 10^5$ yr). The author realized that, even in the crowded bulge of M31, it might be possible to resolve such stars from the ground. Spectroscopy with the 200-inch Hale telescope (Rich et al. 1989) showed that the brightest M31 bulge stars are M giants; the same investigation also discovered an unusual nova-like variable in the bulge of M31, which reached $M_R \sim -10$.

It is also possible to use the infrared to image individual M giants in the bulge of M31 in ground-based seeing (Rich & Mould 1991). The investigation touched off a long controversy. In the early 1990s, Palomar had a 58×62-pixel infrared array. The Rich & Mould (1991) investigation imaged a bulge field ~ 1 kpc from the nucleus using nine separate pointings. Subsequently, Rich, Mould, & Graham (1993) attempted to use the same array to produce mosaics of larger fields in the M31 bulge. The earlier investigation produced the luminosity function in Figure 6 (note that the bright end compares well with the *HST*/NICMOS photometry) but the Rich et al. (1993) photometry finds spuriously

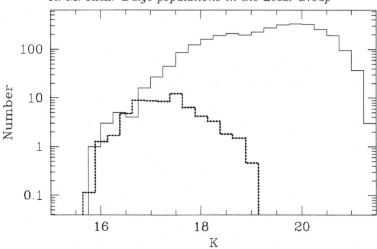

FIGURE 6. Ground-based (dotted) and NICMOS (solid) luminosity functions for a field in the M31 bulge 500 pc from the nucleus. The Palomar 200-inch data are from Rich & Mould (1991) and the *HST* data are from Stephens et al. (2003). Image crowding must affect the ground-based data, yet in this case it gives some useful information at the bright end. Stephens et al. 2003 settled a long standing controversy by showing that the luminosity function of the bulge of M31 resembles that of the Galactic bulge (see the following figure also).

bright stars in the nuclear region. The best guess is that this latter photometry was compromised by the process used to mosaic the images, which broadened and blended the already crowded stellar images.

Rich & Mould (1991) found that the luminosity function of the M31 bulge M giants appeared extended relative to the Frogel & Whitford (1987) luminosity function. They suggested that the bulge of M31 might be younger than the Milky Way bulge. Freedman (1992) imaged M32 in the infrared, and found a population of slightly more luminous AGB stars; as intermediate-age stars were suspected in M32, this observation appeared consistent with expectations. The M31 bulge imaging touched off a debate in the literature with Davies et al. (1993) arguing that the apparent bulge population in the Rich & Mould (1991) field actually suffered from disk contamination. The subsequent survey of Rich, Mould, & Graham (1993) showed largely that disk contamination could not be responsible for the bright stars. De Poy et al. 1993 then argued that photometric crowding was most likely responsible. In the meantime, Guarnieri et al. 1998 found an AGB star with $M_{bol} < -5$ in the old Galactic globular cluster NGC 6553. Old metal-rich populations can spawn very bright AGB stars, and it appears that the bright end of the AGB provides little age constraint at the metal rich end.

The problem has been resolved using high resolution imaging from the ground (Davidge 2001) and from space (Stephens et al. 2003). The luminosity function of AGB stars in the bulge of M31 *is extended* to $M_{bol} < -5$. However, the luminosity function in the bulge of M31 and in Baade's Window in the Milky Way appear identical. Further, the optical imaging of Jablonka et al. (1991) finds no evidence for anomalously bright stars in the I band.

The similarity of the M31 bulge luminosity function illustrated in Figure 7 to that of the Galactic bulge would appear to be compelling evidence that the bulge of M31 is old. However, one can still legitimately worry about the small number of bright stars with $M_{bol} < -4.5$. Might there still be room for an intermediate-age population? A very nice recent study by Rejkuba et al. (2004) on the Mira variables (and infrared

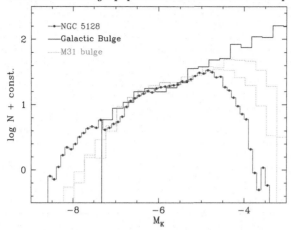

FIGURE 7. Scaled luminosity functions for the Milky Way (solid line), two M31 bulge fields from Stephens et al. (2003; dashed line) and for a halo field in NGC 5128 (beaded histogram); from Rejkuba et al. 2004. The bulge of M31 is similar to the Galactic bulge and we may infer that the bulge of M31 is also old, like the Galactic bulge. The clear extended giant branch (bright stars) found in NGC 5128 corroborate with the presence of a significant number of long-period Mira variable stars which are not found in the Galactic bulge fields. Studies of the luminosity functions of distant galaxies may offer a powerful approach to measuring subtle age differences between stellar populations. The bulge of M31does show some extension relative to the Galactic bulge. The recent discovery of an intermediate-age population in the M31 halo (Brown et al. 2003) raises the interesting possibility that a portion of the bulge population might also harbor intermediate-age metal-rich stars.

luminosity function) of a halo field in the nearby gE NGC 5128 supports the idea that an intermediate age population *would* show an extended giant branch. The presence of Miras with period > 500 days corresponds to a clear 0.5 mag extension of the giant branch, which is not seen in the M31 bulge (Figure 7). The Galactic bulge has few such stars, and Galactic globular clusters have no Miras with periods in excess of 300 days. However, the use of Miras as age indicators is in its infancy and will require extensive additional calibration. However, we conclude then that the case for an old M31 bulge is strengthened.

One may consider the age problem in the bulge to be settled, except for the finding of Brown et al. (2003; and see Brown, this volume) of an intermediate population of metal-rich stars in the M31 halo. At 11 kpc from the nucleus, roughly half the population has age 6–11 Gyr (younger than halo globular cluster age) and [Fe/H] > −0.5. The origin and extent of this stellar population is not known. There is also the long-standing (Burstein et al. 1984) suggestion that the metal-rich globular clusters in M31 might be younger than the oldest halo globular clusters in the Milky Way. This point has been recently sustained by Jiang et al. (2003), although there is no evidence for an extended giant branch in any M31 halo globular clusters (e.g., Stephens et al. 2001). Given the insensitivity of the metal-rich AGB tip luminosity to age, it would be hard to rule out a ∼8–11 Gyr age population in the M31 bulge at this time.

3.1. *M32*

Far less is known about the resolved populations of M32 because it is so difficult to work on. Near the nucleus the high surface brightness makes it very difficult to resolve stars. Davidge et al. (2000) resolves red giants using AO imaging on the Gemini telescope. Near the edge of M32, the disk of M31 dominates. The overall effect is to make it virtually impossible to reach the main sequence turnoff in M32. Nonetheless, some of the earliest WFPC2 photometry in M32 was quite successful. Grillmair et al. (1996) find that M32 has a metal-rich giant branch, much like the halo of M31. We know that an underlying old stellar population is present in the outskirts of M32, as RR Lyrae stars have been discovered by *HST*/WFPC2 imaging (Alonso-Garcia, Mateo, & Worthey 2004). The same study confirms the high metallicity of the red giants in that galaxy. In contrast to the Milky Way bulge and halo, the numbers of RR Lyrae stars in M32 are small and the authors conclude that the implied old stellar population is $\approx 2\%$ (in contrast to the near 100% for the Galactic bulge). We note that M32 is metal rich, so only a small fraction of its population is metal poor enough to land on the blue HB and to become RR Lyrae stars. It would be interesting to use the existing photometry and spectroscopy to constrain further the age distribution in M32.

4. M33, NGC 147, and G1

With these less luminous systems, we reach the boundary of what may properly be called a bulge population. Yet this is also a regime where one gains insight into the formation of bulges.

The infrared photometry of Regan & Vogel (1994) shows that infrared light in the inner five arcmin rises clearly toward the nucleus of M33. The infrared surface photometry of the disk can be decomposed into a clear disk and spheroidal (bulge) population. The extent of this "bulge" is of order one kpc, as might be expected for a central bulge population. Mighell & Rich (1995) image the nuclear region with *HST* and find a centrally concentrated old stellar population which is in contrast to the younger star-forming disk. Although severely affected by image crowding, the nuclear region shows a wide red-giant branch, indicating some metallicity dispersion. It would be of interest to repeat this observation at higher spatial resolution and S/N.

NGC 147 has been imaged using WFPC2 (Han et al. 1997) and has a wide red-giant branch. In contrast to the halo of M31, the giants of NGC 147 range in color from M15 (~ -2 dex) to 47 Tuc (~ -0.7 dex), but not more metal rich.

The globular cluster G1 is the most luminous globular cluster known in the Local Group. Rich et al. (1996) show that the luminosity function and red giant branch of G1 resembles that of 47 Tuc. Meylan et al. (2001) argue that the giant branch shows a clear abundance spread. However, if present, the abundance spread has a relatively subtle effect on the colors of the stars. Gebhardt, Rich, & Ho (2002) find dynamical evidence for a black hole in G1. There have been some suggestions, consequently, that G1 might be the remnant nucleus of a dwarf elliptical galaxy. It is interesting that the nucleus of M33, similar to G1 in its luminosity, lacks a kinematic black hole signature (Gebhardt et al. 2001).

Are these systems bulges? Likely not, for they fall short in total mass. Yet they have a burst history of star formation and clear evidence for a wide abundance spread, indicating that they were able to retain metals from multiple generations of star formation. In a broad sense, the abundance distributions follow the Simple Model of chemical evolution (with upper metallicity limited by wind-driven loss of metals). At the very least, one may safely speculate that these systems share more in common with bulges than they do with globular clusters or dwarf spheroidals, and that much may profitably be learned by studying their formation histories.

5. Conclusions

HST imaging has produced beautiful results for the bulge populations of the Local Group. Hubble's colleague Walter Baade would surely be amazed and delighted if he were able to see an array of the results. In the Galactic bulge, the age of the population is demonstrated to be similar to that of the Galactic halo globular clusters. *HST* imaging has shown that normal hydrogen-burning stars and not some peculiar population of brown dwarfs is responsible for the high rate of microlensing events toward the bulge. Within the coming months, *HST* will search the bulge population for planetary transits.

In the bulge of M31, *HST* imaging has largely settled a long standing question about whether some fraction of the bulge could be intermediate age. Metal rich, perhaps, but age 8–10 Gyr. The infrared luminosity function instead is consistent with that of the Galactic bulge, so the bulge of M31 is almost certainly old.

The spectroscopy of the spheroids undertaken by the NUKER team (see Gebhardt et al. 2003) has all but connected bulges with supermassive black holes, while reaching well beyond the Local Group, is also a critical legacy of *HST*.

The principle legacy of *HST* studies of bulge populations is that we have achieved a clear demonstration that the bulge of the Milky Way is as old as the halo globular clusters. The ages of bulge globular clusters are trivially derived from color-magnitude diagrams and are shown to be halo age. Reaching to M31, the central bulge luminosity function resembles clearly that of the Milky Way bulge, and indicates strongly that the M31 bulge is old.

This evidence regarding the age of bulge populations has broad implications, because it suggests that the formation of bulge populations and supermassive central black holes may be one of the most significant and early stages of the process by which the most massive galaxies, similar to the Milky Way, form.

REFERENCES

ALONSO-GARCÍA, J., MATEO, M., & WORTHEY, G. 2004 *AJ* **127**, 868.
BLANCO, V. M., MCCARTHY, M. F., & BLANCO, B. M. 1984 *AJ* **89**, 636.
BELL, E. F., MCINTOSH, D. H., BARDEN, M., WOLF, C., ET AL. 2004 *ApJ* **600**, L11.
BINNEY, J., GERHARD, O., & SPERGEL, D. 1997 *MNRAS* **288**, 365.
BROWN, T. M., FERGUSON, H. C., ET AL. 2003 *ApJ* **592**, L17.
BURSTEIN, D., FABER, S. M., GASKELL, C. M., & KRUMM, N. 1984 **ApJ 287**, 586.
CATCHPOLE, R. M., WHITELOCK, P. A., & GLASS, I. S. 1990 *MNRAS* **247**, 479.
COHN, H. N., LUGGER, P. M., GRINDLAY, J. E., & EDMONDS, P. D. 2002 *ApJ* **571**, 818.
DAVIDGE, T. J. 2000 *AJ* **119**, 748.
DAVIDGE, T. J. 2001 *AJ* **122**, 1386.
DAVIDGE, T. J., RIGAUT, F., CHUN, M., ET AL. 2000 *ApJ* **545**, L89.
DAVIES, R. L., FROGEL, J. A., & TERNDRUP, D. M. 1991 *AJ* **102**, 1729.
DEPOY, D. L., TERNDRUP, D. M., FROGEL, J. A., ATWOOD, B., & BLUM, R. 1993 *AJ* **105**, 2121.
DWEK, E., ARENDT, R. G., HAUSER, M. G., ET AL. 1995 *ApJ* **451**, 188.
EDVARDSSON, B., ANDERSEN, J., GUSTAFSSON, B., LAMBERT, D. L., NISSEN, P. E., & TOMKIN, J. 1993 *A&A* **275**, 101.
FELTZING, S. & GILMORE, G. 2000 *A&A* **369**, 510.
FELTZING, S. & JOHNSON, R. A. 2002 *A&A* **385**, 67.
FREEDMAN, W. 1992 *AJ* **104**, 1349.
FUKUGITA, M., HOGAN, C. J., & PEEBLES, P. J. E. 1998 *ApJ* **503**, 518.
FIGER, D. F., MORRIS, M., GEBALLE, T. R., RICH, R. M., SERABYN, E., MCLEAN, I. S., PUETTER, R. C., & YAHIL, A. 1999a *ApJ* **530**, L97.

FIGER, D. F., KIM, S. S., MORRIS, M., SERABYN, E., RICH, R. M., & MCLEAN, I. S. 1999b *ApJ* **525**, 750.

FIGER, D. F., RICH, R. M., KIM, S. S., MORRIS, M., & SERABYN, E. 2004 *ApJ* **601**, 319.

FROGEL, J. A. & WHITFORD, A. E. 1987 *ApJ* **320**,199.

GEBHARDT, K., LAUER, T. R., KORMENDY, J., ET AL. 2001 *AJ* **122**, 2469.

GEBHARDT, K., RICH, R. M., & HO, L. C. 2002 *ApJ* **578**, L41.

GEBHARDT, K., ET AL. 2003 *ApJ* **596**, 903.

GRILL MAIR, C. J., LAUER, T. R., WORTHEY, G. ET AL. 1996 *AJ* **112**, 1975.

GUARNIERI, M. D., ORTOLANI, S., MONTEGRIFFO, P., ET AL. 1998 *A&A* **331**, 70.

HAN, M., HOESSEL, J. G., GALLAGHER, J. S. III, HOTZMAN, J., & STETSON, P. B. 1997 *AJ* **113**, 1001.

HAN, C. & GOULD, A. 2003 *ApJ* **592**, 172.

HODGE, P. 1989 *ARA&A* **27**, 139.

HOLTZMAN, J. A., LIGHT, R. M., BAUM, W. A., ET AL. 1993 *AJ* **106**, 1826.

HOLTZMAN, J. A., WATSON, A. M., BAUM, W. A., ET AL. 1998 *AJ* **115**, 1946.

JABLONKA, P., BRIDGES, T. J., SARAJEDINI, A., MEYLAN, G., MADER, A., & MEYNET, G. 1999 *ApJ* **518**, 627.

JIANG, L., ET AL. 2003 *AJ* **125**, 727.

KENT, S. M. 1987 *AJ* **93**, 816.

KUIJKEN, K. & RICH, R. M. 2002 *AJ* **124**, 2054.

LAUNHARDT, R., ZYLKA, R., & MEZGER, P. G. 2002 *A&A* **384**, 112.

MCLEAN, I., ET AL. 1998 *SPIE* **3354**, 566.

MCWILLIAM, A. & RICH, R. M. 1994 *ApJS* **91**, 749.

MCWILLIAM, A. & RICH, R. M. 2003, astro-ph/0312628; Carnegie Astrophysics Series Vol. 4 (eds. A. McWilliam and M. Rauch) http://www.ociw.edu/ociw/symposia/series/symposium4/proceedings.html/.

MEYLAN, G., SARAJEDINI, A., JABLONKA, P., ET AL. 2001 *AJ* **122**, 830.

MIGHELL, K. J. & RICH, R. M. 1995 *AJ* **110**, 1649.

MORRIS, M. & SERABYN, E. 1996 *ARA&A* **34**, 645.

ORTOLANI, S., BARBUY, B., BICA, E., RENZINI, A., ZOCCALI, M., RICH, R. M., & CASSIS, S. 2001 *A&A* **376**, 878.

ORTOLANI, S., RENZINI, A., GILMOZZI, R., MARCONI, G., BARBUY, B., BICA, E., & RICH, R. M. 1995 *Nature* **377**, 701.

REGAN, M. W. & VOGEL, S. N. 1994 *ApJ* **434**, 536.

REJKUBA, M., MINNITI, D., SILVA, D. R., & BEDDING, T. R. 2004 *A&A* **412**, 903.

RICH, R. M. 1988 *AJ* **95**, 828.

RICH, R. M. & MCWILLIAM, A. 2000 *SPIE* **4005**, 150; (astro-ph/0005113).

RICH, R. M., MIGHELL, K. J., FREEDMAN, W. L., & NEILL, J. D. 1996 *AJ* **111**, 768.

RICH, R. M. & MOULD, J. R. 1991 *AJ* **101**, 1286.

RICH, R. M., MOULD, J., PICARD, A., FROGEL, J. A., & DAVIES, R. 1989 *ApJ* **341**, L51.

RICH, R. M., MOULD, J. R., & GRAHAM, J. R. 1993 *AJ* **106**, 2252.

SANDAGE, A. R., BECKLIN, E. E., & NEUGEBAUER, G. 1969 *ApJ* **157**, 55.

STEPHENS, A. W., FROGEL, J. A., DEPOY, D. L., ET AL. 2003 *AJ* **125**, 2473.

TERNDRUP, D. M., FROGEL, J. A., & WHITFORD, A. E. 1991 *ApJ* **357**, 453.

WHITFORD, A. E. 1978 *ApJ* **226**, 777.

ZHAO, H., RICH, R. M., & SPERGEL, D. N. 1996 *MNRAS* **282**, 175.

ZOCCALI, M., CASSISI, S., FROGEL, J. A., ET AL. 2000 *ApJ* **530**, 418.

ZOCCALI, M., RENZINI, A., ORTOLANI, S., BICA, E., & BARBUY, B. 2001 *AJ* **121**, 2638.

ZOCCALI, M., RENZINI, A., ORTOLANI, S., GREGGIO, L., ET AL. 2003 *A&A* **399**, 931.

The Local Group as a laboratory for the chemical evolution of galaxies

By DONALD R. GARNETT

Department of Astronomy and Steward Observatory, University of Arizona, Tucson, AZ 85721,
USA

I review what is known about the general trends of metallicity and element abundance ratios in Local Group galaxies, and some implications of the abundance trends for chemical evolution. The Local Group spirals show radial metallicity gradients and a mean metallicity that increases with luminosity. The composition gradients steepen with decreasing galaxy luminosity, but are roughly similar when the gradients are derived per unit disk scale length. This suggests that the evolution of the metallicity gradient is closely tied to the evolution of the baryon distribution. The M31 and Milky Way bulges appear to have similar metallicity distributions. The high $[\alpha/\mathrm{Fe}]$ in Galactic bulge stars indicates that the bulge formed rapidly. Metallicity distributions for M31 and Galactic halo stars are also similar, except that M31 has more globular clusters that are metal-rich, possibly related to its larger bulge. M33 is anomalous in that its halo clusters may be significantly younger than the Galactic halo. Local Group irregular galaxies are metal-poor, and their mean metallicity correlates with galaxy luminosity. They have low effective yields, as derived from a comparison of mean metallicity with gas fraction, and the effective yield is correlated with galaxy rotation speed (or mass). This is evidence that the irregulars have lost metals to the IGM, either through galactic winds or stripping. Dwarf ellipticals in the Local Group are also metal-poor, and follow a similar metallicity-luminosity relation. The fact that the dEs have no gas also points to loss of metals as a significant factor in their evolution.

1. Introduction

The ultimate aim of the study of element abundances in galaxies is to shed light on the evolution of galaxies. Chemical evolution provides insight into aspects of galaxy evolution that can not be inferred directly from photometry or morphology studies. For example, the ratios of elements that are produced in stars of different masses, and thus ejected into the interstellar medium (ISM) on different time scales, are sensitive to how rapidly star formation has occurred, and thus can distinguish a starburst event from quasi-continuous star formation. The spatial distribution of heavy elements depends on gas flows and timescales for mixing of supernova ejecta into the ISM as well as the local star formation history. Deviations of abundance enrichment derived from star formation histories can thus yield information on gas flows. Comparison of observed abundances with that expected based on the fraction of gas that has been turned into stars can tell us if particular galaxies are ejecting heavy elements into the intergalactic medium.

Abundance studies for Local Group galaxies have special merit because we can determine both abundances and the relevant stellar population in detail. For more distant galaxies, the star formation history must be inferred from either integrated colors or integrated spectra combined with stellar population synthesis, whereas in the Local Group star formation histories can be studied directly by photometry at high spatial resolution with the *Hubble Space Telescope*. We can thus compare the expected production of metals from the known stellar population with the actual inventory. Abundances in Local Group galaxies can be measured down to small spatial scales, yielding information on the homogeneity of the ISM and the efficiency of mixing of SN ejecta into the ISM. In the Local Group, one can combine abundance measurements with the properties of individual objects, such as molecular clouds, to determine how metallicity affects their properties. As a final example, we have a growing capability to measure abundances in stars of different ages to various depths in the Local Group galaxies. This provides information on the time evolution of metallicity in galaxies with different evolution paths—a means of studying galaxy evolution "in our own backyard."

In this review I discuss what we know about trends for abundances and element abundance ratios for the various types of galaxies in the Local Group. Rather than give a detailed account of past results, I will focus on general trends and recent developments where possible, as the literature for Local Group galaxies is vast. The literature on abundances in stars and the ISM for the Galaxy is particularly lengthy, so I plan to discuss the results in only enough detail to place the Galaxy in context with the other members of the Local Group. I will highlight areas where observed abundance trends provide the most interesting information for galaxy evolution.

2. Brief discussion of abundance measurements

This section will briefly review the techniques for deriving abundances in galaxies. Detailed descriptions of abundance measurement methods will not be presented; the main point here is to point out important new developments that have had a significant effect on abundance measurements, and to highlight remaining uncertainties. Interested readers should consult relevant textbooks (Gray 1992, Pagel 1997, and Osterbrock 1989, for example) or primary sources for more in-depth discussions.

2.1. *Stars*

The analysis of the spectra of cool stars ($< 7,000$ K) is fairly well established. Spectrum synthesis using LTE stellar atmospheres has been the standard practice, with non-LTE analyses investigated for special cases. As UV opacities and oscillator strength values have improved, so have the analyses.

One remarkable aspect has been the new developments in stellar atmosphere models, in particular, 3D time-dependent hydrodynamical models (Asplund 2003a,b). These models incorporate the effects of granulation (convection) in the stellar atmosphere and its effects on radiative transfer. Temperature gradients in the warm upflows and cool downflows are very different, and the presence of pockets of cool gas enhance the lines from molecules such as OH, so that one derives lower abundances from such species.

Application of these models to the Sun yields O abundances from OH line that are reduced by about 0.25 dex compared to standard 1D stellar atmospheres, in much better agreement with abundances derived from [O I]. The abundance from [O I] was itself reduced after taking into account the blending of the [O I] 6300 Å line with a Ni I line (Allende Prieto, Lambert, & Asplund 2001). The result is that C, N, and O abundances in the Sun now appear to be 0.2–0.3 dex smaller than previously thought. This has reduced the paradoxical discrepancy that the Sun appeared to be more metal-rich than nearby young stars and the ISM.

For metal-poor halo dwarf stars, the impact of 3D atmosphere models is also significant. Temperatures in the photospheres of metal-poor stars can reach very low values, again leading to enhancement of molecular lines. The models indicate that this effect can account for the very large discrepancies in [O/Fe] between analyses of OH lines and those using [O I], in the sense that the OH lines yield much lower O abundances with 3D atmospheres. These results may resolve the debate over whether [O/Fe] is essentially constant in Galactic halo stars or increases linearly as [Fe/H] decreases.

For hot stars, the intense radiation field dominates over collisions in populating levels, so the statistical equilibrium must be computed explicitly. This is a nonlinear, computationally intensive problem. A major difficulty is dealing with the strong stellar winds that typically accompany such strong radiation fields, which affect the line blanketing in a manner that depends on the mass loss rate and the velocity field. Non-planar geometry is also an important complication.

In main-sequence B stars with $T_{\text{eff}} < 30,000$ K, LTE and non-LTE analyses tend to give similar abundances, although the non-LTE analysis reduces trends in derived abundance with excitation potential somewhat (e.g., see Daflon et al. 2001). For hotter O stars and supergiants, however, non-LTE calculations are crucial, and the effects of wind blanketing increase.

2.2. *Interstellar absorption lines*

Although the study and analysis of interstellar absorption lines is a classic astronomical subject, *HST* has dramatically expanded this area by allowing sensitive ultraviolet spectroscopy, first with GHRS, and now with STIS. Most absorption line abundance studies still target sightlines with the Galaxy, but STIS is sensitive enough to study interstellar lines in the Magellanic Clouds (e.g., Welty et al. 1999, 2001). In addition to the more abundant light elements and iron-peak elements, absorption line spectroscopy has been used to determine the interstellar abundances of even *s*-process elements such as cadmium and tin (Sofia, Meyer & Cardelli 1999).

2.3. *Photoionized nebulae*

The emission-line spectra of H II regions and planetary nebulae (PNs) have long been used to probe gas-phase abundances. Luminous H II regions can be studied in distant galaxies and provide information on ambient interstellar material. Planetary nebulae are fainter, and can also be significantly enriched (or depleted) in C, N, He, and *s*-process elements by stellar processes, but certain populations of PNs have relatively old stellar progenitors and can be used to study abundances in a galaxy at ages of a few Gyr in the past.

Two standard techniques are in common use for deriving abundances from collisionally-excited emission lines in photoionized nebulae:

(1) If the nebula is warm enough, and the spectrum deep enough, one can measure the electron temperature and density from certain diagnostic emission line intensity ratios, e.g., [O III] $\lambda5007/\lambda4363$ (for T_e), or [S II] $\lambda6717/\lambda6731$ (for n_e). With these in hand, it is straightforward to calculate the abundance of any ionic species by solving the rate equations for the low-lying levels or by using an analytic approximation if the density is very low ($\approx$100 cm^{-3} or less). The abundance of any element is then the sum of the abundances of all its ions that are present within the H$^+$ zone.

Uncertainties for abundances derived this way include the at-the-present debated effects of temperature and density inhomogeneities on the derived physical conditions, and the uncertain corrections for unobserved ions, all of which can introduce systematic errors. Forbidden lines in the visible spectrum are sensitive to temperature fluctuations, while many fine-structure lines, such as [O III] 88 μm and [N III] 57 μm, are sensitive to density fluctuations. Combined optical-IR spectroscopy could mitigate these issues in principle, but few studies with sufficient sensitivity and accurate beam matching have been done. Nevertheless, abundances derived from H II regions in the solar neighborhood are in good agreement with those derived from interstellar absorption lines and nearby B stars, suggesting that the systematic uncertainties are probably small (Deharveng et al. 2000, Meyer et al. 1998).

(2) If the nebula is too faint or too cool to measure T_e directly, one can use the intensities of the stronger forbidden lines with a suitable calibration (such as, for example,

[O II]+[O III]/Hβ; Pagel et al. 1980) to estimate the abundances, or else compute a tailored photoionization model to match the emission line strengths.

Strong-line abundance analyses dominate the literature for spiral galaxies, because of the expense in telescope time needed to measure electron temperatures for the relatively metal-rich H II regions in spirals. This is less of an issue for metal-poor irregular galaxies; their H II regions are hotter and the $\lambda 4363/\lambda 5007$ ratio is commonly measured.

Difficulties with the strong-line method arise in metal-rich regions, because the calibration of line strengths versus abundance has been based on photoionization models. Ionization models suffer from degeneracy because predicted forbidden line strengths depend in similar ways on the inferred stellar radiation field, the metallicity (which controls the cooling), and the ionization parameter (roughly speaking, the ratio of photons to gas density). This is particularly the case in metal-rich nebulae, where the cooling is very sensitive to the amount of O^{+2} and the electron density (Oey & Kennicutt 1993), as well as grain heating/cooling processes and depletion (Shields & Kennicutt 1995, Henry 1993).

Deep measurements of T_e in metal-rich H II regions should resolve some of these difficulties. Recent studies (Kinkel & Rosa 1994; Castellanos et al. 2002; Kennicutt, Bresolin, & Garnett 2003) have now measured T_e in H II regions with O/H as high as the solar value for the first time. These new measurements suggest that the commonly-used calibrations of strong nebular lines may overestimate O/H by a factor of two or so. If confirmed, the metallicity scale in galaxies would be shifted systematically downward, and metallicity dependences for certain phenomena, such as the Cepheid period-luminosity relation, would become significantly steeper.

2.4. *Other techniques*

Abundances for a variety of elements can be derived from high-resolution spectroscopy of interstellar absorption lines. The technique is straightforward, either directly determining column densities from unsaturated lines, or by using the curve of growth for stronger lines. The sources of uncertainty are relatively few: oscillator strengths, separating multiple velocity components, and the effects of ionized gas on column densities for species that straddle the hydrogen ionization limit. The primary difficulty is that most of the interesting absorption lines are found in the vacuum ultraviolet, necessitating spacecraft observations. The *Hubble Space Telescope* has greatly expanded this field by allowing observation of fainter stars at sufficiently high spectral resolution to sample sightlines toward OB stars across the Galaxy and in the Magellanic Clouds, as well as toward moderately faint QSOs to sample halo gas in the Milky Way and other galaxies.

Bulk metallicities for resolved stellar populations can be estimated from photometry. The color of the main-sequence turnoff and the slope and color of the red giant branch are both sensitive to the stellar metallicity, and are often used to estimate the metallicity of stellar populations. However, these features are also sensitive to the age of the stellar population as well, and so there is some degeneracy in solutions as a result. One possible way to break this degeneracy (for stars too faint to do traditional stellar spectroscopy) is to measure the equivalent width of the Ca II triplet near 8500 Å. This has been shown to be sensitive to metallicity in globular cluster giants (Armandroff & Zinn 1988), and requires only moderate resolution spectroscopy. One potential source of systematic error that needs to be investigated is the effect of varying [Ca/Fe] ratios on the Ca II index.

Spectrum synthesis is commonly used to estimate ages and metallicities for unresolved stellar populations, particularly for elliptical galaxies and spiral bulges. This technique suffers from degeneracies in age and metallicity in a similar way as the color-magnitude diagram (Worthey 1998; Henry & Worthey 1999). The effects of element abundance

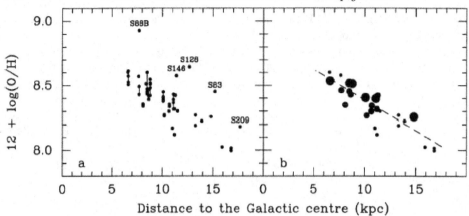

FIGURE 1. Oxygen abundance gradient in the Galaxy from H II regions (Deharveng et al. 2000). The left panel shows all data, while the right panel shows the points weighted by data quality (larger points have higher quality), with the fitted gradient. The regions labeled in the left panel are excluded from the fit.

ratios (in particular that of alpha-capture elements to iron, α/Fe) must also be taken into account.

3. Abundances in Local Group spirals

Here I discuss what we have learned about element abundance trends in the Local Group spirals: our own Galaxy, M31, and M33. This discussion will focus on comparing and contrasting what is observed in the three galaxies, connecting the general trends with models for the formation and evolution of disk galaxies. This discussion can be broken down to deal with disks, bulges, and halos separately, although there is some overlap.

3.1. *Disk abundances*

There is a great deal of information on abundances in both stars and gas in the solar neighborhood, and this has provided the most constraints on models for the evolution of the Milky Way. Detailed abundance measurements for dwarf stars and giants have yielded a great deal of information on the time evolution of metal enrichment in the Galaxy.

The age-metallicity relation for dwarf stars in the solar neighborhood shows the rate of increase of metallicity with time. Various studies have shown that the mean metallicity of stars has increased relatively steadily with time in the solar neighborhood (Twarog 1980; Edvardsson et al. 1993; Rocha-Pinto et al. 2000), from [Fe/H] $\approx$ -0.6 (in dex relative to the Sun) for the oldest disk stars up to [Fe/H] $\approx$ $+0.1$ at the present time. The fact that the oldest disk stars have metallicities that are a significant fraction of the solar value is noteworthy, and is also reflected in the metallicity distribution of disk stars, which shows a distinct lack of metal-poor stars. This has been popularly known as the "G-dwarf problem," as a simple, closed-box model for the evolution of abundances predicts many more metal-poor dwarf stars than observed. A number of solutions have been proposed to solve this problem; the two most promising explanations are:

(1) Infall of metal-poor gas. Infall of gas delays much of the star formation into the disk to later times, after stars that have formed early have significantly enriched only a small

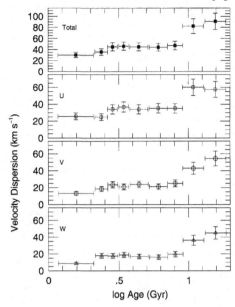

FIGURE 2. Variation in velocity dispersion of the U, V, and W components for stars from Edvardsson et al. (1993). The vertical error bars show the square root of the number of stars in each bin. The horizontal error bars reflect the width of each age bin.

fraction of the present-day mass of the galaxy (Lynden-Bell 1975). Thus, metal-poor stars would be quite rare today.

(2) Initial enrichment of the gas. If the disk formed later than the halo, the disk gas could have been enriched by supernova ejecta from massive halo stars, although it is not yet clear if enough stars were produced in the halo to enrich the disk to an initial [Fe/H] of at least −0.6.

Element abundance ratios provide further insight into chemical evolution. Numerous studies have been made of element abundance ratios in dwarf stars in the solar neighborhood. The most comprehensive study to date is that of Edvardsson et al. (1993; hereafter EAGLNT), who presented spectroscopy for 13 elements, as well as age and kinematic data, for 189 slightly-evolved field stars within 80 pc. EAGLNT confirmed a number of abundance trends from previous smaller studies, and provided a number of interesting new trends:

(1) [O/Fe], [Si/Fe], [Mg/Fe], [Ca/Fe], and [Ti/Fe] all decline steadily from a value of ≈ +0.2–0.3 down to about 0.0 as [Fe/H] increases from −1 to the solar value. The elements, excluding Fe, are generally lumped together as 'α' elements, are thought to be produced mainly in non-explosive nucleosynthesis in massive stars (although there is some question for Ca and Ti), while Fe is expected to come mainly from the explosion of white dwarfs in Type Ia supernovae. The observed trend in [α/Fe] is interpreted to mean that the early solar neighborhood was enriched mainly by the products of massive stars, with [α/Fe] declining as a result of the cumulative enrichment of the disk by Type Ia SNe with increasing age of the disk.

(2) [O/Fe] appears to be higher at [Fe/H] ≈ − 0.4 to − 0.2 for stars with mean orbit radii < 7 kpc than for those with radii >9 kpc. This can be interpreted as evidence that the inner parts of the disk formed more rapidly than the outer parts. However, the

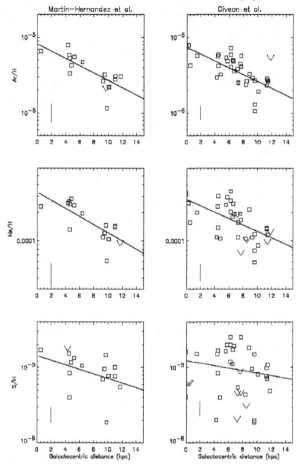

FIGURE 3. Abundance gradients for Ne, S, and Ar from ISO spectroscopy of Galactic H II regions (Martín-Hernandez et al. 2002; Giveon et al. 2002).

number of stars observed with large and small orbit radii may not be large enough yet to consider this a firm result.

(3) [Ba/Fe] appears to be smaller in old disk stars than in young disk stars. This suggests that the source of s-process elements is stars that have longer lives than SN Ia progenitors. The masses for these progenitor stars are not well-known, but the implication is that the production of s-process elements is dominated by relatively low mass AGB stars.

(4) The velocity dispersion of the disk star sample increases sharply for [Fe/H] < -0.4 and ages $>$ 9–10 Gyr (Fig. 2). It is likely that this represents the onset of the Galactic thick disk, and implies that the thick disk is an ancient feature, possibly a result of a minor merger with a dwarf satellite galaxy (Freeman 1991).

(5) EAGLNT noted a significant scatter in their age-metallicity relation, with σ([Fe/H]) > 0.2 dex. This is puzzling as the earlier relation from Twarog (1980) showed much less scatter with less accurate metallicities. However, one must treat the scatter in the EAGLNT relation with some skepticism, as the star sample was selected by metallicity (EAGLNT; Garnett & Kobulnicky 1999). The more recent study of Rocha-Pinto et al. (2000) shows much less scatter in the age-metallicity relation.

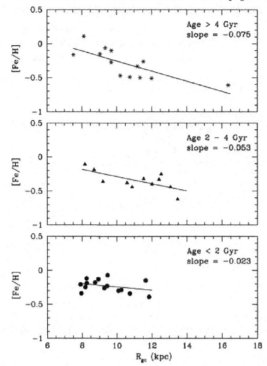

FIGURE 4. The time variation of the radial gradient of [Fe/H] based on observations of open clusters, from Friel et al. (2002).

Abundances for low- and intermediate-mass stars are not available yet for M31 and M33. It should be possible to obtain some data for red giants in these galaxies in the foreseeable future.

The spatial distribution of abundances in the Galaxy, M31, and M33 have been roughly established for young populations (OB stars and H II regions). All three galaxies show radial gradients in O/H. This was inferred first for the Galaxy from a radial variation in T_e for H II regions, and confirmed by direct visible-light spectroscopy of H II regions (Churchwell et al. 1978; Shaver et al. 1983; Deharveng et al. 2000; see Fig. 1). Recent studies of B stars in the Galaxy yield similar result (e.g., Rolleston et al. 2000; Daflon et al. 2001). Much of the uncertainty in determining the Galactic abundance gradient involves uncertain distances and reddening for both H II regions and stars. Recent *ISO* spectroscopy of relatively large samples of compact H II regions have also shown Ne/H and Ar/H abundance gradients of about −0.05 dex/kpc, in good agreement with that for O/H (Martín-Hernandez et al. 2002; Giveon et al. 2002; Fig. 3). Sulfur abundances from these studies appear to show a shallower gradient, but the scatter in the S/H data is quite large, likely because of the relatively poor performance of the *ISO* SWS spectrometer in the 18–40 μm band, where the [S III] lines are located.

In the Galaxy, it is possible to study the time evolution of the composition gradient through suitable choice of targets. This has been done for open clusters (Friel et al. 2002) and planetary nebulae (PNs; Maciel, Costa, & Uchida 2003). Friel et al. measured [Fe/H] for stars in open clusters. Since the clusters have well-defined ages, it is easy to sort out the time evolution (Fig. 4). The main limitations of this work are the currently limited distance range for the cluster sample, 8–12 kpc, and the lack of very old clusters. Planetary nebulae, on the other hand, have been observed over a much larger range

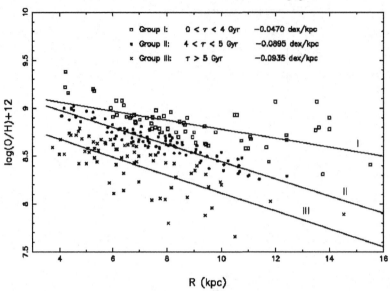

FIGURE 5. Time variation of the O/H gradient in the Galaxy based on observations of planetary nebulae (Maciel et al. 2003). In this plot the open squares represent the youngest PNs, while the crosses represent the oldest PNs. The solid lines show the fitted gradients for each group.

of distances. The major difficulties here are in determining ages and distances, both of which must be derived in general by statistical methods. Maciel et al. separated the PN sample into rough age groups based on kinematics: Type I PNs, which are He- and N-rich and believed to come from relatively massive, and thus young (<4 Gyr) progenitors; a second group of 'old disk' PNs with ages of 4–5 Gyr; and a third group with 'thick disk' kinematics and ages > 5 Gyr. Both studies find evidence that the composition gradient has flattened as the Galaxy has aged, with surprisingly similar results: -0.07 to -0.09 dex/kpc for objects with ages > 4–5 Gyr, and -0.03 to -0.05 dex/kpc for objects <4 Gyr old (Fig. 5). If confirmed, this could rule out whole swaths of chemical evolution models for the Galaxy, which predict that either the gradient steepens with time or flattens with time.

Disk abundances in M31 and M33 have been determined mainly from H II regions. These galaxies have not been as well studied as some more distant galaxies; few T_e measurements have been made, so abundances have been derived mainly via photoionization modelling and strong-line calibrations. Nevertheless, various studies generally agree on the interstellar abundances in M31 and M33 (Kwitter & Aller 1983; Vílchez et al. 1988; Blair, Kirshner & Chevalier 1982; Galarza, Walterbos, & Braun 1999). These studies have found O/H gradient slopes of approximately -0.1 dex/kpc for M33 and -0.01 to -0.02 dex/kpc for M31, compared to -0.07 to -0.04 dex/kpc for the Galaxy. Abundances for luminous OBA stars in M31 and M33 agree quite well with the H II region values, although only a handful of stars have been observed to date (Monteverde, Herrero, & Lennon 2000; Venn et al. 2000).

Although no direct measurements of abundances in older stars exist in M33, Kim et al. (2002) indirectly estimated mean metallicities across the M33 disk from the color of the red giant branch based on *HST* imaging. They found a well-defined gradient in $<$[Fe/H]$>$ with a slope of -0.05 dex/kpc, shallower than the O/H gradient discussed above. If confirmed, this would imply that the composition gradient has *steepened* with time, in contrast to what is seen in our Galaxy.

The difference in composition gradient slopes for these three galaxies are consistent with a general trend of slope decreasing with luminosity for spiral galaxies. At the same time it has been noted that spirals all have roughly the same slope of about -0.2 dex per unit (B band) disk scalelength (Garnett et al. 1997). It is also observed that the disk abundances are generally higher in M31 than in the Galaxy, while those in M33 appear to be smaller. The reasons for these differences are not well understood. The differences are consistent with a picture in which M31 has evolved more rapidly than the Galaxy, which would naturally lead to a higher metallicity and flatter composition gradient, while by similar argument M33 would have evolved more slowly. On the other hand, it is not known if significant quantities of metals are escaping the disks as a result of supernova-driven outflows. A proper comparison of the spirals awaits improved data on disk abundances for M31 and M33, as well as detailed studies of their stellar populations.

3.2. *Bulge abundances*

Some of the more interesting questions about bulges are whether they formed separately or together with disks, when they formed, and over what timescale they formed. Study of bulges in the Local Group is still in relative infancy; the Galactic bulge is difficult to study because of high obscuration by dust, while the much more distant M31 bulge suffers from high crowding of stars.

Studies of the Galactic bulge got a big boost with the first Keck spectroscopy of bulge giants by McWilliam & Rich (1994). This study showed that, surprisingly, the mean [Fe/H] of the bulge is slightly lower than in the solar neighborhood, and that [α/Fe] and [Ti/Fe] were enhanced in the bulge, compared to the Sun. The [Ti/Fe] enhancement proved to crucial as the TiO molecule is an important source of opacity in these giants. McWilliam & Rich showed that because of the high Ti previous [Fe/H] measurements for bulge giants were overestimated by a factor of two. This result has important implications for the interpretation of population synthesis models for luminous elliptical galaxies, where [α/Fe] may also be enhanced.

The high [α/Fe] in the bulge giants (discussed also by Rich in his review at this symposium) implies that the bulge formed rapidly: the α elements are produced mainly in massive stars with lifetimes shorter than 10 Myrs, while Fe is expected to come mainly from exploding white dwarfs as Type Ia SNe. In addition, the bulge does not appear to have a G-dwarf problem, indicating that any infall into the bulge must have occured rapidly. The lifetimes for SN Ia progenitors is not very well known, but is generally taken to be <1–2 Gyr. Indeed, Ferreras, Wyse, & Silk (2003) have inferred from modeling that the Galactic bulge collapsed on a timescale < 0.5 Gyr.

The inner bulge can be studied only in the infrared because of high extinction. Recent improvements in IR detector capabilities have made it possible to study inner bulge giants at moderate resolution. From the slope of the red giant branch in J versus $J - K$ a gradient in metallicity is seen along the minor axis of the bulge (Frogel et al. 1999); however, this gradient disappears in the innermost part of the bulge (Ramírez et al. 2000). The mean metallicity of the inner bulge is also subsolar. The lack of gradient in the inner bulge appears to be a problem for models of bulge formation by dissipative collapse.

No spectroscopy of stars in the M31 bulge is available, so direct comparison with the Galactic bulge is not possible. However, PNs in both bulges have been observed, providing an alternative comparison. Jacoby & Ciardullo (1999) observed 40 PNs in the M31 bulge. The mean value and spread of O/H in the M31 PNs is very similar to that of the Milky

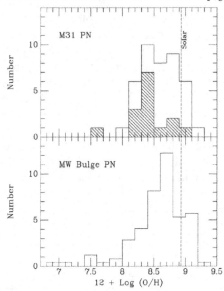

FIGURE 6. Distributions of O/H for planetary nebulae in the bulges of M31 (top panel) and the Galaxy (bottom panel), from Jacoby & Ciardullo (1999). The range and distribution of PN abundances appear to be similar in both bulges.

Way bulge PNs (Stasińska et al. 1998), although the M31 PNs may have a broader peak (Fig. 6). Unfortunately, it is not possible to measure $[\alpha/\text{Fe}]$ or an equivalent in PNs because of depletion of Fe onto grains. Thus, while the similarity between the M31 and Galactic bulge PNs suggests similar histories, it does not demonstrate this. Population synthesis of M31 bulge spectra (Trager et al. 2000) suggest that the M31 bulge does not have enhanced $[\alpha/\text{Fe}]$, implying a longer formation timescale (or perhaps a merger). Such a difference would be important for bulge formation models and warrants closer examination.

3.3. *Halo abundances*

An enormous amount of research has been done on abundances in Galactic halo stars—too much to review here. I refer the interested reader to the excellent review by McWilliam (1997) for more information on abundances in halo stars up to that time. Here I review the general abundance trends observed in halo stars and recent developments. The general trends have not changed significantly since McWilliam's review, although a few notable issues have arisen (and in some cases, have been resolved).

The Galactic halo is quite metal-poor, with $<[\text{Fe/H}]> \approx -1.6$. The metallicity distribution is still being refined, but appears to be consistent with the predictions of closed-box chemical evolution, albeit with a very low yield, about 1/20th the solar metallicity. It is not yet known how low metallicities can be in halo stars. For many years no stars with $[\text{Fe/H}] < -4$ were known. However, the Hamburg/ESO survey has turned up the current record-holder for low metallicity: HE0107−5240, with an apparent $[\text{Fe/H}] \approx -5.3$. This discovery keeps open the hope that one day zero-metallicity ('Population III') stars may be found. It also highlights a possible paradox in that the halo has a significant number of field stars that are more metal-poor than any of the globular clusters.

It has been known for some time that $[\alpha/\text{Fe}]$ is enhanced in halo stars compared to the sun, by at least a factor of two. There has been some question recently about whether $[\text{O/Fe}]$ was essentially constant in the halo stars, or whether it increased steeply as

[Fe/H] goes down (cf. Boesgaard et al. 1999, Israelian et al. 1998 versus results reviewed by McWilliam 1997). The derived trend depends on which oxygen lines were analyzed. Although this issue does not have much bearing on the evolution of the halo, it is relevant to understanding the stellar IMF in the early Galaxy. This discrepancy may be resolved in favor of flatter [O/Fe] with the revised temperature structure of metal-poor stars in 3D atmosphere models (Asplund 2003a,b).

A third important result regards neutron-capture elements. In halo stars the [Ba/Eu] ratio is lower than in the Sun, and increases to the solar ratio for [Fe/H] > -1 in the disk. Barium is produced mainly via slow neutron captures ('*s*-process') in red giants, while Eu comes mainly from rapid neutron capture processes ('*r*-process') in a site that is not yet known, but is believed to be associated with massive stars or supernovae. The low [s/r] ratio in halo stars thus suggest enrichment primarily from massive stars. Indeed, the detailed analyses that have been done so far indicate that all halo stars have heavy element abundances consistent with pure scaled solar *r*-process distribution (Sneden et al. 2003).

Taken together, the properties listed above describe a picture in which the stars in the halo formed over a short time scale, shorter than the time for significant enrichment by SN Ia. Although the metallicity distribution is consistent with closed-box chemical evolution, the low mean metallicity indicates that the halo lost much of its gas before it could be turned into stars, perhaps through a superwind. This gas may have become part of the disk or bulge at a later time. Another issue that is in doubt is whether the halo formed by monolithic collapse or by merger of smaller objects.

At the same time, there is a question of whether the halo formed as a single event, or whether much of the halo could have been accreted through tidal disruption of dwarf galaxies. We know that the globular clusters are ancient, but, lacking parallaxes, we do not know much about the age distribution of halo *field* stars. The discovery of the Sagittarius dwarf demonstrated that such accretion has occurred at least once. The relatively small scatter of [α/Fe] in halo stars has been used as an argument against most halo stars having been accreted from dwarf galaxies at late times (very early accretion events would be difficult to distinguish). However, a new study of 150 metal-poor dwarf stars with *Hipparcos* parallaxes and kinematics by Gratton et al. (2003) re-opens this possibility. Gratton et al. looked at stars in three different kinematic groups: (1) the rapidly rotating thin disk; (2) a second component with rotation speed greater than 40 km s^{-1}, moderate eccentricities, and apogalactic distances less than 15 kpc, which they identify with the thick disk; and (3) a non-rotating or counter-rotating stellar component with predominantly radial orbits, which they identify with the halo population. Looking at the [α/Fe] ratio, Gratton et al. found that their 'halo' sample showed a larger scatter in [α/Fe] at fixed metallicity than the 'thick disk' or 'thin disk' samples. Since [α/Fe] is sensitive to the star formation timescale, the increased scatter implies that the 'halo' stars have a larger range of ages than the 'thick disk' stars. This situation could occur if the halo field stars were accreted from metal-poor dwarf galaxies that formed stars at different times in their histories (see discussion of dwarf ellipticals in Sect. 5 below). The combination of abundance measurements with kinematics for much larger samples of stars should yield important clues to the evolution of all of the components of the Galaxy.

Another interesting possibility is that there was a distinct gap between the end of star formation in the halo and the onset of formation of the disk. This idea was posited by Fuhrmann (1998) from an analysis of [Mg/Fe] ratios in halo and disk stars. Fuhrmann found an apparently abrupt transition from halo-like [Mg/Fe] to disk-like [Mg/Fe] at [Mg/H] between -0.5 and 0.0, with a possible gap in [Mg/Fe] between the halo and disk

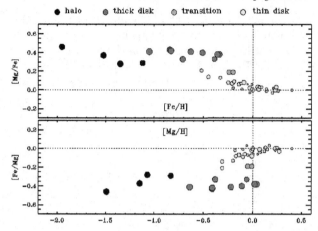

FIGURE 7. [Mg/Fe] versus [Fe/H] (top panel) and [Fe/Mg] versus [Mg/H] (bottom panel), for stars in the halo and disk of the Galaxy. The abrupt change in [Fe/Mg] at [Mg/H] ≈ −0.2, and the significant overlap in [Mg/H] for thick disk and thin disk stars, is interpreted as a gap between the end of star formation in the thick disk and the start of star formation of the thin disk.

samples (Fig. 7). This has been interpreted as the result of a time gap of 1–2 Gyr between the end of star formation in the halo and the formation of the halo, during which time the gas that fell into the disk had time to be enriched in Fe from Type Ia SNe (Chiappini et al. 1997). The evidence for this is tenuous at present, but worth further exploration.

One startling observational result is the enormous increase in the scatter of the abundances for r-process elements (ratioed to Fe) for [Fe/H] < −2.5. These very metal-poor stars show essentially the same r-process pattern for neutron-rich elements (Sneden et al. 2002). However, the amount of r-process material in any star varies wildly. This phenomenon suggest that we are seeing inhomogeneous enrichment in the early galaxy from individual events in whatever source is responsible for the r-process. This source is not yet understood, but is likely associated with Type II SNe or perhaps nucleosynthesis on the surface of neutron stars (see Truran et al. 2002; Qian & Wasserburg 2003).

Another amazing development has been the recent measurements of thorium and uranium abundances in halo stars (e.g., Sneden et al. 2000; Cayrel et al. 2001). These measurements allow the derivation of ages for stars independently of color-magnitude diagram methods. Although the uncertainties in the age for individual stars based on this method are still fairly large, measurements of several stars yield ages which are quite consistent with recent estimates based on the cosmic microwave background.

The halos of our companion spirals are not as well known. Most metallicity estimates have come from photometry of the globular clusters or the field stars, so informative abundance ratios such as [α/Fe] are not yet available. Nevertheless, enough is known about the bulk metallicity distributions to provide some interesting comparisons with the Galactic halo.

Recent *HST* imaging studies have resolved the globular clusters of M31 and M33, as well as the halo field stars populations, while the Keck observatory has provided the first moderate-resolution spectroscopy of red giants in the M31 halo. Reitzel & Guhathakurta (2002) presented Ca II triplet measurements for field giants in the M31 halo, and compared the metallicity distributions for field stars and globular clusters in the M31 and Milky Way halos. The M31 field stars appear to have similar mean metallicity and metallicity distribution as the Galactic halo stars. A few stars with roughly solar metallicity

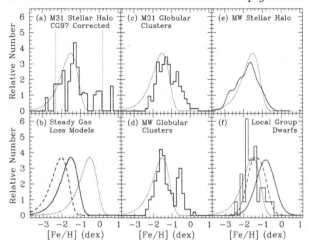

FIGURE 8. Comparison of metallicity distributions for globular clusters and field stars in the M31 and Galactic halos, from Reitzel & Guhathakurta (2000).

appear in the M31 distribution, but contamination from the disk is still an unresolved problem (see also Brown's contribution to this symposium). For the globular clusters, however, the M31 system has a higher mean metallicity than the Galactic clusters. From the distribution shown by Reitzel & Guhathakurta, it appears that this difference is a result of M31 having more metal-rich clusters, rather than a systematic shift of the whole distribution (Fig. 8). The M31 GC system shows a hint of a bimodal metallicity distribution, not as pronounced as that seen for the Galactic GCs. This may indicate that the M31 bulge may have an associated metal-rich GC system like that seen in the Galaxy (Barmby et al. 2000).

The M33 globular cluster system exhibits some peculiarities compared to those of the larger spirals. From the slope of the red giant branches, Sarajedini et al. (2000) derived metallicities for nine M33 GCs, finding a mean metallicity of -1.3, similar to the means for M31 and the Galaxy. The observed metallicity spread is only 1 dex, compared to the 2 dex spread for the larger spirals, although with such a small sample this comparison may not be conclusive yet. No radial gradient in metallicity is seen for the M33 clusters, implying that no bulge/disk GC population is present, consistent with the lack of a large bulge. Most striking, however, is that the M33 GC population appears to be much younger than the Galactic halo, with a mean age of only 7 Gyr. This could be an important clue to the formation of M33 and demands closer examination and confirmation.

4. Local Group irregular galaxies

4.1. *The Magellanic Clouds*

The Magellanic Clouds are our closest neighbor galaxies (not counting the Sagittarius dwarf spheroidal, whose boundaries are ill-defined), and thus are the best-studied irregular galaxies. Thanks to *HST*, we now have photometry of stars in the LMC and SMC that reach well below the oldest main-sequence turnoffs, providing detailed star formation histories. Meanwhile, new high-efficiency spectrographs also permit detailed abundance analyses for red giants and luminous young stars. The combination of star formation history and abundance measurements (along with orbital analysis enabled by the first extragalactic proper motion measurements) provide an opportunity to build comprehensive models for the evolution of the Magellanic Clouds.

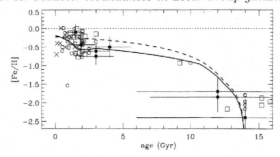

FIGURE 9. Age-[Fe/H] relation for star clusters in the LMC, from Pagel & Tautvaišienė (1998). The solid line shows a chemical evolution model with a reduced star formation rate between 3 and 12 Gyrs ago and a sharp increase in the SFR 3 Gyr ago. The dashed line shows a model with a smoothly declining SFR over the past 10 Gyr.

Abundances measured in H II regions and young stars (B stars, supergiants) give generally the same results for the present-day metallicities of the Clouds: the LMC is down by −0.2 dex relative to the solar neighborhood, while the SMC is down by −0.6 dex in O, Ne, S, Ar, Si, and Fe (as reviewed by Garnett 1999). Supergiants tend to have higher C and N than the H II regions, but main sequence B stars agree with the H II regions (Korn et al. 2000; Rolleston et al. 2003). This most likely signifies the mixing of hydrogen and helium burning products to the surface of the supergiants.

A metallicity distribution for red giants in the LMC (which should reflect the metallicity distribution of the older stellar populations) has been derived from Ca II triplet measurements by Cole et al. (2000). Their derived [Fe/H] distribution for the LMC inner disk a mean [Fe/H] = −0.64, but the peak is at −0.57 and has a long tail down to [Fe/H] = −1.6. The distribution appears truncated at the upper end at [Fe/H] = −0.25, consistent with the present-day metallicity derived from the young stars and H II regions. The outer disk [Fe/H] distribution appears to be similar at the metal-rich end, but has an additional peak between [Fe/H] = −0.8 and [Fe/H] = −1.6. Neither of these metallicity distributions agrees with that derived from LMC clusters (Olszewski et al. 1996), which exhibits a pronounced gap in clusters with −2 < [Fe/H] < −0.8 and 3 Gyr < age < 10 Gyr. The reason for this gap is a profound mystery.

The author is not aware of a similar spectroscopy study of the metallicity distribution for the SMC. Clusters in the SMC appear to have a relatively uniform distribution of metallicities (Olszewski et al. 1996), although only eleven clusters have been studied so far. There is some debate at the present time whether the SMC has had a relatively constant star formation rate or whether there have been periods of enhanced activity (Mighell, Sarajedini, & French 1998; Rich et al. 2000).

Age-[Fe/H] relations are available for star clusters in both the LMC and SMC, and for α-elements from planetary nebulae in the LMC (Dopita et al. 1997). The age-metallicity relations for the two galaxies, although still fairly uncertain and not well sampled, share some common characteristics. The oldest clusters are quite metal-poor, although the SMC appear to have been more enriched at early times ([Fe/H] $\approx$ −1.6, versus −2 for the oldest LMC clusters). Both galaxies may have experienced an intermediate period with little metal enrichment, followed by a steep increase in metallicity within the past 2–3 Gyr (Figs. 9, 10). Such behavior would be consistent with a star formation rate that is relatively high in the first few Gyrs and the last 2–3 Gyrs, with a period of several Gyrs in between in which the star formation rate is rather low (Pagel & Tautvaišienė 1998). The recent increase could be related to close encounters between the LMC and

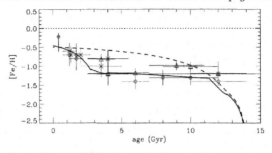

FIGURE 10. Similar to Fig. 9 for the SMC. The solid line in this case shows a chemical evolution model with a reduced star formation rate between 4 and 12 Gyrs ago and a sharp increase in the SFR 4 Gyr ago.

SMC approximately 2 Gyrs ago as inferred from dynamical models (Gardiner & Noguchi 1996).

[O/Fe] has been derived for LMC giants only recently. A study of giants in four populous clusters (Hill et al. 2000) found a relatively flat trend of [O/Fe] versus [Fe/H], but the addition of measurements for twelve LMC field red giants (Smith et al. 2002) revealed a pattern of [O/Fe] versus [Fe/H] that is similar to that seen in the solar neighborhood: high [O/Fe] for metal-poor stars, and declining [O/Fe] for the most metal-rich LMC giants. At the same time, it was noted that [O/Fe] starts to decline at lower [Fe/H] than in the Galaxy. Smith et al. interpret this in terms of a reduced SN II rate in the LMC; it may have taken longer for the LMC to reach [Fe/H] $= -1$ than the solar neighborhood, allowing time for SNe Ia to 'catch up' and enrich the LMC with iron at a lower metallicity. These studies are in their infancy, however, and more stars should be studied before reaching any conclusions regarding the enrichment history of the Magellanic Clouds.

4.2. *Other Local Group irregulars*

Besides the Magellanic Clouds, there are several dwarf irregular galaxies with blue absolute magnitudes fainter than $M_B = -16$. There is some ambiguity as to the Local Group membership for some nearby irregulars. The certain members include IC 10, NGC 6822, IC 1613, WLM, and Leo A. The less certain members are Sextans A, Sextans B, and the Sagittarius dwarf irregular (to be distinguished from the Sagittarius dwarf elliptical).

Abundances for the dwarf irregulars have been obtained largely from spectroscopy of ionized nebulae, and all have at least one measurement (Lequeux et al. 1979; Skillman, Kennicutt, & Hodge 1989). Because of their greater distances, stellar abundances in these galaxies are rare. Spectra for a handful of supergiants have been obtained in NGC 6822 and WLM (Venn et al. 2001,2003). The NGC 6822 supergiants have abundances in agreement with nebular results, but for WLM (Venn et al. 2003) the supergiants are systematically more metal-rich than the H II regions. On the other hand, looking at the WLM measurements one sees that Mg and Si have the same underabundances in the supergiants as O in the H II regions, and only the single O measurement in one of the supergiants is far off. Such comparisons are valuable and merit greater effort.

O/H in the Local Group irregulars range from 0.3 solar down the 0.04 solar. No spatial variation in the nebular abundances are seen across any of the irregulars, implying that the gas is well-mixed. This in turn suggests that supernova ejecta travel significant distances before they are mixed with the ambient ISM; another possibility is that the SN products are ejected into the halo of the galaxy, and enrich the ISM of a dwarf galaxy after falling back onto the disk from the halo.

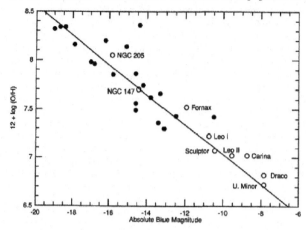

FIGURE 11. Mean O/H for nearby dwarf irregulars and dwarf ellipticals versus M_B, from Skillman, Kennicutt, & Hodge (1989). Dwarf irregulars are plotted as filled circles, ellipticals as open circles. O/H for the ellipticals has been estimated assuming O/H = [Fe/H] + 0.3.

The present-day abundances of irregular galaxies correlate with their luminosities (masses) (Lequeux et al. 1979; Skillman et al. 1989; Fig. 11). In fact, the metallicity-luminosity relation extends over 11 magnitudes to include massive spiral galaxies as well (Garnett & Shields 1987). This correlation is remarkable and suggests a key physical process. One idea is that low-mass galaxies lose significant quantities of metals to the IGM in supernova-driven winds. On the other hand, a variation in gas fraction, which indicates how much gas has been turned into stars, could also produce a metallicity-luminosity relation. In the closed-box (no gas inflows or outflows) chemical evolution model with the instantaneous recycling of elements (a good approximation for massive star products like oxygen), the metallicity Z scales linearly with the log of the gas mass fraction μ_{gas} as

$$Z = -y \, ln(\mu_{gas}), \tag{4.1}$$

where y is the yield of metals.

One can then define an effective yield y_{eff} in terms of observables

$$y_{eff} = -ln(\mu_{gas})/Z. \tag{4.2}$$

Edmunds (1990) showed that y_{eff} is always reduced with respect to the true yield y in the presence of metal-poor infall into a system and any outflow of gas. Thus, variations of y_{eff} among galaxies can be used to gauge the importance of gas flows on metallicity evolution.

Garnett (2002) explored this, compiling data on abundances and gas fractions for nearby spiral and irregular galaxies, deriving y_{eff} for each system. It was found that the y_{eff} correlates strongly with galaxy rotation speed (mass), increasing sharply with V_{rot} for low-mass galaxies, and flattening to approximately constant y_{eff} for the more massive spirals (Fig. 12).

The simplest interpretation of this trend is that low-mass dwarf galaxies lose significant fractions of their metals, perhaps in supernova-driven winds into the IGM, while in more massive galaxies metallicity evolution is driven mainly by evolution in the gas fraction. If the true yield is roughly constant, then the variation in y_{eff} is related to the fraction of metals lost. The low effective yields for dwarf galaxies imply that very low-mass galaxies such as Leo A lose as much as 80–90% of their metals to the IGM.

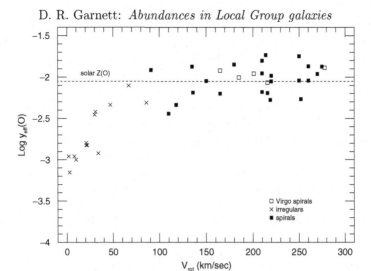

FIGURE 12. Effective yields for nearby irregular and spiral galaxies as derived using equation (4.2), from Garnett (2002). Note that the dwarf irregulars (crosses) have lower effective yields than spirals, and that y_{eff} correlates strongly with rotation speed (that is, mass) for the irregulars.

5. The dwarf ellipticals/spheroidals

The dwarf ellipticals in the Local Group including some of the most intrinsically faint galaxies known in the universe, with Carina, Draco and Ursa Minor less luminous than many globular clusters. Their smooth ellipsoidal structure and lack of gas make them appear deceptively simple. However, detailed studies of their stellar populations show that the dwarf ellipticals have had complex histories.

Photometric studies indicated that the dwarf ellipticals are all metal-poor, with mean metallicities ranging from [Fe/H] = -0.8 for NGC 205 down to [Fe/H] ≈ -2.2 for Ursa Minor (Mateo 1998). The dEs follow a metallicity-luminosity relation similar to the dwarf irregulars (Fig. 12); whether they follow the same relation depends on how one compares [Fe/H] in stellar populations of the dEs to O/H in the gas of the irregulars. Thus, some of the processes that govern the metallicity evolution of irregular galaxies, including, possibly, outflows of metals, probably operate in the dwarf ellipticals as well.

Despite their apparently simple structures, it has been known for some time that most of the dEs are not simple, single-aged stellar populations. The discovery of carbon stars (Mould & Aaronson 1983) in several dEs demonstrated the presence of at least intermediate-age (few Gyr old) populations. This has been strikingly confirmed by high-resolution deep *HST* imaging of the Local Group dEs, which show that they exhibit a wide variety of star formation histories (see Mateo 1998). Two interesting questions that arise are: (1) Does the element abundance evolution reflect the star formation history, and if not, why? (2) Are the element abundance ratios in the dEs consistent with those observed for Galactic halo stars? If the halo was formed mainly by late accretion of dwarf galaxies, element abundance ratios, such as [O/Fe] should be similar in stars in both the halo and the dEs.

These questions are starting to be addressed by a second striking recent development, the first detailed spectroscopic analyses of red giant stars in the nearest dEs, thanks to the powerful capabilities of the Keck, VLT, and Gemini telescopes. Shetrone et al. (2001, 2003) have made the first high-resolution, high signal/noise measurements of red giants in

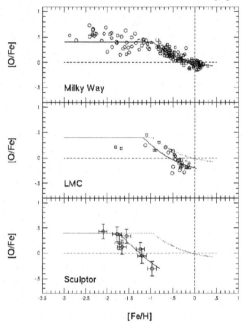

FIGURE 13. [O/Fe] versus [Fe/H] for stars in the solar neighborhood (top panel), LMC (middle panel), and Sculptor (bottom panel). The Milky Way pattern is repeated as the dotted line in the lower two panels. [O/Fe] in Sculptor declines at a lower value of [Fe/H] than in the Galaxy or the LMC, implying that it took longer for Sculptor to reach [Fe/H] $= -1.7$ than in the other galaxies.

Carina, Sculptor, Draco, Ursa Minor, and Fornax, measuring a variety of α and neutron-capture elements. These studies have found some contradictory results, based on a small number of stars observed. On one hand, they found that [r/Fe], [Cu/Fe], and [Mn/Fe] in the dEs were similar to Galactic halo stars; on the other hand, [α/Fe] appeared to be lower on average than in the Galactic halo. Largely on the basis of [α/Fe], Shetrone et al. 2003 argued that the stars in dEs did not share the Galactic halo abundance pattern, and so the bulk of the halo could not consist of stars accreted from dwarf satellite galaxies at late times (cf. Searle & Zinn 1978).

However, more recent work suggest that these conclusions might have been premature. Shetrone (2003), at the Fourth Carnegie Centennial symposium, reported on spectra for additional giants in dEs that show that the average [α/Fe] in dEs is higher than previously reported. In addition, Smith (2003) has new data for nine red giants in Sculptor, showing that [α/Fe] versus [Fe/H] shows a pattern similar to that seen in the Galactic halo + disk, except that [α/Fe] in Sculptor starts to decline at [Fe/H] ≈ -1.7, rather than at -1 as in the Galaxy (Fig. 13). This indicates that it took longer for Sculptor to reach [Fe/H] $= -1.5$ than did the Galaxy. Thus, it may be that most stars in the dEs have enhanced [α/Fe]. Finally, I recall the finding by Gratton et al. (2003) that stars with halo kinematics may have a relatively large scatter in [α/Fe], indicating a possibly large range in ages for halo field stars (Sect. 3.3). This remains a very interesting issue for the formation of the Galactic halo.

The availability of detailed star formation histories from deep color-magnitude diagrams from *HST* imaging and element abundances from spectroscopy of red giants using 8–10m class telescopes raise the exciting possibility of constructing self-consistent models of the evolution of star formation and metallicity in the dE companions of the Milky Way.

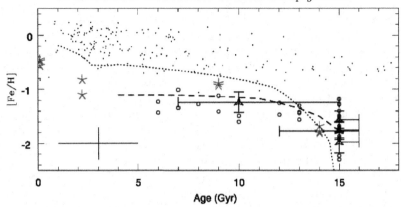

FIGURE 14. The age-metallicity relation for Sculptor (Tolstoy et al. 2003). The triangles represent [Fe/H] measurements for Sculptor giants from high-resolution spectroscopy (Shetrone et al. 2003), while the open circles show [Fe/H] for Sculptor giants from Ca II triplet measurements. The small dots show the age-metallicity relation for the solar neighborhood. in the other galaxies. The dashed line is the result of chemical evolution model for Sculptor. Other symbols are defined in Tolstoy et al. (2003).

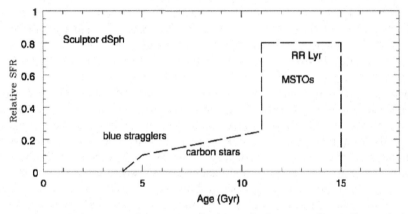

FIGURE 15. The star formation history for Sculptor (Tolstoy et al. 2003), as derived from deep color-magnitude diagrams, and adjusted for metallicities of stars in the CMD. The stellar population components which dominate age determinations in the CMD are labeled.

These efforts are still in their infancy, but an example of what can be done is provided by Tolstoy et al. (2003). Tolstoy et al. collected color-magnitude diagrams, [Fe/H] estimates for individual red giants in several dEs based on Ca II triplet measurements, and the abundance data from Shetrone et al. (2003). Using the CMDs and [Fe/H] measurements, they were able to assign an age to each red giant and construct an age-metallicity relation (Fig. 14), while the metallicity measurements allowed them to refine the star formation history for each galaxy (Fig. 15). With this information it became possible to derive a chemical evolution model for each galaxy. Based on these models, Tolstoy et al. argued that the metallicity evolution and the low [α/Fe] of the dEs could be reproduced with a simple, closed-box evolution model, but only with mean star formation rates so low that the upper end of the stellar IMF is truncated due to stochastic effects. The metallicity evolution of a dE is then governed by the slow star formation rate, minus the contribution of Type II SNe.

There are several flaws with this picture, however: (1) as mentioned above, there is growing evidence that $[\alpha/\text{Fe}]$ is not as low in the dEs as assumed by Tolstoy et al.; (2) the mean $[\text{Fe/H}]$ of the dEs is only about 0.03 times the solar value, but they have no interstellar gas. The effective yield one derived from closed-box chemical evolution is thus about 1/30th solar. This is so low that in order to explain it the chemical evolution model would require that both Type Ia and Type II SNe are suppressed. This seems highly unlikely. A more likely explanation is that the dEs have lost their metals, either by galactic winds or stripping (Lynden-Bell 1992). (3) Tolstoy et al. argue that the Type II SN rate is suppressed by stochastic star formation effects when the average star formation rate is about 2.4×10^{-4} solar masses per year. However, based on their derived star formation histories, the peak star formation rates for Fornax and Sculptor are of order 10^{-3} to 10^{-2} solar masses per year, much larger than the threshold proposed by Tolstoy et al. It is therefore unlikely that the dEs follow closed-box chemical evolution.

6. Conclusions

Despite reservations about the conclusions, the work by Tolstoy et al. nicely illustrates how combined stellar metallicity and age data can be used to constrain galaxy evolution. The improving resolution and sensitivity of *HST*'s imaging systems, and the deep spectroscopic capabilities of the new 8–10m class telescopes, means that the potential exists to determine in detail the history of star formation and metal enrichment for nearly every galaxy in the Local Group. Such data will provide important constraints on galaxy evolution models.

This work is just beginning to bear fruits. While deep imaging of all of the nearby dwarf ellipticals has been done, and a few of the dwarf irregulars, the disks of M31 and M33 have yet to be imaged at the same level. The crowding of stellar images is much more severe in the spirals, but ACS should provide some improvement, and *JWST* will be able to highly resolve the stars in these galaxies in the IR. Meanwhile, Mike Rich showed a remarkable simulation at the Hubble Science Legacy workshop (Rich 2003) in which an imager on a proposed 6–8m class optical/UV space telescope could resolve the bulge of M31 down to the nucleus. It is no stretch to imagine that within the next 10 years or so we will have an understanding of the star formation history of the disks of M31 and M33 comparable to what we now know for the Galaxy and its satellite galaxies. Deep imaging of the remaining Local Group irregulars is also needed; such observations will resolve some of the questions regarding the nature of star formation in dwarf galaxies— for example, are starbursts a frequently important mode of star formation, and, was the first major epoch of star formation in these galaxies early or delayed, and just when did that first star formation occur?

There is more room for discovery in the area of stellar spectroscopy for Local Group galaxies. First, better metallicity distributions and ages for stars are needed for all of the Local Group galaxies. This goes hand in hand with photometry from imaging, as ages derived from photometry depend somewhat on metallicity, while photometry is needed to identify good candidates for spectroscopy (red giants not on the asymptotic giant branch). The metallicity distribution and the age-metallicity relation both provide important constraints on galaxy evolution models. It is important to recognize that these can not be derived accurately from observations of 10–20 stars—one can miss important stellar populations (such as a weak, ancient, metal-poor population). Ideally, metallicities for a few hundred stars should be obtained; metallicities combined with kinematics is even more valuable. This is well within the realm of possibility for, say, Ca II triplet measurements of red giants with fiber spectrographs on large ground-based telescopes, as demonstrated by Tolstoy et al. (2003).

High-resolution spectroscopy of both red giants and luminous young stars in Local Group galaxies is also very important. These observations provide the element abundance ratios that tell us much about the timescales for star formation in galaxies. This has been demonstrated for the dwarf elliptical satellites of the Milky Way. However, we have also seen that it is dangerous to base conclusions about galaxy evolution on a handful of stars; again, it is necessary to measure a larger number of stars per galaxy (of order 20 or more) to obtain a fair picture of the evolution of element abundance ratios in galaxies.

There is even room for improvement in abundances for ionized nebulae in Local Group galaxies. The dwarf irregulars have been studied in great detail, but M33 and M31 are relatively unexplored with modern spectroscopy. Most data for these galaxies dates back to the 1980s, with little CCD spectroscopy. Only a handful of H II regions have observed been with CCD spectrographs in M33, while in M31 no H II regions have been observed deeply enough to measure the T_e-sensitive diagnostic lines. The issues that can be addressed with improved observations include:

• Is the radial distribution of abundances consistent with a pure exponential dependence? If the slope changes, is the change related to disk dynamics?

• What is the intrinsic scatter in disk abundances? Spatial fluctuations provide information on the timescales for mixing of supernova ejecta. We have little information on the abundance scatter in galaxies, other than known radial gradients.

• How do composition gradients vary with time, and do they vary the same way in all spirals? This can be addressed by studies of both stars and PNs in all three Local Group spirals. Chemical evolution models tend to predict that gradients either flatten with time or steepen with time. New observations on this could rule out a wide variety of models.

It is clear that we can learn a great deal about metallicity evolution, and galaxy evolution in general, from further measurements of abundances in stars and the interstellar medium in Local Group galaxies. I look forward to some very exciting developments over the next five to ten years.

I thank the Institute for the opportunity to review this fascinating research area, and for partial travel support. Additional support from NASA LTSA grant NAG5-7734 is also gratefully acknowledged. Thanks also to Verne Smith for providing Fig. 13, and Eline Tolstoy for electronic versions of Figs. 14 and 15.

REFERENCES

ALLENDE PRIETO, C., LAMBERT, D. L., & ASPLUND, M. 2001 *ApJ* **556**, L63.

ARMANDROFF, T. E. & ZINN, R. 1988 *AJ* **96**, 92.

ASPLUND, M. 2003a, in *IAU Symposium 210: Modelling of Stellar Atmospheres* (eds. N. E. Piskunov, W. W. Weiss, & D. F. Gray), in press; (astro-ph/0302407).

ASPLUND, M. 2003b, *in CNO in the Universe* (eds. C. Charbonnel, D. Schaerer, & G. Meynet), in press; (astro-ph/0302409).

BARMBY, P., HUCHRA, J. P., BRODIE, J. P., FORBES, D. A., SCHRODER, L. L., & GRILLMAIR, C. J. 2000 *AJ* **119**, 727.

BLAIR, W. P., KIRSHNER, R. P., & CHEVALIER, R. A. 1982 *ApJ* **254**, 50.

BOESGAARD, A. M., KING, J. R., DELIYANNIS, C. P., & VOGT, S. S. 1999 *AJ* **117**, 492.

CASTELLANOS, M., DÍAZ, A. I., & TERLEVICH, E. 2002 *MNRAS* **329**, 315.

CAYREL, R., ET AL. 2001 *Nature* **409**, 691.

CHIAPPINI, C., MATTEUCCI, F., & GRATTON, R. 1997 *ApJ* **477**, 765.

CHURCHWELL, E., SMITH, L. F., MATHIS, J., MEZGER, P. G., & HUCHTMEIER, W. 1978 *A&A* **70**, 719.

COLE, A. A., SMECKER-HANE, T. A., & GALLAGHER, J. S. III 2000 *AJ* **120**, 1808.

DAFLON, S. L., CUNHA, K., BECKER, S. R., & SMITH, V. V. 2001 *ApJ* **552**, 309.

DEHARVENG, L., PEÑA, M., CAPLAN, J., & COSTERO, R. 2000 *MNRAS* **311**, 329.

DOPITA, M. A., VASSILIADIS, E., & WOOD, P. R. 1997 *ApJ* **474**, 188.

EDMUNDS, M. G. 1990 *MNRAS* **246**, 678.

EDVARDSSON, B., ANDERSSON, J., GUSTAFSSON, B., LAMBERT, D. L., NISSEN, P. E., & TOMKIN, J. 1993 *A&A* **275**, 101 (EAGLNT).

FERRERAS, I., WYSE, R. F. G., & SILK, J. 2003 *MNRAS*, **345**, 1381.

FRIEL, E. D., ET AL. 2002 *AJ* **124**, 2693.

FROGEL, J. A., TIEDE, G. P., & KUCHINSKI, L. E. 1999 *AJ* **117**, 2296.

FUHRMANN, K. 1998 *A&A* **338**, 161.

GALARZA, V. C., WALTERBOS, R. A., & BRAUN, R. 1999 *AJ* **118**, 2775.

GARDINER, L. T. & NOGUCHI, M. 1996 *MNRAS* **278**, 191.

GARNETT, D. R. 1999 in *IAU Symposium 190: New Views of the Magellanic Clouds*, (eds. Y.-H. Chu, N. Suntzeff, J. Hesser, & D. Bohlender). p. 266. ASP.

GARNETT, D. R. 2002 *ApJ* **581**, 1019.

GARNETT, D. R. & KOBULNICKY, H. A. 2000 *ApJ* **532**, 1192.

GARNETT, D. R. & SHIELDS, G. A. 1987 *ApJ* **317**, 82.

GARNETT, D. R., SHIELDS, G. A., SKILLMAN, E. D., SAGAN, S. P., & DUFOUR, R. J. 1997 *ApJ* **489**, 63.

GIVEON, U., STERNBERG, A., LUTZ, D., FEUCHTGRUBER, H., & PAULDRACH, A. W. A. 2002 *ApJ* **566**, 880.

GRATTON, R. G., CARRETTA, E., DESIDERA, S., LUCATELLO, S., MAZZEI, P., & BARBIERI, M. 2003 *A&A* **406**, 131.

GRAY, D. F. 1992, *The observation and analysis of stellar photospheres*. Cambridge University Press.

HENRY, R. B. C. 1993 *MNRAS* **261**, 306.

HENRY, R. B. C. & WORTHEY, G. 1999 *PASP* **111**, 919.

HILL, V., FRANÇOIS, P., SPITE, M., PRIMA, F., & SPITE, F. 2000 *A&A* **364**, 19.

ISRAELIAN, G., GARCÍA LÓPEZ, R. J., & REBOLO, R. 1998 *ApJ* **507**, 805.

JACOBY, G. H. & CIARDULLO, R. 1999 *ApJ* **515**, 169.

KENNICUTT, R. C., JR., BRESOLIN, F., & GARNETT, D. R. 2003 *ApJ* **591**, 801.

KIM, M., KIM, E., LEE, M. G., SARAJEDINI, A., & GEISLER, D. 2001 *AJ* **123**, 244.

KINKEL, U. & ROSA, M. 1994 *A&A* **282**, 37.

KORN, A. J., BECKER, S. R., GUMMERSBACH, C. A., & WOLF, B. 2000 *A&A* **353**, 655.

KWITTER, K. B. & ALLER, L. H. 1983 *MNRAS* **195**, 939.

LEQUEUX, J., RAYO, J. F., SERRANO, A., PEIMBERT, M., & TORRES-PEIMBERT, S. 1979 *A&A* **80**, 155.

LYNDEN-BELL, D. 1975 *Vist. Ast.* **19**, 299.

LYNDEN-BELL, D. 1992, in *Elements and the Cosmos*, (eds. M. G. Edmunds & R. J. Terlevich). p. 270. Cambridge University Press.

MACIEL, W. J., COSTA, R. D. D., & UCHIDA, M. M. M. 2003 *A&A* **397**, 667.

MARTÍN-HERNÁNDEZ, N. L., ET AL. 2002 *A&A* **381**, 606.

MATEO, M. 1998 *ARAA* **36**, 435.

MCWILLIAM, A. 1997 *ARAA* **35**, 503.

MCWILLIAM, A. & RICH, R. M. 1994 *ApJS* **91**, 749.

MEYER, D. M., JURA, M., & CARDELLI, J. A. 1998 *ApJ* **493**, 222.

MIGHELL, K. J., SARAJEDINI, A., & FRENCH, R. S. 1998 *AJ* **116**, 2395.

MONTEVERDE, M. I., HERRERO, A., & LENNON, D. J. 2000 *ApJ* **545**, 813.

MOULD, J. & AARONSON, M. 1983 *ApJ* **273**, 530.

OEY, M. S. & KENNICUTT R. C., JR. 1993 *ApJ* **411**, 137.

OLSZEWSKI, E. W., SUNTZEFF, N. B., & MATEO, M. 1996 *ARAA* **34**, 511.

OSTERBROCK, D. E. 1989, in *Astrophysics of Gaseous Nebulae and Active Galactic Nuclei.* University Science Books.

PAGEL, B. E. J. 1997, in *Nucleosynthesis and chemical evolution of galaxies.* Cambridge University Press.

PAGEL, B. E. J., EDMUNDS, M. G., & SMITH, G. 1980 *MNRAS* **193**, 219.

PAGEL, B. E. J. & TAUTVAIŠIENĖ, G. 1998 *MNRAS* **299**, 535.

QIAN, Y.-Z. & WASSERBURG, G. J. 2003 *ApJ* **588**, 1099.

RAMÍREZ, S. V., STEPHENS, A. W., FROGEL, J. A. & DEPOY, D. L. 2000 *AJ* **120**, 833.

REITZEL, D. B. & GUHATHAKURTA, P. 2002 *AJ* **124**, 234.

RICH, R. M. 2003 in *Hubble's Science Legacy: Future Optical/Ultraviolet Astronomy from Space,* (eds. J. C. Blades, G. Illingworth, & R. C. Kennicutt, Jr.). ASP Conference Series, v. 291, p. 132. ASP.

RICH, R. M., SHARA, M., FALL, S. M., & ZUREK, D. 2000 *AJ* **119**, 197.

ROCHA-PINTO, H. J., MACIEL, W. J., SCALO, J., & FLYNN, C. 2000 *A&A* **358**, 850.

ROLLESTON, W. R. J., SMARTT, S. J., DUFTON, P. L., & RYANS, R. S. I. 2000 *A&A* **363**, 537.

ROLLESTON, W. R. J., VENN, K., TOLSTOY, E., & DUFTON, P. L. 2003 *A&A* **400**, 21.

SARAJEDINI, A., GEISLER, D., SCHOMMER, R., & HARDING, P. 2000 *AJ* **120**, 2437.

SEARLE, L., & ZINN, R. 1978 *ApJ* **225**, 357.

SHAVER, P., MCGEE, R. X., NEWTON, L. M., DANKS, A. C., & POTTASCH, S. R. 1983 *MNRAS* **204**, 53

SHETRONE, M. D. 2003, in *Carnegie Observatories Astrophysics Series, Vol. 4: Origin and Evolution of the Elements,* (eds. A. McWilliam & M. Rauch). Cambridge University Press, in press.

SHETRONE, M. D., CÔTÉ, P., & SARGENT, W. L. W. 2001 *ApJ* **548**, 592.

SHETRONE, M., VENN, K. A., TOLSTOY, E., PRIMAS, F., HILL, V., & KAUFER, A. 2003 *AJ* **125**, 684.

SHIELDS, J. C. & KENNICUTT, R. C., JR. 1995 *ApJ* **454**, 807.

SKILLMAN, E. D., KENNICUTT, R. C., JR. & HODGE, P. W. 1989 *ApJ* **347**, 875.

SMITH, V. V. 2003 in *CNO in the Universe,* (eds. C. Charbonnel, D. Schaerer, & G. Meynet). ASP, in press.

SMITH, V. V., ET AL. 2002 *AJ* **124**, 3241.

SNEDEN, C., JOHNSON, J., KRAFT, R. P., SMITH, G. H., COWAN, J. J., & BOLTE, M. S. 2000 *ApJ* **536**, L85.

SNEDEN, C., ET AL. 2003 *ApJ* **591**, 936.

SOFIA, U. J., MEYER, D. M., & CARDELLI, J. A. 1999 *ApJ* **522**, L137.

STASIŃSKA, G., RICHER, M. G., & MCCALL, M. L. 1998 *ApJ* **522**, L137.

TOLSTOY, E., VENN, K., SHETRONE, M., PRIMAS, F., HILL, V., KAUFER, A., & SZEIFERT, T. 2003 *AJ* **125**, 707.

TRAGER, S. C., FABER, S. M., WORTHEY, G., & GONZÁLEZ, J. J. 2000 *AJ* **119**, 1645.

TRURAN, J. W., COWAN, J. J., PILACHOWSKI, C. A., & SNEDEN, C. 2002 *PASP* **114**, 1293.

TWAROG, B. A. 1980 *ApJ* **242**, 242.

VENN, K. A., LENNON, D. J., KAUFER, A., MCCARTHY, J. K., PRYZBILLA, N., KUDRITZKI, R.-P., LEMKE, M., SKILLMAN, E. D., & SMARTT, S. J. 2001 *ApJ* **547**, 765.

VENN, K. A., MCCARTHY, J. K., LENNON, D. J., PRYZBILLA, N., KUDRITZKI, R.-P., & LEMKE, M. 2000 *ApJ* **541**, 610.

VENN, K. A., TOLSTOY, E., KAUFER, A., SKILLMAN, E. D., CLARKSON, S. M., LENNON, D. J., SMARTT, S. J., & KUDRITZKI, R.-P. 2003 *AJ* **126**, 1326.

VÍLCHEZ, J. M., PAGEL, B. E. J., DÍAZ, A. I., TERLEVICH, E., & EDMUNDS, M. G. 1988 *MNRAS* **235**, 633.

WELTY, D. E., FRISCH, P. C., SONNEBORN, G., & YORK, D. G. 1999 *ApJ* **512**, 636.

WELTY, D. E., LAUROESCH, J. T., BLADES, J. C., HOBBS, L. M., & YORK, D. G. 2001 *ApJ* **554**, L75.

WORTHEY, G. 1998 *PASP* **110**, 888.

Massive stars in the Local Group: Star formation and stellar evolution

By PHILIP MASSEY

Lowell Observatory, 1400 W. Mars Hill Road, Flagstaff, AZ 86001, USA

The galaxies of the Local Group that are currently forming stars can serve as our laboratories for understanding star formation and the evolution of massive stars. In this talk I will summarize what I think we've learned about these topics over the past few decades of research, and briefly mention what I think needs to happen next.

1. Introduction

My talk today will be restricted to giving a brief introduction to the study of massive stars in the Local Group; I'll begin by discussing why I think the subject is important, and giving you a few of the complications and caveats. I'll spend most of my time then talking about what I think we've learned, first about star formation (stories of star formation, the initial mass function, and the upper mass cut-off), and second about the evolution of massive stars (including Luminous Blue Variables, Wolf-Rayet stars, and red supergiants). Finally I'll conclude with a brief discussion of what I think we need to do next. This talk is based in large part on an *Annual Reviews of Astronomy & Astrophysics* paper that I have coming out in October (Massey 2003), and the reader is referred there for a more in-depth analysis. I have used this opportunity to update some of the figures and thoughts from that, so hopefully the two will be somewhat complementary.

Massive stars are extremely rare. If we integrate a Salpeter (1955) initial mass function (IMF) and allow for the relative lifetimes involved, we expect that there are about a hundred thousand solar-type stars (i.e., stars of $1(\pm 0.1)$ $M_\odot$ for every $20(\pm 0.2)$ $M_\odot$ in the Galaxy. There should be over a million solar-type stars for every $100(\pm 10)$ $M_\odot$ star. Yet, these stars have a disproportionate effect upon their environment: they provide most of the mechanical energy input into the interstellar medium through stellar winds and supernova (Abbott 1982); they provide most of the UV ionizing radiation of galaxies, plus power the far-IR luminosities of galaxies through the heating of dust; and they serve as the primary source of CNO enrichment of the interstellar medium (Maeder 1981).

Why study massive stars in the more distant galaxies of the Local Group, when there are examples closer to home? The primary reason is that the metallicity of the gas out of which these stars are born differs by a factor of nearly 20 amongst the galaxies currently actively forming stars (Massey 2003 and references therein). And metallicity should matter! First, we expect that the initial mass function will depend upon metallicity, as the Jeans length depends upon temperature, and the temperature depends upon the metallicity of the star-forming cloud. Understanding how star-formation processes (and the IMF) varies with metallicity is important for interpreting the integrated properties of galaxies at large look-back times. Secondly, the stellar winds of massive stars are driven by radiation pressure in highly ionized metal lines, and the evolution of massive stars is dominated by the effects of radiatively-driven stellar winds. Thus we expect to see large differences in the evolved massive star populations depending upon the metallicity of the host galaxy.

And this is a wonderful time for such studies. Increasing sophisticated stellar evolutionary models are available through the kindness of both the Geneva and Padova teams.

At the same time we have excellent observational capabilities, such as the high resolution optical imaging and UV spectroscopic facilities of *HST*. There are large-format CCD cameras available on 4-m class telescopes around the world, and my own "Local Group Survey" team is busy producing *UBVRI* photometry of 300 million stars in nine nearby galaxies. At the same time there are high through-put spectrographs on large telescopes, such as the Blue Channel on the MMT, GMOS on Gemini, and DEIMOS on Keck.

1.1. *Some difficulties and caveats*

Because massive stars have very high effective temperatures (30,000–50,000° K) while on the main-sequence, there are some unique problems in characterizing the massive star population of a nearby galaxy. Only a small fraction ($<10\%$) of their light leaks out into the visible range, and because what we see is so far down on the *tail* of the Rayleigh-Jeans distribution that we may not recognize the actual beast (Figure 1). The bolometric luminosity depends critically on the effective temperature, and the deduced mass depends critically on the bolometric luminosity. Massey (1998a) gives explicit equations, namely:

$$L \propto m^{2.0} \ ,$$

$$\Delta \log m = -0.2 \times \Delta M_{\rm bol} \ ,$$

$$\Delta {\rm BC} = -6.84 \times \Delta \log T_{\rm eff} \ .$$

A reddening-free index, such as Q, will have a dependence $\Delta {\rm BC}/\Delta Q = 33$ (note that the ratio is mistakenly inverted in Massey 1998a); i.e., a 0.10 mag uncertainty in Q will lead to an uncertainty in the BC (and $M_{\rm bol}$ of 3.3 mag and hence to an uncertainty of 0.7 dex in $\log m$. This is not going to lead to finding a meaningful IMF slope.

Another implication of the high effective temperatures, and corresponding large bolometric corrections, is that *the visually brightest stars are not the most massive*. A young 85 $M_\odot$ star O-type star will be about 3 mags (a factor of 15) fainter than a 25 $M_\odot$ A-type supergiant, although the later is about 1.5 mag (factor of 4) less luminous bolometrically. The reader is referred to Figure 1 in Massey et al. (1995) and the corresponding discussion.

In both of these cases spectroscopy provides the answer to such ambiguities, allowing accurate temperatures (and hence BCs, luminosities, and thus masses) to be determined, and allowing the construction of meaningful H-R diagrams.

1.2. *Tests*

If indeed the Local Group is our "astrophysical laboratory," then we must be good scientists and design our experiments with some care. Generally there are two types of tests possible.

(*a*) *Mixed-age tests*, where we compare the relative number of one thing to another, such as the relative number of red supergiants (RSGs) to Wolf-Rayet stars (WRs). The implicit assumption in such tests—not often stated, and possibly not always realized—is that we are averaging over a sample population to be have a completely heterogeneous mix of ages. If this assumption is invalid—if we are instead looking at a population with a strong bias of 10 Myr (typical for a RSG) but for which 4 Myr old stars (typical for WRs) are under-represented, then we are going to have some problems in interpreting the results.

(*b*) *Coeval tests*, where we study the main-sequence stars in a cluster in order to derive a stellar IMF or determine the turn-off mass in a cluster containing evolved massive stars. We must test if indeed all the stars were "born on a particular Tuesday" (Hillenbrand et al. 1993), or if instead the age spread is significant.

PROBLEMS WITH CONTINUUM MEASURES

OF HOT STARS ...

(with apologies to H.J.G.L.M.L.)

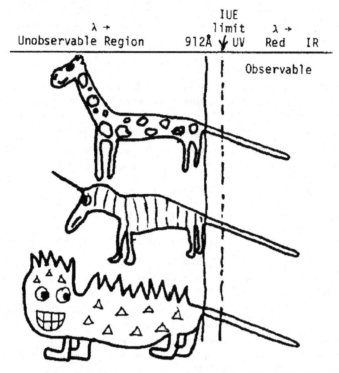

FIGURE 1. For stars with high effective temperatures, the *tails* of the Rayleigh-Jeans distribution gives little information about the effective temperatures, and hence the bolometric luminosities or masses of hot stars. The figure is from Conti (1986) and is used with permission.

2. What have we learned?

2.1. *Star formation, the IMF, and the Upper-Mass Limit*

2.1.1. *Star Formation*

Detailed spectroscopic and photometric studies of young clusters reveal very different histories of star formation. Herbig (1962) first suggested that low- and intermediate-mass stars might form over a prolonged time in a cluster, followed by the formation of high mass stars, which halts all star formation. But not all clusters follow this paradigm.

Back before *HST* made M16 famous, Hillenbrand et al. (1993) studied the stellar content of the cluster and found that there were intermediate-mass stars, pre main-sequence stars with ages as young as a few times 10^5 years. Yet M16's massive star population has an age of two million years. As Hillenbrand et al. (1993) conclude, "...Thus the formation of O stars neither ushered in nor concluded the star-formation process in this young complex."

By contrast, the prototype "super star cluster" R136 in the 30 Doradus region of the LMC does fit the standard model. *HST* photometry and spectroscopy reveals that the intermediate-mass population began forming 6 Myr ago, and stopped about 2 Myr years

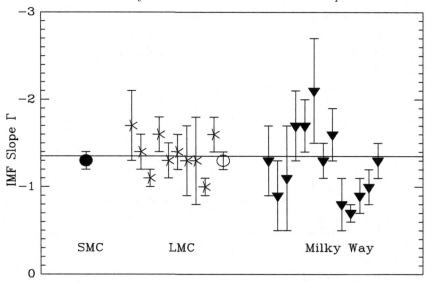

FIGURE 2. The slope (Γ) of the initial mass function of OB associations and clusters in the SMC, LMC, and Milky Way is constant over a range of 4 in metallicity and 200 in stellar density. The circle shows the IMF of the R136 cluster in the LMC. The Salpeter (1955) $\Gamma = -1.35$ value is denoted by the solid line. Based upon data in Massey (2003) and references therein.

ago. This cluster contains a very large number of extremely massive stars that formed 1–2 Myr ago (Massey & Hunter 1998).

These two clusters provide an interesting contrast in their stories of star formation. In very rich, dense clusters, such as R136, we find that the formation of intermediate-mass stars is stopped shortly after the *formation* of the high-mass stars, due presumably to the effects that their stellar winds have on the surrounding gas. In a less rich cluster such as M16, production of intermediate-mass stars is not halted by the formation of the high-mass stars. One might speculate that there the intermediate-mass stars will continue to form until the first Wolf-Rayet stars (with their strong stellar winds) are produced, or possibly will continue until the first SNe. Nevertheless, in both M16 and R136 the massive stars themselves were formed over a very short period of time (<1 Myr), a fact that can prove very useful, as we'll see in Section 2.3.

2.1.2. *The IMF of massive stars*

Studies of the stellar IMF of OB associations and clusters in the Milky Way and Magellanic Clouds show no evidence for any effect with metallicity. The variations that are seen are observational and/or statistical in nature (Massey 1998a; Kroupa 2001). In Figure 2 we see that the *slope* if the IMF, Γ, is approximately Salpeter ($\Gamma = -1.35$) over a factor of 4 in metallicity and 200 in stellar density.

This is actually somewhat surprising given that metals provide the primary cooling mechanism in molecular clouds, and hence cloud temperatures (and thus the Jeans mass) should depend upon metallicity. But so far we have probed only a small range in metallicity: from the SMC to the Milky Way, the metal content changes by only a factor of 3.7. We could extend this to cover a range of 17 if we were to push such studies further out in the Local Group (WLM to Andromeda). Our Local Group survey project is producing *UBVRI* photometry of 300 million stars, but will require spectroscopy if we are to determine IMFs and star formation histories.

2.1.3. *The upper mass limit*

One of the other important lessons that spectroscopy of the R136 stars taught us (Massey & Hunter 1998) is concerns the "upper mass limit" seen in clusters. The highest mass stars in R136 were an unprecedented 150 $M_\odot$. However, this is just what we would have expected from extrapolating the IMF, as this is just where the IMF would peter down to a single star. This suggests that the "upper mass limits" we've so far encountered are statistical, and not physical. Whatever it is that limits the ultimate mass of a star, we have yet to encounter it in nature.

2.2. *Evolved Massive Stars*

The evolved (He-burning) massive stars include:
- Luminous Blue Variables (LBVs): very luminous stars that undergo occasional "eruptions" (S Doradus, η Carinae, the Hubble-Sandage variables).
- Wolf-Rayet stars (WRs): stars with broad emission lines whose surface compositions are consistent with the equilibrium products of H-burning (WN types) or He-burning (WC types).
- Red supergiants (RSGs).

2.2.1. *The Conti scenario*

Conti (1976) first proposed that stellar winds (mass-loss) would result in some of the chemical peculiarities observed for Wolf-Rayet stars. The most luminous stars (at a given metallicity) will have the highest mass-loss rates and suffer the greatest fraction of mass loss. A modern version of the "Conti scenario" might look something like this:

$$m > 85\ M_\odot: \quad \text{O} \longrightarrow \text{LBV} \longrightarrow \text{WN} \longrightarrow \text{WC} \longrightarrow \text{SN}$$

$$40 > m > 85\ M_\odot: \quad \text{O} \longrightarrow \text{WN} \longrightarrow \text{WC} \longrightarrow \text{SN}$$

$$25 > m > 40\ M_\odot: \quad \text{O} \longrightarrow \text{RSG} \longrightarrow \text{WN} \longrightarrow \text{WC} \longrightarrow \text{SN}$$

$$20 > m > 25\ M_\odot: \quad \text{O} \longrightarrow \text{RSG} \longrightarrow \text{WN} \longrightarrow \text{SN}$$

$$10 > m > 20\ M_\odot: \quad \text{OB} \longrightarrow \text{RSG} \longrightarrow \text{BSG} \longrightarrow \text{SN}$$

Thus, an 85 $M_\odot$ star starts out as an O-type star, and then becomes a luminous blue variable. Extreme mass-loss during the LBV stage then helps lead first to a WN-type Wolf-Rayet star (with the H-burning products revealed at the surface), while subsequent evolution (and mass-loss) results in a WC-type WR (with the more advanced He-burning products revealed at the surface. Stars of lower mass (20–25 $M_\odot$) might not have enough mass loss for a star to evolve beyond the WN stage, while stars of even lower mass might never go through a WR phase at all, but rather spend their He-burning lives first as red supergiant (RSG) and then as a "second generation" blue supergiant (BSG), similar to the precursor of SN1987A.

Of course, we don't know how right this overall "cartoon" version of massive star evolution is, and in particular we don't know what the corresponding mass ranges are for the various evolutionary paths. We especially don't know how these ranges vary with metallicity!

2.2.2. *LBVs*

Hubble & Sandage (1953) called attention to five irregular variables in M31 and M33 that were (at times) among the brightest resolved stellar objects in these galaxies. Photographic plates from Mt. Wilson extended back to 1916, which revealed episodic visual brightenings of several magnitudes. These objects were extremely luminous and blue, and had spectra that were of intermediate F-type during maxima. A footnote in their paper suggested that the LMC star S Doradus might be similar. The connection to the Galactic stars η Car and P Cyg came later. Conti (1984) coined the term "luminous blue variables" to describe these objects.

LBVs are extremely luminous bolometrically, and are found near the edge of the observed upper luminosity limit in the H-R diagram. They undergo episodic bouts of high mass loss, during which they brighten visually, while remaining roughly constant in bolometric luminosity.

In a series of papers in the late 1970s and early 1980s, Roberta Humphreys demonstrated that there was an observed limit to the luminosities of stars (see, for example, Humphreys & Davidson 1979). This limit decreases with decreasing effective temperatures until $T_{\rm eff} = 10,000°$ K, after which it is nearly constant at $\log L/L_\odot \sim 5.7 (M_{\rm bol} = -9.5$, corresponding roughly to 50 $M_\odot$. It was understood that this upper luminosity limit was *somehow* related to LBVs, since these star occupied a narrow band near this limit.

It was very tempting to try to interpret the observed upper luminosity limit in terms of some fundamental physics, such as the Eddington limit. In the classic Eddington limit,

$$\frac{L}{M} = \frac{4\pi Gc}{\kappa} \quad ,$$

where κ is the flux-mean opacity (0.347 for electron scattering if $T_{\rm eff}$ is high). In that case

$$\frac{L/L_\odot}{M/M\odot} = 3.8 \times 10^4 \quad .$$

This is similar to, but considerably greater than, the *observed* ratio of the Humphreys-Davidson limit, $\frac{L/L_\odot}{M/M_\odot}$:

$$\frac{10^{5.7} L_\odot}{50\ M_\odot} = 1.0 \times 10^4 \quad .$$

This is tantalizingly close, but the *classical* Eddington limit is thus factors of several times higher than what is observed, and doesn't show the same $T_{\rm eff}$ dependence.

Lamers & Fitzpatrick (1988) used model atmospheres to explore where radiation pressure and gravity are balanced if the full effects of metal-line opacities are included, rather than just the effects of electron scattering. This turned out to completely explain the Humphreys-Davidson limit. The actual limit is a trough, descending to lower luminosity with decreasing effective temperatures until a temperature of 10,000° K is reached, after which the opacity decreases and hence the allowed luminosity increases again. However, since stars evolve from higher temperatures to lower, the effect is to produce a decreasing envelope followed by a line at constant luminosity for $T_{\rm eff} < 10,000°$ K.

This suggests then that LBVs are a normal phase in the lives of the most massive stars. As such stars try to evolve to cooler effective temperatures, their radii expand, their surface gravities get lower, and most importantly their atmospheric opacities increase and radiation pressure overcomes gravity, leading to vastly increased mass loss. These stars are stopped dead in their (evolutionary) tracks! Such a star will dump a lot of material, heat up, and then try again.

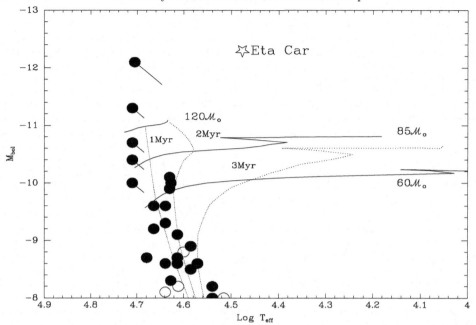

FIGURE 3. The location of η Car in the H-R diagram is just what we expect if it happens to be a highly evolved massive star, as its bolometric luminosity is slightly higher than the highest luminosity unevolved massive star. The highest mass stars are quite coeval, with an age of about 1 Myr. This figure is based upon Figure 4 in Massey, DeGioia-Eastwood, & Waterhouse (2001).

This may neatly wrap up the whole issue of LBVs, but possibly not. Kenyon & Gallagher (1985) suggested that some LBVs were the result of evolution in close binary systems, an argument partially supported by the relative isolation of some LBVs in M31. This has been given some boast by the controversial claim that η Car itself is a binary.

The problem with fully resolving the question of the origins of LBVs is complicated by the fact that the bolometric luminosities of LBVs are poorly known, making their placement on the H-R diagram uncertain. The real exception to this is η Car itself, where the surrounding gas and dust reprocesses the radiation to the IR, where it can be measured. η Car is a member of the Tr14/16 complex, and when we place it on the H-R diagram (Figure 3) we find something quite comforting: it is exactly where it should be, in the sense of being just slightly more bolometrically luminous than the highest luminosity unevolved star (Massey, DeGioia-Eastwood, & Waterhouse 2001). One is forced to conclude that even if η Car is a binary, its binary nature may be irrelevant to its LBV nature.

Another complication in understanding the origins of LBVs is that it is very hard to gather meaningful statistics. η Car underwent its last major outburst in 1830, and P Cyg underwent its last major outburst in 1655. If these stars were located in the Magellanic Clouds, would we even consider them LBVs? One such Galactic example of an LBV-in-waiting may be the star VI Cyg No. 12 (Massey & Thompson 1991). This star is one of the visually most luminous stars known in the Milky Way, it is surrounded by circumstellar material, but it has not had a photometric "episode" during the few decades it has been known. Given that Hubble & Sandage (1953) relied upon 40 years' worth of observations just to identify the first five such stars in M31 and M33 (one of these 5 stars, Var A, is no longer even considered an LBV), it may require centuries to obtain firm statistics on the number of LBVs in nearby galaxies.

2.2.3. *Wolf-Rayet stars*

In 1867 two French astronomers, Wolf and Rayet, came across three stars whose spectra showed broad, strong emission lines rather than the absorption spectra that character- ized other stars. Beals (1930) correctly identified the emitting atoms as ionized helium, nitrogen, and carbon.

WN-type Wolf-Rayets show lines of helium and nitrogen. Abundance studies show that for WN stars we are seeing an atmospheric composition that reflects what the composition that the stellar core would have during the main-sequence CNO hydrogen-burning cycle. (See discussion and references in Massey 2003.) WC stars show helium, carbon, and oxygen; the latter two are the products of the triple-α helium-burning cycle. Somehow, then, the outer layers of these stars have been stripped away, revealing the nuclear burning products at the surface. Prior to 1976, the prevailing notion was that Wolf-Rayet stars were the product of evolution in close binaries, with Roche-lobe overflow being responsible for this stripping. However, Conti (1976) noted that the ubiquitous presence of stellar winds in hot stars provided an alternative explanation, and proposed that O-type stars would naturally evolve first to WN-type and then to WC-type Wolf-Rayets.

Since mass-loss depends upon metallicity, one expects that the relative number of WCs and WNs should vary from place to place. Indeed, roughly equal number of WCs and WNs were known in the vicinity of the sun, while in the lower metallicity Magellanic Clouds most of the WRs are of WN type. Quantitative numbers provide the means for comparison with evolutionary models.

One complication in finding unbiased samples of Wolf-Rayet stars is that the WCs are much easier to find than are the WNs. The strongest emission line in WCs (C IIIλ4650) have a median line strength that is a factor of four greater than that of the strongest emission line in WNs (He II λ4686). My colleagues and I designed a set of three inter- ference filters in order to facilitate identification of WRs in nearby galaxies, and have used this with some success in regions of nearby galaxies (see Massey & Johnson 1998 for a recent summary of such studies). Candidate stars are identified on the basis of having a magnitude difference from an emission-band to continuum-band filter, if that magnitude difference is significant compared to the photometric errors. Candidates are then observed spectroscopically.

Such studies have revealed a strong trend in the relative number of WCs and WNs with respect to metallicity. We show the data in Figure 4. Two points deserve comments. First, the point for the Milky Way is higher than the trend based upon the other galaxies, and one has to wonder if the data for the region within 2 kpc is complete. No systemic surveys for WRs in the Milky Way have been carried out; instead, such stars were found either as part of the HD survey (which would be very incomplete at a distance of 2 kpc for WRs given typical reddenings, as noted by Massey & Johnson 1998), or by "accidental" spectroscopic discovery as part of a study of stars in, say, a specific cluster. IC 10 poses a far more interesting discrepancy. Massey & Armandroff (1995) discovered 22 Wolf-Rayet candidates in this small galaxy (by contrast, the SMC, which is twice the size, contains only 11 known WRs), and argued that IC 10 is the nearest example of a starburst galaxy. One peculiarity, though, is that the number of WCs was about two times greater than the number of WNs, while a ratio of 0.1 would be more in keeping with its metallicity. Various explanations have been advanced to explain this: it's possible the IMF is top- heavy, or alternatively, massive stars might have formed within the same short time interval all across the galaxy and are evolving in lock-step. A recent deeper survey by

Massey & Holmes (2002) suggests another explanation: they found a large number of additional candidates, confirming two by spectroscopy. If spectroscopic confirmation of the remainder turns out the way they expect, then the number ratio of WCs to WNs may be normal in IC 10, but the galaxy may have a total WR content of a hundred WRs, a surface density that would be about 20 times higher than that of the LMC. In terms of its overall massive star population, IC 10 must be more like an OB association, but on a kpc scale.

How well do the stellar models do in reproducing this trend? We see in the bottom panel of Figure 4 that the normal-mass loss models of the Geneva (Schaller et al. 1992; Schaerer et al. 1993; Charbonnel et al. 1993) and Padova (Fagotto et al. 1994; Bressan et al. 1993) groups come relatively close to reproducing the trends, although both predict too low a WC/WN at higher metallicities. (One should note, though, that the statistics for M31 are not well established.) The "enhanced" mass-loss tracks of the Geneva group (Meynet et al. 1994) do not match the data at all. Massey (2003) has examined the historical basis for the introduction of these "enhanced" mass-loss models, and with the advantage of 20–20 hindsight, suggests that the problems these were trying to solve were, to some extent, unreal. Unfortunately, it is the enhanced mass-loss models that are commonly used in starburst models (e.g., Schaerer & Vacca 1998; Leitherer et al. 1999; Smith, Norris, & Crowther 2003).

2.2.4. *Red supergiants*

A long-standing problem in identifying the RSG content of nearby galaxies has been the confusion by foreground Galactic dwarfs. For instance, Massey (1998b) found that roughly half of the sample of red stars seen towards M33 by Humphreys & Sandage (1980) were likely foreground Galactic objects. BVR two-color diagrams partially resolve this ambiguity, as RSGs will have redder B-V colors at a given V-R color (Massey 1998b). Spectroscopy of the Ca II triplet lines in the NIR is a powerful way of distinguishing Galactic disk dwarfs from bona-fide extra-galactic RSGs given the heliocentric radial velocities of most Local Group galaxies.

Massey (1998b) investigated the RSG content of three Local Group galaxies (NGC 6822, M33, and M31), which span a range of 0.8 dex in metallicity. He found that as the metallicity goes up, the fraction of high luminosity RSGs goes down, but that there is not a sharp cut-off in either M_V or M_{bol}. This suggests that moderately high-mass stars (40 $M_\odot$) become RSGs even at higher metallicities, but what changes is that the RSG phase is shorter at higher metallicities, in accord with the suggestion of Maeder, Lequeuex, & Azzopardi (1980). A recent study (Massey & Olsen 2003) has extended this work to that of the Magellanic Clouds.

This work has revealed a potentially serious drawback to stellar evolutionary models, namely that none of the "modern" stellar evolutionary models produce RSGs as cool and luminous as what is actually observed. This is illustrated in Figure 5. We see too that there is a tight relationship between luminosity and effective temperature of the RSGs. Matching this will provide a challenge to future stellar evolutionary models.

Finally, we note that the relative number of RSGs and WRs show a very strong trend with metallicity, changing by two orders of magnitude over just 0.8 dex in metallicity (Figure 6). Again, none of the stellar evolutionary models reproduce this trend.

2.3. *Coeval tests*

So far the tests described have involved mixed-aged populations, where we compare the relative number of WCs to WNs, or RSGs to WRs. But, in principle, a far more sensitive

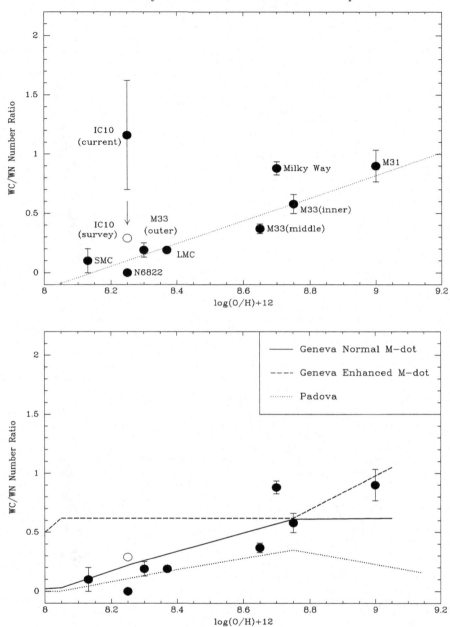

FIGURE 4. The relative number of WCs and WNs show a strong trend with metallicity among the Local Group galaxies (top). The normal-mass loss models of the Geneva and Padova groups do a reasonable job at matching the data for all but the highest metallicities (bottom). The "enhanced" mass-loss models of the Geneva group do not match the data at all.

test is possible, which involves using coeval regions to determine the progenitor masses as a function of metallicity.

This is a classical method first used by Sandage (1953) to determine the masses of RR Lyrae stars using globular clusters, and subsequently used by Anthony-Twarog (1982) to find the progenitor masses of white dwarfs in intermediate-age clusters. However, it

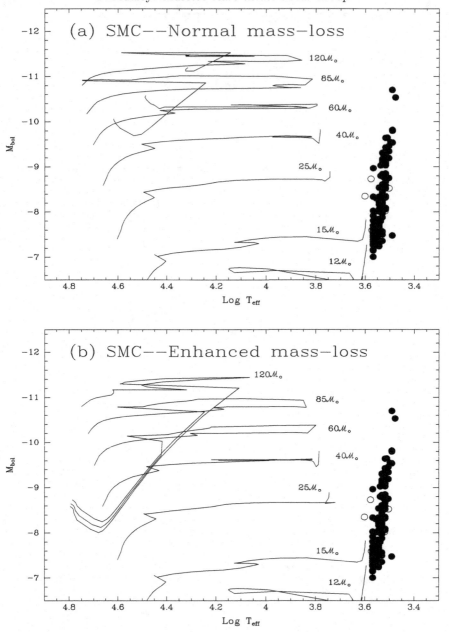

FIGURE 5. The evolutionary tracks do not actually go sufficiently far to the right to produce RSGs as cool and as luminous as what is observed. Note too that the tight relation of RSGs in these diagrams, with stars of lower luminosities being of slightly earlier (hotter) spectral types. The figure is based upon data from Massey & Olsen (2003).

is one thing do do this for clusters that are 10^{10} yr old, or even 40–70 Myr old, and quite another to do it for clusters that are only a few million years old: the degree of coevality required is quite high. However, recall that the Hillenbrand et al. (1993) study of NGC 6611 (discussed in Section 2.1.1) found that most of the massive stars were born

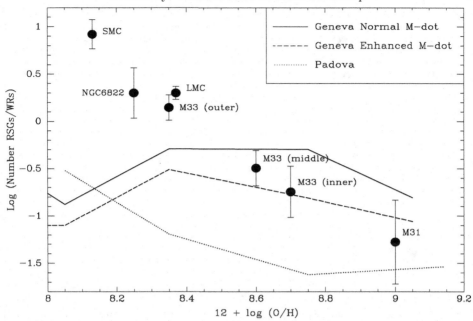

FIGURE 6. Although the relative number of RSGs and WRs show a very strong trend with metallicity, none of the current (non-rotation) evolutionary models reproduce this trend.

over a very short period of time (<1 Myr). The data used to construct the H-R diagrams themselves can be used to answer the degree of coevality.

Massey, Waterhouse, & DeGioia-Eastwood (2000) and Massey, DeGioia-Eastwood, & Waterhouse (2001) selected 19 clusters in the Magellanic Clouds and 12 clusters in the Milky Way that contained evolved massive stars, and obtained photometry and spectroscopy of the most luminous unevolved cluster members. About half of these met their stringent criteria for coevality, and hence could be used to determine the progenitor masses of the evolved massive stars using the masses of as-yet unevolved massive stars. The results are shown in Figure 7.

What we've learned from this is that LBVs all come from the highest mass stars, as we expect. The Ofpe/WN9s, which have sometimes been linked to the LBVs, do not come from the highest mass stars—instead, their progenitor masses are quite low. Interestingly, there are strong trends in the progenitor masses of WN stars in the three galaxies: apparently, in the SMC only the highest mass stars become WRs, which is consistent with the SMC's low metallicity and the expectations of the Conti scenario— that the amount of mass loss will determine the evolution of massive stars. In the LMC, a larger range of progenitor masses yield WNs. And, in the Milky Way, stars from masses as low as 20 $M_\odot$, up to that of $\gg 120$ $M_\odot$ become WRs.

3. What's next?

Studies of the massive star content of nearby galaxies (including our own!) have shed important light on star-formation and the evolution of massive stars. What comes next?

• We need to move beyond the Magellanic Cloud and complete galaxy-wide surveys for WRs, RSGs, and BSGs in M31, M33, NGC 6822, IC 10, and other galaxies of the Local Group.

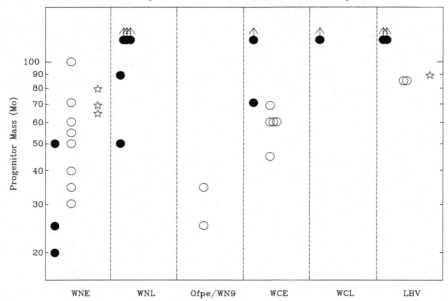

FIGURE 7. The progenitor masses are shown for various types of evolved massive stars for the SMC (stars), LMC (open circles), and Milky Way (filled circles). We see that LBVs come from only the most massive stars. The Ofpe/WN9s come from lower mass stars. Of the WNEs, the progenitor masses are highest in the SMC, span a larger range to lower masses in the LMC, and are even lower in the Milky Way, as one might expect. The figure is modified from one that appears in Massey et al. (2001).

- Follow-up spectroscopy with 8-m-class telescopes will allow meaningful H-R diagrams to be constructed, allowing careful tests of models as a function of metallicity.
- Coeval studies need to be extended to higher metallicity systems, such as M31.

This contribution has been supported by the National Science Foundation, through grant AST0093060.

REFERENCES

ABBOTT, D. C. 1982 *ApJ* **163**, 723.

ANTHONY-TWAROG, B. J. 1982 *ApJ* **255**, 245.

BEALS, C. S. 1930 *Publ. Dominion Astrophys. Obs.* **4**, 271.

BRESSAN, A., FAGOTTO, F., BERTELLI, G., & CHIOSI, C. 1993 *A&AS* **100**, 647.

CHARBONNEL, C., MEYNET, G., MAEDER, A., SCHALLER, G., & SCHAERER, D. 1993 *A&AS* **101**, 415.

CONTI, P. S. 1976 *Mem. Soc. R. Sci. Liege* **9**, 193.

CONTI, P. S. 1984. In *Observational Tests of the Stellar Evolutionary Theory* (eds. A. Maeder & A. Renzini). IAU Symp. 105, p. 253. Reidel.

CONTI, P. S. 1986. In *Luminous Stars and Associations in Galaxies* (eds. C. W. H. de Loore, A. J. Willis, & P. Laskarides). IAU Symp. 116, p. 199. Reidel.

FAGOTTO, F., BRESSAN, A., BERTELLI, G., & CHIOSI, C. 1994 *A&AS* **105**, 29.

HERBIG, G. H. 1962 *ApJ* **135**, 736.

HILLENBRAND, L. A., MASSEY, P., STROM, S. E., & MERRILL, M. K. 1993 *AJ* **106**, 1906.

HUBBLE, E. & SANDAGE, A. 1953 *ApJ* **118**, 353.

HUMPHREYS, R. M. & DAVIDSON, K. 1979 *ApJ* **232**, 409.

HUMPHREYS, R. M. & SANDAGE, A. 1980 *ApJS* **44**, 319.

KENYON, S. J. & GALLAGHER, J. S. 1985 *ApJ* **290**, 542.

KROUPA, P. 2001 *MNRAS* **322**, 231.

LAMERS, H. J. G. L. M. & FITZPATRICK, E. L. 1988 *ApJ* **324**, 279.

LEITHERER, C., SCHAERER, D., GOLDADER, J. D., DELGADO, R. M. G., ROBERT, C., ET AL. 1999 *ApJS* **123**, 3.

MAEDER, A. 1981 *A&A* **101**, 385.

MAEDER, A., LEQUEUEX, J., & AZZOPARDI, M. 1980 *A&A* **90**, L17.

MASSEY, P. 1998a. In *The Stellar Initial Mass Function* (eds. G. Gilmore & D. Howell). 38th Herstmonceux Conference, p. 17. ASP.

MASSEY, P. 1998b *ApJ* **501**, 153.

MASSEY, P. 2003 *ARA&A* **41**, 15.

MASSEY, P. & ARMANDROFF, T. E. 1995 *AJ* **109**, 2470.

MASSEY, P., DEGIOIA-EASTWOOD, K., & WATERHOUSE, E. 2001 *AJ* **121**, 1050.

MASSEY, P. & HOLMES, S. 2002 *ApJ* **580**, L35.

MASSEY, P. & HUNTER, D. A. 1998 *ApJ* **493**, 180.

MASSEY, P. & JOHNSON, O. 1998 *ApJ* **505**, 793.

MASSEY, P., LANG, C. D., DEGIOIA-EASTWOOD, K., & GARMANY, C. D. 1995 *ApJ* **438**, 188.

MASSEY, P. & OLSEN, K. A. G. 2003 *AJ*, submitted.

MASSEY, P. & THOMPSON, A. B. 1991 *AJ* **101**, 1408.

MASSEY, P., WATERHOUSE, E., & DEGIOIA-EASTWOOD, K. 2000 *AJ* **119**, 2214.

MEYNET, G., MAEDER, A., SCHALLER, G., SCHAERER, D., & CHARBONNEL, C. 1994 *A&AS* **103**, 97.

SALPETER, E. E. 1955 *ApJ* **121**, 161.

SANDAGE, A. 1953 *Mem. Soc. R. Sci. Liege* **4**, 254.

SCHAERER, D., MEYNET, G., MAEDER, A., & SCHALLER, G. 1993 *A&AS* **98**, 523.

SCHAERER, D. & VACCA, W. D. 1998 *ApJ*, **497**, 618.

SCHALLER, G., SCHAERER, D., MAYNET, G., & MAEDER, A. 1992 *A&AS* **96**, 269.

SMITH, L. J., NORRIS, R. P. F., & CROWTHER, P. 2002 *MNRAS* **337**, 1309.

Massive Young Clusters in the Local Group

By JESÚS MAÍZ-APELLÁNIZ[1,2]

[1]Space Telescope Science Institute, 3700 San Martin Drive, Baltimore, MD 21218, USA

[2]ESA Space Telescope Division

We analyze the properties of the Massive Young Clusters in the Local Group, concentrating on the youngest segment of this population and, more specifically, on the two best studied cases: 30 Doradus and NGC 604. 30 Doradus is a Super Star Cluster and will likely evolve to become a Globular Cluster in the future. NGC 604, on the other hand, is a Scaled OB Association that will be torn apart by the tidal effects of its host galaxy, M33. Given their extreme youth, each cluster is surrounded by a Giant H II Region produced by the high ionizing fluxes from O and WR stars. The two Giant H II Regions are found to be rather thin structures located on the surfaces of Giant Molecular Clouds, and their geometry turns out to be not too different from that of classical H II regions such as the Orion or Eagle Nebulae.

1. Introduction

We have all learned in textbooks that the stellar clusters of the Milky Way can be classified as either globular or open. Globular clusters are old (~ 10 Ga), massive ($3 \times 10^4 - 3 \times 10^6$ $M_\odot$), metal-poor, and spherically-symmetric members of the Galactic halo. Open clusters are young ($\lesssim 1$ Ga), low-mass ($< 5 \times 10^3$ $M_\odot$), metal-rich, and non-spherically-symmetric members of the Galactic disk. Thanks to *HST* and other telescopes, in the last decade we have learned that such classification is not valid for other galaxies and that even for the Milky Way it is not completely correct. Some clusters can be both young and massive and some galaxies can have large numbers of such Massive Young Clusters (or MYCs, see Maíz-Apellániz 2001; Whitmore et al. 1999; Larsen & Richtler 1999).

The youngest MYCs are associated with Giant H II Regions (GHRs), one of the most conspicuous type of objects in spiral and irregular galaxies at optical wavelengths. GHRs are the clearest manifestation of the violent interaction of MYCs with their nearest environment by means of gravity, ultraviolet light, stellar winds, and supernova explosions. The study of GHRs in our neighborhood is a necessary step to understand the properties of their relatives at high redshift, where they may appear as bright point sources against the background of their host galaxies.

2. The sample

In this presentation I will concentrate on the two youngest of the most massive stellar clusters located in external galaxies of the Local Group, 30 Doradus and NGC 604. For completeness, I also discuss briefly at the end of this section other clusters in the Local Group which are somewhat similar to 30 Doradus and NGC 604, but which are not as massive or as young or which are heavily obscured by Galactic extinction.

2.1. *30 Doradus*

30 Doradus is certainly the most studied MYC, since its location in the LMC makes it the closest unobscured example. Its proximity, combined with the extreme youth of some of its stellar populations, has prompted the denomination of "Starburst Rosetta" (Walborn 1991). It is located close to the SE tip of the LMC bar, an area of intense star formation of which it is the brightest exponent. It has an integrated M_V (corrected for extinction)

of -11.6, which yields a stellar mass of $\sim 10^5 \ M_\odot$ for a Salpeter IMF (Hunter et al. 1995; Sirianni et al. 2000; Maíz-Apellániz 2001). It has a complex star-formation history, with the main star formation episode taking place ~ 2 Ma ago (Walborn & Blades 1997; Massey & Hunter 1998). The GHR powered by the MYC extends over 200 pc and has a kinematic structure with many fast-expanding shells (Chu & Kennicutt 1994), a result of the star-formation history of the region over the past ~ 20 Ma. The high-excitation, high-intensity part of the nebula (see Fig. 1) is much smaller and extends to the N and W of R136, the MYC core, at a distance of 10–20 pc (Maíz-Apellániz et al. 2002). The kinematic profile there can be well characterized by a single gaussian plus a few weak components (Melnick et al. 1999).

2.2. NGC 604

NGC 604 is a GHR in M33 ($d = 840$ kpc; Freedman et al. 2001), located along a direction with low foreground extinction (González Delgado & Pérez 2000). It is the brightest FIR source in M33 (Hippelein et al. 2003) and is powered by a MYC without a central core which contains $\gtrsim 200$ O and WR stars (Hunter et al. 1996; Maíz-Apellániz et al. 2004b). The age of NGC 604 is estimated to be 3.0–3.5 Ma; its stellar mass (as derived from its luminosity) is 0.5–$2.0 \times 10^5 \ M_\odot$, approximately the same as 30 Doradus. (Yang et al. 1996; Maíz-Apellániz 2000; González Delgado & Pérez 2000). The MYC is located inside a cavity (see Fig. 2) which has the kinematic profile of an incomplete expanding bubble. The majority of the nebular photons (especially those from high-excitation species such as O III) originate at the borders of that central cavity and in a nearby high-intensity knot (Maíz-Apellániz et al. 2004a). The kinematic profiles in those areas can be well characterized by single gaussians (Sabalisck et al. 1995; Maíz-Apellániz 2000), a strong indication that the overall dynamics of the GHR is not yet dominated by the kinetic energy deposition by stellar winds and SN explosions (Tenorio-Tagle et al. 1996). Other cavities are visible surrounding the central one: some have multiple components detected in their $H\alpha$ kinematic profiles and are likely to be secondary expanding bubbles created where the primary bubble has punctured the surrounding medium (Maíz-Apellániz 2000; Tenorio-Tagle et al. 2000).

2.3. Other interesting clusters in the LG

• N11 is the second largest H II region in the LMC after 30 Doradus in terms of $H\alpha$ luminosity (Kennicutt & Hodge 1986). It includes several OB associations, of which the largest are LH9 and LH10, some of them still in the process of emerging from their parent molecular clouds (Parker et al. 1992; Rosado et al. 1996; Barbá et al. 2003). Its stellar mass is $\approx 1/4$ that of 30 Doradus.

• NGC 346 is the largest H II region in the SMC (Kennicutt & Hodge 1986) and it contains at least 33 O stars (Massey et al. 1989; Rubio et al. 2000). It has received less attention recently compared to other objects (especially with *HST*) and it appears to be smaller than N11.

• NGC 595 is the second largest H II region in M33 and contains a significant WR star population (Drissen et al. 1993). Its stellar mass was calculated by Malamuth et al. (1996) to be in the range 1–$2 \times 10^4 \ M_\odot$, significantly smaller than that of 30 Doradus or NGC 604.

• Some clusters in the LG appear to be quite massive but have lost most of their ionizing power already due to their greater age ($\gtrsim 10$ Ma) in comparison to 30 Doradus or NGC 604. Among them, we can include NGC 1805 and NGC 1818 in the LMC (de Grijs et al. 2002) and NGC 330 in the SMC (Sirianni et al. 2002).

• We should finally mention some of the young Galactic clusters: NGC 3603 (Moffat et al. 1994; Drissen et al. 1995); Central, Arches and Quintuplet (Figer et al. 1999; Figer

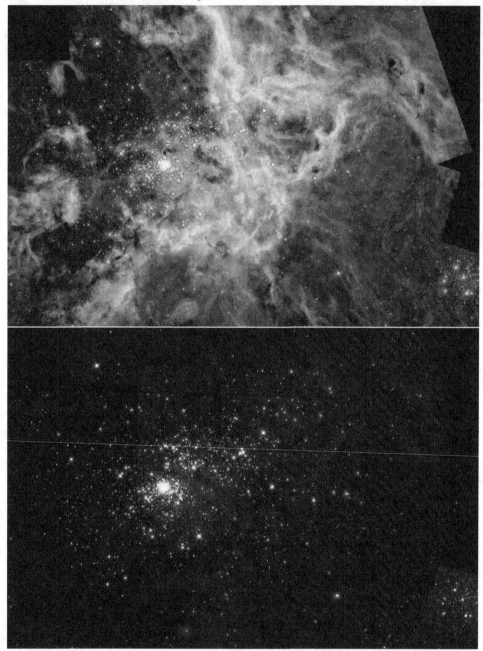

FIGURE 1. Black and white reproduction of two color mosaics of 30 Doradus comprising 5 WFPC2 fields. The top one has F814W and F673N in the red channel, F555W and F656N in the green channel and F336W in the blue channel. The bottom one only has the three continuum filters (F814W, F555W, and F336W). Each field has a size of 68 pc × 44.5 pc, with N towards the upper right corner. See http://www.stsci.edu/~jmaiz for a color version of this figure.

2003); W43 (Blum et al. 1999); W49A (Alves & Homeier 2003); Cyg OB2 (Knödlseder 2000; Comerón et al. 2002); and Westerlund-1 (Neguerela & Clark 2003). These objects have ages ≲10 Ma but they are all less massive than 30 Doradus or NGC 604 and are harder to study due to their higher extinctions and to confusion with Galactic fore-

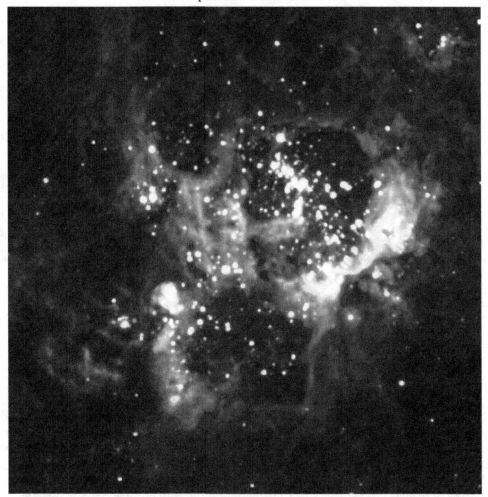

FIGURE 2. Black and white reproduction of a color mosaic of NGC 604 produced with WFPC2 data using F673N in the red channel, F555W and F656N in the green channel and F336W in the blue channel. The field has a size of 160 pc × 160 pc with N towards the top. See http://www.stsci.edu/~jmaiz for a color version of this figure.

ground objects. Furthermore, those close to the Galactic Center may be easily destroyed (Portegies Zwart et al. 2001; McMillan & Portegies Zwart 2003).

3. The structure of MYCs

3.1. *SSCs, SOBAs, cores, and halos*

Stars have a tendency to be born in groups. In some cases, the group is compact enough to be bound, at least for a period much longer than a typical orbital time for a given star, and we have a (real) cluster. In others, the group is too extended and, although the stars have approximately the same velocity (which differentiates them from nearby non-group members), the tidal field of the galaxy easily disrupts the group within one galactic rotation. In that case, the group is called an (OB) association. It is not uncommon to have clusters (bound groups) within more extended associations, with both originating from the same progenitor molecular cloud (de Zeeuw et al. 1999, and references therein).

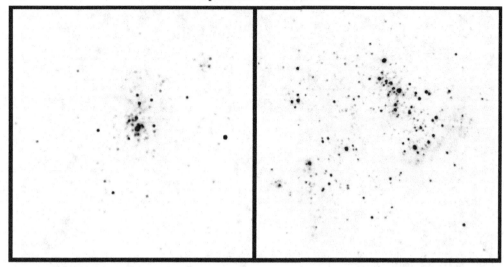

FIGURE 3. A comparison between 30 Doradus (left) and NGC 604 (right). Both images were obtained using similar filters (continuum at Hβ for 30 Doradus, WFPC2 F547M for NGC 604), and have the same physical size (120 pc × 120 pc), orientation (N at top), and resolution (the ground-based 30 Doradus image was degraded to attain this objective). R136 is the bright object at the center of the 30 Doradus image.

The above description was originally applied to the relatively young low-mass groups in the solar neighborhood. In the past decade we have found out that pretty much the same classification applies to the upper end of the young cluster mass spectrum. Thus, MYCs can be divided into two types: Super Star Clusters (SSCs) are organized around a compact (half-light radius, $r_{1/2} = 1$–3 pc) core while Scaled OB Associations (SOBAs) lack such structure and are more extended objects, with $r_{1/2} > 10$ pc (Hunter 1995; Maíz-Apellániz 2001). SSCs are bound and represent the high mass end of young stellar clusters while SOBAs are (at least from a global point of view) unbound and are the massive relatives of regular OB associations.† Furthermore, the core of some SSCs is surrounded by extended halos which are themselves similar to SOBAs in terms of structure and number of stars, thus representing the high-mass equivalent of those associations with clusters inside (Maíz-Apellániz 2001).

30 Doradus and NGC 604 provide a good example of the differences between SSCs and SOBAs, as shown in Fig. 3. 30 Doradus is an SSC with a very well defined compact core, R136, surrounded by an extended halo. The core alone contains at least 40 O2-3 stars (Massey & Hunter 1998), a large fraction of the stars of those spectral types so far identified anywhere (Walborn et al. 2002). On the other hand, NGC 604 is a prototypical SOBA, without a core and with quarter- and half-light radii ($r_{1/4}$ and $r_{1/2}$) as measured with WFPC2 F336W of 18.4 and 28.4 pc (Maíz-Apellániz 2001). As a comparison, 30 Doradus has $r_{1/4} = 3.3$ pc and $r_{1/2} = 9.2$ pc (the presence of a halo influences the value of $r_{1/2}$ more strongly than that of $r_{1/4}$.)

We have retrieved archival UIT data and combined it with WFPC2 F170W images to produce Fig. 4, the radial profile of 30 Doradus in the FUV. The presence of an extended

† Note that by calling a SOBA a MYC we are introducing a slight terminological inconsistency with respect to the low-mass end of the spectrum, since the classical definition of a stellar cluster implies a bound object. Here we are applying a less restrictive definition of cluster to include associations and SOBAs, i.e., to mean a group of stars born from the same molecular cloud within a short (~ 10 Ma) period of time.

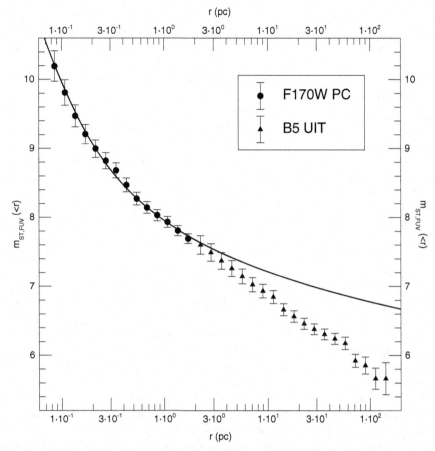

FIGURE 4. The radial profile of 30 Doradus in the FUV as obtained from F170W PC data for the region around R136 and B5 UIT data for the halo. The line shows the best fit to the PC data using an Elson et al. (1987) model, which yields $r_c = 0.096$ pc. The two data sets were joined artificially by applying a 1.15 magnitudes shift to the UIT data to account for a variety of effects regarding absolute calibration (contamination in the the PC, R136 saturation in the UIT data, and passband differences). Note that the error bars should not be interpreted as independent, since the plotted magnitude is integrated in the radial direction.

halo reveals itself quite clearly at radii larger than 5 pc from R136. At longer wavelengths, the presence of the halo is even more obvious, since R136 and its vicinity are especially rich in O stars, the dominant contributors to the FUV flux. Fitting an Elson et al. (1987) model to the inner regions we obtain a value for the core radius $r_c = 0.096$ pc, which agrees with previous equivalent determinations. The relative weight of the inner halo with respect to the core as a function of wavelength is the likely cause of the differences in r_c as a function of stellar mass (Brandl et al. 1996). R136 is an extremely compact object, when we compare it either with other SSCs (Maíz-Apellániz 2001) or with other clusters in the LMC (Elson et al. 1989). As we will see later when we analyze the evolution of stellar clusters, this is probably due to its extreme youth: R136 is the youngest SSC among those with reliable age and radius determinations.

3.2. Stellar generations

How long does it take for a stellar cluster to form? A compact low-mass cluster can be formed on a short time scale (few 10^5 years), according to recent numerical simulations

(Bonnell et al. 2003), and an SSC core is likely to form on similar time scales. Therefore, given our current inability to measure stellar ages with much better precision than 1 Ma, single SSC cores can be treated as single-age populations (see, however, Maoz et al. 2001; Origlia et al. 2001, for the case of an SSC with a double core). But what about SOBAs and SSC halos? Given their much larger physical sizes, one would not expect them to form on such fast time scales and indeed that is what it is found.

We show in Fig. 1 a WFPC2 mosaic of the central region of the 30 Doradus Nebula. Four of the five populations described by Walborn & Blades (1997) can be seen there. The core (R136), located at center left, has an age of ~ 2 Ma (see also Massey & Hunter 1998). A second younger population, partially embedded in the nebulosity and with some of its members easily detectable only in the IR, has an age of less than 1 Ma and is located at distances of 10–20 pc to the N and W of R136 (Rubio et al. 1998a; Walborn et al. 1999, 2002). A third population whose brightest components are O and B supergiants is scattered throughout the field and has an age of 4–6 Ma. Finally, Hodge 301, a 20–25 Ma smaller cluster (Grebel & Chu 2000), can be seen at the lower right corner (the fifth missing population is located outside the mosaic towards the left and has an age similar to that of the third one). We can use those populations to divide the star-forming history of 30 Doradus into three episodes: (1) an initial phase lasting up to 20 Ma in which part of the halo (including the low-mass cluster Hodge 301) was formed; (2) the main phase, in which R136 was formed almost instantaneously ~ 2 Ma ago. (3) A third ongoing phase that is completing the halo population by using up the last remaining molecular gas.

The most important episode so far in terms of numbers of stars formed is undoubtedly the second one. However, we should not forget that the third one is not finished and that it is also producing massive stars, including at least one of spectral type O3-4, and that several IR sources are found inside the molecular cloud, including one cluster (Walborn et al. 2002). This "second generation" of star formation is induced by the compression of the molecular gas produced by the energy input of the "first generation" stars (Walborn et al. 1999) and could lead to a full-blown "two-stage starburst" once $\gtrsim 200$ stars explode as SNe in R136 within a few Ma of each other.† An example of such an event can be seen in N11, ~ 3 Ma after the formation of the central cluster (LH9), a second cluster (LH10) has been recently formed within the SOBA, likely triggered by the multiple SN explosions in LH9 (Meaburn et al. 1989; Parker et al. 1992; Walborn & Parker 1992; Rosado et al. 1996).

The data for NGC 604 also shows that the SOBA is not completely coeval. A RSG is present, something unexpected in a single-population cluster with ≈ 10 WR stars, since RSGs are expected to appear only at a later stage (Drissen et al. 1993; Terlevich et al. 1996, note, however, that this could be a foreground object). Also, some of the massive stars are partially embedded in the H II region that is sandwiched between the cavity and the molecular cloud, with some of them showing a higher than average extinction, properties which in 30 Doradus correspond to the younger generation (Maíz-Apellániz et al. 2004a,b). It would be interesting to obtain high-spatial-resolution IR observations of that region in NGC 604 in order to find out whether there is a massive star population deeply embedded in the molecular cloud.

How does this fit into our current understanding of ISM evolution and star formation? Recent numerical simulations of galactic disks show that molecular clouds are high-density, high-pressure regions that form mainly by turbulent ram pressure (as opposed

---

† Note that, according to this terminology, the stars formed before R136 would belong to a "zeroth" generation. They may belong to a larger field of similar age, in which 30 Doradus is immersed.

to by self-gravity), which has its ultimate origin mostly in SN explosions (Mac Low 2004). In this scenario, molecular clouds are very transient features that are easily created and destroyed. Gravity would play a role only after turbulent pressure creates filamentary structures dense enough to start collapsing (Bate et al. 2003). The final stages of cluster formation appear to be a hierarchical process (Bonnell et al. 2003), with subclusters forming first, which then merge to form larger structures until the formation process is ended when there is no more gas available or when the gas is dispersed by the stellar winds from the newly-born massive stars.

The models and simulations in the above paragraph refer to "normal" (i.e., low-mass) clusters but the similarities that we have found between the structural properties of low-mass and high-mass clusters and associations suggest that MYCs may form in the same way. The most important difference would be the need for a large initial amount of gas and for an extremely high pressure in order to compress it into a relatively small volume. The extra high pressure could be caused e.g., by a head-on encounter between high-density regions due to the disordered velocity field of an irregular galaxy or by a global gravitational instability in the galaxy. In any case, the hierarchical nature of the process indicates that the same type of filamentary structures should form in the dense molecular gas during the early stages of formation of a MYC. Subclusters would then form along those structures and, if a region is dense enough to produce a large number of them within a small volume, a core would be formed when they merge. The rest of the subclusters would form the halo or SOBA part of the MYC and in a time scale of the order of 10–30 Ma (the typical orbital periods around the center of the cluster for stars located at radii of 10–20 pc) the relative positions of the stars there would bear little resemblance to their original ones. Given the size of several tens of pc of the cloud, the whole MYC formation process could take ~ 10 Ma from the time when the first stars are born until the time when the molecular material is dispersed, a value consistent with the observed properties of 30 Doradus, NGC 604, and N11.

Is there a way to detect those filamentary structures in 30 Doradus or NGC 604 in order to test our understanding of the physics of MYC formation? The problem here is that our current capabilities do not allow us to measure with good enough resolution the spatial distribution of the dense molecular gas (though the situation should change when ALMA becomes available). Furthermore, a large fraction of the original molecular gas has already been transformed into stars or photoevaporated and dispersed by the action of the newly-born massive stars, so we should look for the alleged structures in the spatial distribution of the stars themselves. Such "star chains" structures are visible in 30 Doradus and in other MYCs (Fig. 5), which would indeed suggest the persistence of filamentary structures for a few Ma. However, no firm link can be yet established due to the lack of reliable kinematical stellar data. Furthermore, some of the stars that delineate the "star chains" may actually be of different ages (Walborn & Blades 1997), thus weakening the case for a causal connection between the current stellar spatial distribution and the original density distribution of the molecular gas. Therefore, we cannot provide a final answer to the question at this stage.

3.3. *Dynamical evolution*

A number of processes influence the dynamical evolution of a stellar cluster: mass loss through stellar winds and SNe, a process that heats the cluster; two-body interactions that lead to relaxation, energy redistribution as a function of stellar mass, stellar ejections, evaporation, and possible core-collapse; binary star formation, another cluster-heating process; and tidal interactions in the form of static tidal fields, tidal shocks, and

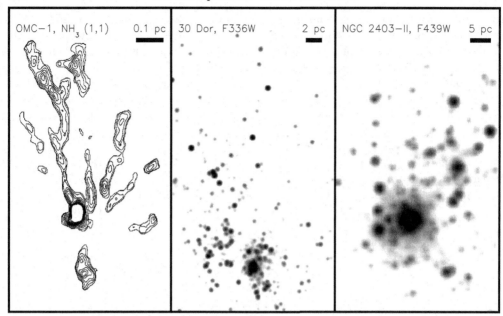

FIGURE 5. A comparison between the structures observed in the dense material of the Galactic star-forming molecular cloud OMC-1 (adapted from Wiseman & Ho 1998) and two SSCs
with strong halos. Even though the scales are quite different, in both cases the same core +
quasiradial filamentary structure is observed. Since structures in molecular clouds are apparently
hierarchical, it is possible that the "star chains" in the cluster halos are a consequence of the
mass distribution of the parent cloud.

dynamical friction (all leading towards an increase in the speed of destruction for the
cluster) and tidal merging (Gerhard 2000).

Reviewing those processes, we see that the distinction between SSCs and SOBAs is
important for the long-term evolution of the cluster, since the extended character (and
likely unboundness) of the latter make them highly vulnerable to disruption by tidal
forces (Fall & Rees 1977). A SOBA like NGC 604 in M33 is not likely to survive more
than a galactic rotation. Thus, SOBAs should follow an evolution similar to that of their
low-mass relatives, OB associations, and end up dispersing their components among the
field population of their galaxy.

A more interesting question is the destiny of SSCs. Will they become the Globular
Clusters (GCs) of the future? Fall & Rees (1977) determined that in order for a cluster
to survive for a Hubble time, a large initial mass ($\approx 3 \times 10^4$ $M_\odot$) was required. For
lower-mass clusters the combined effects of two-body interactions and tidal forces was
too strong for the cluster to last that long. More recent studies (Fall & Zhang 2001)
confirm this conclusion. The second condition required for long-term survival is having
the right size. Clusters which are too compact are easily affected by two-body interactions
(though their immediate destiny is probably not destruction but only expansion) while
clusters that are too extended get disrupted by tides, as we have already mentioned.
SSCs have the right intermediate size to ensure survival (Maíz-Apellániz 2001) and one
would expect them to become GCs in the future. Furthermore, it is probably a good idea
to define an SSC as a cluster which has the right size and enough mass to become a GC
in the future. The only problem with this definition is its practicality. On the one hand,
SSC masses cannot always be measured directly and easily and, on the other hand, the
minimum mass required for survival is a strong function of the environment and the same

cluster which may be able to last for 10 Ga in the outskirts of an irregular galaxy may be destroyed in less than 1 Ga near the center of a spiral galaxy (Boutloukos & Lamers 2003).

30 Doradus is a good example of the difficulties in measuring SSC masses directly (i.e., by means of velocity dispersions). SSCs that have had their masses measured in that way with some precision are all older than 30 Doradus and have a large number of RSGs, allowing the Ca triplet or the NIR lines which originate in those stars as tracers of the velocity dispersion of the cluster (Ho & Filippenko 1996; Maíz-Apellániz 2004, and references therein). Only one such star is present in the vicinity of R136 (see Fig. 1) and it is likely a field interloper, so a similar analysis is not possible there. An alternative is to use the He lines of massive stars but the analyses up until now have been hampered by the presence of undetected binaries (Bosch et al. 2001). Furthermore, it is likely that many of the stars analyzed belong not to the R136 population (i.e., the core of the SSC) but to the halo and, therefore, an analysis of their motions would tell us little about the mass of R136 (given their youth and distances from R136, their velocities are likely to be determined mostly by the initial conditions of their formation and not by their supposed orbital motion around R136). Therefore, the only way we have of measuring the mass of 30 Doradus is by converting its luminosity into mass by means of an IMF. This has been attempted by a number of authors (Sirianni et al. 2000, and references therein) to conclude that it appears to have a Salpeter slope at higher masses with a flattening at lower masses. A few words of caution have to be said here: the IMF has been measured only in the periphery of R136 (the core itself is too crowded); differential extinction could play a role, especially at lower masses; the IMF has not been measured for masses below 0.6 $M_\odot$, where a significant fraction of the total mass could be "hidden"; and it is not clear what fraction of its halo (if any) will be retained by R136 in the long term. With all those caveats, the total stellar mass of 30 Doradus appear to be in the vicinity of 10^5 $M_\odot$, with maybe one third or one half of that belonging to a future R136 cluster once the halo has disappeared from its surroundings. If those values prove to be correct, R136 has a good chance of becoming a low-mass GC in the future.

The possible evolution of R136 described in the above paragraph refers to the long term ($\gtrsim 1$ Ga). In the short term (few to tens of Ma), the most important evolutionary process is likely to be tied to mass loss through stellar winds and SNe and to binary heating. As previously mentioned, R136 is both the most compact and the youngest of all well-measured SSCs and of all massive LMC clusters. This correlation between age and size is not unexpected, since R136 is still to experience its first SN, and numerical models predict that it should expand by a factor of 2 within the first 7 Ma of its life time due to the above mentioned processes (Portegies Zwart et al. 1999). After that, the expansion is expected to slow down but not to halt completely, since tidal effects become important in the long term (Wilkinson et al. 2003)

4. GHRs and MYCs

4.1. *GHR structure*

The structures visible in the optical nebular images of 30 Doradus and NGC 604 (Figs. 1 and 2) reveal that, despite the obvious causal connection between the massive stars and the ionized gas in the GHR, there is no clear-cut correlation between the positions of one and the other. If anything, a weak anticorrelation is present. In 30 Doradus, R136 sits on a region with little nebular emission, most of it of rather low-excitation, with most of the high-excitation, high-intensity emission coming from an area 10–20 pc to the N and W

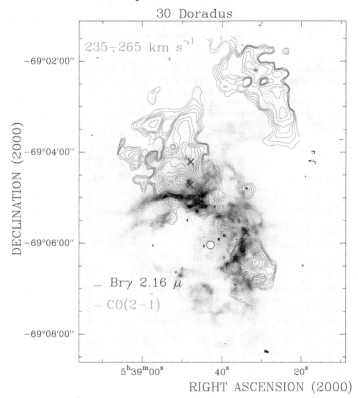

FIGURE 6. CO (2-1) contour plot of 30 Doradus superimposed on a Br-γ greyscale image. The circle marks the position of R136 and the 'X's indicate the positions of IR sources. Figure provided by Mónica Rubio.

which overlaps with the massive-star halo. In NGC 604, most of the SOBA is located again in a region of low-intensity, low-excitation nebular gas but some of the massive stars to the S and to the E appear to be in regions of high-intensity nebular gas.

A much better correlation can be observed in Fig. 6 between the Brackett-γ image (once again tracing the GHR) and the CO contours. There are two large molecular clouds to the W and NE of R136 joined by a lower-intensity CO bridge. The high-intensity, high-excitation part of the GHR (darker in Fig. 6) is located at the sides of the molecular clouds directly exposed to the ionizing radiation of R136 and the inner halo of 30 Doradus. The velocity data reveal that the two large molecular clouds have the same (approximately stationary) velocity while in the bridge at least two velocity components are present. One is blueshifted and corresponds to the location of the "stapler" nebula, the dark reddish structure ~ 7 pc to the right of R136 in Fig. 1, and the other is slightly redshifted and extends to the NW of the first one (Rubio et al. 1998b). One explanation of these CO observations would be to consider the molecular gas in 30 Doradus as shaped originally like a disk with an inclination of $\sim 70°$ with respect to the plane of the sky and with its axis in the NW-SE direction. R136 would have been formed near the center of the disk and would have carved it into its present form by means of the UV radiation and stellar winds of its massive stars. The foreground part of the disk would have been almost completely destroyed, leaving only a few pieces accelerated towards us, such as the stapler nebula. The background part would be shaped like a doughnut cut in half, with the two observed main molecular clouds corresponding to the two points where the

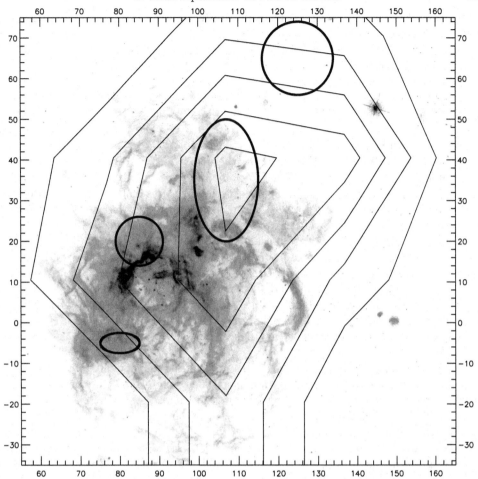

FIGURE 7. CO contour map of NGC 604 (Wilson & Scoville 1992, from) superimposed on a F656N WFPC2 image. The ellipses mark the positions of individual CO clouds from the higher-resolution data of Engargiola et al. (2000). The axes are labeled in arcseconds ($1'' = 4$ pc) with N towards the bottom.

optical depth would be largest, and the molecular bridge to the central part of the half-doughnut, which would be located ~ 10 pc in the background and would have started to move away from us due to the influence of R136 and the stars in its inner halo.

The spatial resolution of the CO observations of NGC 604 (see Fig. 7) is lower than in 30 Doradus but the observed structures are similar. The MYC sits in a cavity with little or no CO emission and the molecular gas is located in several clouds in its vicinity. Furthermore, the brightest parts of the GHR are located at the borders of the molecular clouds directly exposed to the radiation of the MYC. In N11 (Israel et al. 2003), we find a similar disposition. The CO clouds there have been swept to larger distances from the central and older association LH9 to form an incomplete 140 pc × 200 pc ring. Once again, the GHR is brightest on those sides of the CO gas directly exposed to the UV radiation from the massive stars, especially in those areas where secondary star formation has been produced, such as towards the N in LH10 and towards the E in LH13.

The general picture that emerges from the study of GHRs in the LG is the following: MYCs are formed from Giant Molecular Clouds (GMCs) and the UV radiation from the massive stars (with some help from stellar winds) carves an initial cavity in the molecular gas of a few tens of pc in size in the first 3.5 Ma by first dissociating and later ionizing the molecular gas. The GHR is formed as a highly-stratified, thin (1–2 pc) region on the surface of the GMC directly exposed to the UV radiation that can extend for several tens of pc and that, in many senses, is nothing but a scaled-up version of what we observe in lower-mass nearby H II regions (Scowen et al. 1998; Ferland 2001). After 3.5 Ma, the first SNe start exploding and the molecular gas is swept to larger distances, such as those observed in N11; in a few Ma more probably there is little trace of the original molecular gas around the MYC. It is interesting to point out that at all times the sizes of the cavities created by MYCs and their expansion velocities are smaller than those predicted by the models of Weaver et al. (1977), as noted by Maíz-Apellániz & Walborn (2001) for a sample of nearby MYCs. The explanation for this is two-fold: (a) the creation of the cavities is not driven (at least initially) by stellar winds and SNe but mainly by ionizing radiation and (b) the surrounding GMCs are massive, pressurized, and dynamic objects that offer a higher resistance to expansion than the low-density, unpressurized, and static medium of the ISM around single stars in the Weaver et al. (1977) models. The kinematic profiles of the emission lines support this secondary role for the kinetic energy input from stellar winds and SNe in the early stages of GHRs (see references in Section 2 and also Maíz-Apellániz et al. 1999).

For a ~ 2 Ma old GHR such as 30 Doradus, the situation may look like what is depicted in Fig. 8. Some aspects that we have not mentioned before but which appear in that diagram are treated here. A second thin layer, the Photo Dissociation Region (PDR), where the non-ionizing UV light transforms H_2 into atomic hydrogen, exists between the H II gas and the molecular cloud (Hollenbach & Tielens 1997). This structure has been observed in the NIR in 30 Doradus by Rubio et al. (1998b) and its spatial extent fits the description in the above paragraphs quite well. Also, some small cloudlets are seen accelerating away from the MYC. Some of them are shells produced by early SNe in the halo of the MYC which, as we have seen for 30 Doradus, can start to form a few Ma in advance of the main starburst episode. Also, note that the sides of the molecular cloud not directly exposed to the ionizing radiation from the core of the MYC as well as distant gas cloudlets can still be ionized by the low-intensity ionizing radiation coming from halo stars and from scattered light. This gas will shine in low excitation species such as S II but will produce negligible emission from higher excitation species such as O III. Another interesting point to make about the diagram in Fig. 8 is the importance of the orientation. Observers along different directions can obtain very different pictures of the GHR depending on whether one of the GMCs blocks the view towards the central regions of the MYC or not (Walborn 2002). Regarding the "empty space" around the MYC, the classical answer is that it is filled with hot coronal gas (at this early stage, from stellar winds; later, from SNe) and X-ray observations support this idea, at least partially (Wang 1999). However, the S/N ratio provided even by *Chandra* or *XMM/Newton* is not conclusive to support that all the empty space is filled up in this manner. Some of it could indeed be filled with very hot gas escaping from a punctured dense ISM and of such low density as to be very hard to detect in X rays. An alternative would be a non-equilibrium $\sim 10^4$ K Diffuse Ionized Gas (DIG) originating in the nearby (denser) H II gas through champagne-like flows. The latter alternative is favored by current numerical models of the ISM, which show that the pressure distribution is quite broad and that there is very strong mixing (Mac Low 2004).

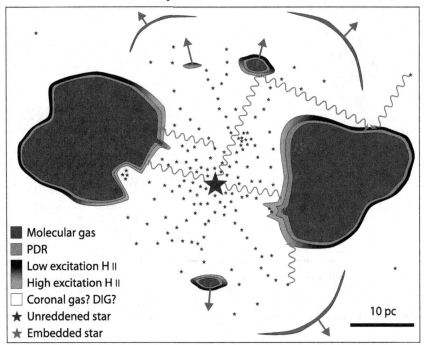

FIGURE 8. Model of a ~ 2 Ma old Giant H II Region. The PDR is the region sandwiched between the H II gas and the molecular cloud. Wiggly arrows indicate ionizing radiation while straight ones signal gas motions. The big star indicates the position of the core of an SSC. See http://www.stsci.edu/~jmaiz for a color version of this figure.

A final point should be made about the high excitation part of the GHR. The high excitation there is not due only to the direct exposure to the ionizing radiation from the central regions of the MYC but also to the ionization from within by the second generation of massive stars being produced there. The existence of this second generation of stars in 30 Doradus is revealed by the detection of IR sources (Rubio et al. 1998a; Walborn et al. 1999) and of pillars similar to those in M16 (Scowen et al. 1998; Walborn et al. 1999). Spectroscopy of some of those sources reveal them to be O stars (Walborn et al. 2002). For NGC 604 there are no IR data available and even at *HST* resolution we would not be able to resolve pillars but, still, we see massive stars embedded in the bright regions of the nebulosity, some of them with high values of extinction and located at excitation peaks (Maíz-Apellániz et al. 2004a,b). Those stars would contribute substantailly to increasing the excitation of the nearby gas by producing high fluxes of ionizing radiation in situ.

4.2. *The effect of extinction*

Most analyses of GHRs assume a constant foreground extinction. We can see in Fig. 9 that this approximation is too rough, at least for 30 Doradus. Extinction is very patchy, with complex filamentary structures producing strong variations on pc and sub-pc scales. A comparison between the nebular image on the left and the color excess map on the right reveals that some low-intensity areas (e.g., the one ~ 20 pc to the S of R136) are indeed caused by extinction while others (e.g., the one ~ 20 pc to the E of R136) are caused by the absence or weakness of the H II gas there.

Furthermore, an $E(B-V)$ map like the one shown here is only part of the story behind extinction (and, what is more important, correcting for it):

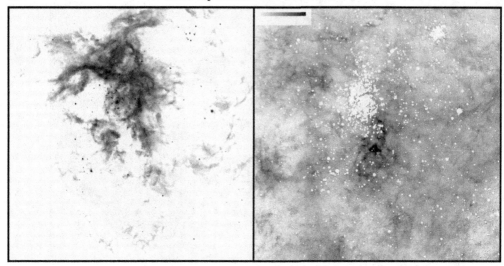

FIGURE 9. Hα + continuum image (left) and $E(B-V)$ map (right) of 30 Doradus. $E(B-V)$ is calculated from the Hα/Hβ ratio using a foreground screen model with $R_V = 3.2$ and the scale shown in the upper left corner goes from 0.0 to 0.75. White areas indicate the presence of stars ($E(B-V)$ cannot be calculated accurately there). The field size is 112 pc × 112 pc and N is up.

• Most simple extinction models assume a foreground screen. However, as we have seen, in most instances there is molecular gas located near the H II gas, so one would expect more complex geometries involving unseen highly-extincted gas and scattered radiation, which makes an Hα/Hβ map only a first-order correction.

• The extinction law in the optical-IR is poorly known for GHRs (previous studies have concentrated on the UV). In the Galaxy it is known that normal H II regions show high values of R_V (Cardelli et al. 1989) but no similar study exists for GHRs and most models actually assume $R_V = 3.2$.

• The spatial distribution (not only in the plane of the sky but also along the line of sight) can be very different for stars and gas and the measured extinction can also be very different. When possible, one should measure stellar and gas extinctions independently.

We are currently undertaking an in-depth study of NGC 604 using a variety of data, including radio observations, to quantify more precisely some of the above effects. Our preliminary results indicate that the patchy character of the dust distribution plays a significant role there as well and that some of the material producing the thermal radio emission is almost completely hidden in either Hα or Hβ due to the large values of A_V.

5. Summary

• Massive Young Clusters (MYCs) can be either Super Star Clusters (SSCs) or Scaled OB Associations (SOBAs). SSCs have a compact core and possibly an extended halo while SOBAs lack such a core.

• There can be multiple generations in a MYC separated by a few Ma. Some of them precede the main episode of star formation while others take place at a later stage and are likely induced by it.

• SSCs are the likely predecessors of Globular Clusters. SOBAs are not expected to last for long periods of time but instead they should dissolve rather rapidly and their stars become part of the field population.

- ∼ 2–4 Ma old MYCs still retain a considerable fraction of their primordial molecular gas around them.
- The GHR around a MYC is located on the wall of the molecular cloud facing the MYC. Thus, the H II gas tends to form basically 2-D instead of 3-D structures, pretty much like in low-mass Galactic H II regions.
- The extinction around a MYC is not uniform and the extinction law is poorly known.

I would like to thank Nolan Walborn, Enrique Pérez, Rodolfo Barbá, Miguel Mas-Hesse, Casiana Muñoz-Tuñón, and Mónica Rubio for sharing their knowledge of 30 Doradus and NGC 604 with me, and Mario Livio for comments on this paper. Go to http://www.stsci.edu/~jmaiz to retrieve color figures and animations. Support for this work was provided by NASA through grant GO-09096.01-A from the Space Telescope Science Institute, Inc., under NASA contract NAS5-26555, and by the Spanish Government grant AYA-2001-3939.

REFERENCES

ALVES, J. & HOMEIER, N. 2003 *ApJ* **589**, L45.

BARBÁ, R. H., RUBIO, M., ROTH, M. R., & GARCÍA, J. 2003 *AJ* **125**, 1940.

BATE, M. R., BONNELL, I. A., & BROMM, V. 2003 *MNRAS* **339**, 577.

BLUM, R. D., DAMINELI, A., & CONTI, P. S. 1999 *AJ* **117**, 1392.

BONNELL, I. A., BATE, M. R., & VINE, S. G. 2003 *MNRAS* **343**, 413.

BOSCH, G., SELMAN, F., MELNICK, J. & TERLEVICH, R. 2001 *A&A* **380**, 137.

BOUTLOUKOS, S. G. & LAMERS, H. J. G. L. M. 2003 *MNRAS* **338**, 717.

BRANDL, B., SAMS, B. J., BERTOLDI, F., ECKART, A., GENZEL, R., DRAPATZ, S., HOFMANN, R., LOEWE, M., & QUIRRENBACH, A. 1996 *ApJ* **466**, 254.

CARDELLI, J. A., CLAYTON, G. C., & MATHIS, J. S. 1989 *ApJ* **345**, 245.

CHU, Y.-H. & KENNICUTT, JR., R. C. 1994 *ApJ* **425**, 720.

COMERÓN, F., PASQUALI, A., RODIGHIERO, G., STANISHEV, V., DE FILIPPIS, E., LÓPEZ MARTÍ, B., GÁLVEZ ORTIZ, M. C., STANKOV, A., & GREDEL, R. 2002 *A&A* **389**, 874.

DE GRIJS, R., JOHNSON, R. A., GILMORE, G. F., & FRAYN, C. M. 2002 *MNRAS* **331**, 228.

DRISSEN, L., MOFFAT, A. F. J., & SHARA, M. M. 1993 *AJ* **105**, 1400.

DRISSEN, L., MOFFAT, A. F. J., WALBORN, N. R., & SHARA, M. M. 1995 *AJ* **110**, 2235.

ELSON, R. A. W., FALL, S. M., & FREEMAN, K. C. 1987 *ApJ* **323**, 54.

ELSON, R. A. W., FREEMAN, K. C., & LAUER, T. R. 1989 *ApJ* **347**, L69.

ENGARGIOLA, G., PLAMBECK, R. L., & BLITZ, L. 2000 In *The Interstellar Medium in M31 and M33* (eds. E. M. Berkhuijsen, R. Beck, & R. A. M. Walterbos). WE-Heraeus Seminar, Proc. 232, p. 49. Shaker.

FALL, S. M. & REES, M. J. 1977 *MNRAS* **181**, 37P.

FALL, S. M. & ZHANG, Q. 2001 *ApJ* **561**, 751.

FERLAND, G. J. 2001 *PASP* **113**, 41.

FIGER, D. F. 2003 In *A Massive Star Odyssey, from Main Sequence to Supernova* (Eds. K. A. van der Hucht, A. Herrero, & C. Esteban). Proc. IAU Symposium No. 212, p. 487. ASP.

FIGER, D. F., KIM, S. S., MORRIS, M., SERABYM, E., RICH, R. M., & McLEAN, I. S. 1999 Hubble Space Telescope/NICMOS observations of massive stellar clusters near the Galactic Center. *ApJ* **525**, 750.

FREEDMAN, W. L., MADORE, B. F., GIBSON, B. K., FERRARESE, L., KELSON, D. D., SAKAI, S., MOULD, J. R., KENNICUTT, R. C., FORD, H. C., GRAHAM, J. A., HUCHRA, J. P., HUGHES, S. M. G., ILLINGWORTH, G. D., MACRI, L. M., & STETSON, P. B. 2001 *ApJ* **553**, 47.

GERHARD, O. 2000 In *Massive Stellar Clusters* (eds. A. Lançon & C. M. Boily). ASP Conf. Ser. 211, p. 12. ASP.

GONZÁLEZ DELGADO, R. M. & PÉREZ, E. 2000 *MNRAS* **317**, 64.

GREBEL, E. K. & CHU, Y. 2000 *AJ* **119**, 787.

HIPPELEIN, H., HAAS, M., TUFFS, R. J., LEMKE, D., STICKEL, M., KLAAS, U., & VÖLK, H. J. 2003 *A&A* **407**, 137.

HO, L. C. & FILIPPENKO, A. V. 1996 *ApJ* **472**, 600.

HOLLENBACH, D. J. & TIELENS, A. G. G. M. 1997 *ARA&A* **35**, 179.

HUNTER, D. A. 1995 *Rev. Mex. Astron. Astrofís. (conference series)* **3**, 1.

HUNTER, D. A., BAUM, W. A., O'NEIL, JR., E. J., & LYNDS, R. 1996 *ApJ* **456**, 174.

HUNTER, D. A., SHAYA, E. J., HOLTZMAN, J. A., LIGHT, R. M., O'NEIL, JR., E. J., & LYNDS, R. 1995 *ApJ* **448**, 179.

ISRAEL, F. P., DE GRAAUW, T., JOHANSSON, L. E. B., BOOTH, R. S., BOULANGER, F., GARAY, G., KUTNER, M. L., LEQUEUX, J., NYMAN, L.-A., & RUBIO, M. 2003 *A&A* **401**, 99.

KENNICUTT, R. C. & HODGE, P. W. 1986 *ApJ* **306**, 130.

KNÖDLSEDER, J. 2000 *A&A* **360**, 539.

LARSEN, S. S. & RICHTLER, T. 1999 *A&A* **345**, 59.

MAC LOW, M.-M. 2004 In *How does the Galaxy work? A Galactic tertulia with Don Cox and Ron Reynolds* (eds. E. Alfaro, E. Pérez, & J. Franco. Kluwer Academic Publishers, in press.

MAÍZ-APELLÁNIZ, J. 2000 *PASP* **112**, 1138.

MAÍZ-APELLÁNIZ, J. 2001 *ApJ* **563**, 151.

MAÍZ-APELLÁNIZ, J. 2004 In *How does the Galaxy work? A Galactic tertulia with Don Cox and Ron Reynolds* (eds. E. Alfaro, E. Pérez, & J. Franco). Kluwer Academic Publishers, in press.

MAÍZ-APELLÁNIZ, J., MUÑOZ-TUÑÓN, C., TENORIO-TAGLE, G., & MAS-HESSE, J. M. 1999 *A&A* **343**, 64.

MAÍZ-APELLÁNIZ, J., PÉREZ, E., & MAS-HESSE, J. M. 2004*a* in preparation.

MAÍZ-APELLÁNIZ, J. & WALBORN, N. R. 2001 In *Galaxies and their constituents at the Highest Angular Resolution* (ed. R. T. Schilizzi). Proc. IAU Symposium No. 205, p. 222. ASP.

MAÍZ-APELLÁNIZ, J., WALBORN, N. R., & BARBÁ, R. H. 2002 In *Extragalactic Star Clusters* (eds. E. Grebel, D. Geisler, & D. Minniti). Proc. IAU Symposium No. 207, p. 691. ASP.

MAÍZ-APELLÁNIZ, J. ET AL. 2004*b* in preparation.

MALAMUTH, E. M., WALLER, W. H., & PARKER, J. W. 1996 *AJ* **111**, 1128.

MAOZ, D., HO, L. C., & STERNBERG, A. 2001 *ApJ* **554**, L139.

MASSEY, P. & HUNTER, D. A. 1998 *ApJ* **493**, 180.

MASSEY, P., PARKER, J. W., & GARMANY, C. D. 1989 *AJ* **98**, 1305.

MCMILLAN, S. L. W. & PORTEGIES ZWART, S. F. 2003 *astro-ph/0304022*.

MEABURN, J., SOLOMOS, N., LASPIAS, V., & GOUDIS, C. 1989 *A&A* **225**, 497.

MELNICK, J., TENORIO-TAGLE, G., & TERLEVICH, R. 1999 *MNRAS* **302**, 677.

MOFFAT, A. F. J., DRISSEN, L., & SHARA, M. M. 1994 *ApJ* **436**, 183.

NEGUERELA, I. & CLARK, J. S. 2003 In *A Massive Star Odyssey, from Main Sequence to Supernova* (eds. K. A. van der Hucht, A. Herrero, & C. Esteban). Proc. IAU Symposium No. 212, p. 531. ASP.

ORIGLIA, L., LEITHERER, C., ALOISI, A., GREGGIO, L., & TOSI, M. 2001 *AJ* **122**, 815.

PARKER, J. W., GARMANY, C. D., MASSEY, P., & WALBORN, N. R. 1992 *AJ* **103**, 1205.

PORTEGIES ZWART, S. F., MAKINO, J., MCMILLAN, S. L. W., & HUT, P. 1999 *A&A* **348**, 117.

PORTEGIES ZWART, S. F., MAKINO, J., MCMILLAN, S. L. W., & HUT, P. 2001 *ApJ* **546**, L101.

ROSADO, M., LAVAL, A., LE COARER, E., GEORGELIN, Y. P., AMRAM, P., MARCELIN, M., GOLDES, G., & GACH, J. L. 1996 *A&A* **308**, 588.

RUBIO, M., BARBÁ, R. H., WALBORN, N. R., PROBST, R., GARCÍA, J., & ROTH, M. R. 1998*a* *AJ* **116**, 1708.

RUBIO, M., CONTURSI, A., LEQUEUX, J., PROBST, R., BARBÁ, R., BOULANGER, F., CESARSKY, D., & MAOLI, R. 2000 *A&A* **359**, 1139.

RUBIO, M., GARAY, G., & PROBST, R. 1998*b* *The Messenger* **93**, 38.

SABALISCK, N. S. P., TENORIO-TAGLE, G., CASTAÑEDA, H. O., & MUÑOZ-TUÑÓN, C. 1995 *ApJ* **444**, 200.

SCOWEN, P. A., HESTER, J. J., SANKRIT, R., GALLAGHER, III, J. S., BALLESTER, G. E., BURROWS, C. J., CLARKE, J. T., CRISP, D., EVANS, R. W., GRIFFITHS, R. E., HOESSEL,

J. G., HOLTZMAN, J. A., KRIST, J., MOULD, J. R., STAPELFELDT, K. R., TRAUGER, J. T., WATSON, A. M., & WESTPHAL, J. A. 1998 *AJ* **116**, 163.

SIRIANNI, M., NOTA, A., DE MARCHI, G., LEITHERER, C., & CLAMPIN, M. 2002 *ApJ* **579**, 275.

SIRIANNI, M., NOTA, A., LEITHERER, C., DE MARCHI, G., & CLAMPIN, M. 2000

TENORIO-TAGLE, G., MUÑOZ-TUÑÓN, C., & CID-FERNANDES, R. 1996 *ApJ* **456**, 264.

TENORIO-TAGLE, G., MUÑOZ-TUÑÓN, C., PÉREZ, E., MAÍZ-APELLÁNIZ, J., & MEDINA-TANCO, G. 2000 *ApJ* **541**, 720.

TERLEVICH, E., DÍAZ, A. I., TERLEVICH, R., GONZÁLEZ DELGADO, R. M., PÉREZ, E., & GARCÍA-VARGAS, M. L. 1996 *MNRAS* **279**, 1219.

WALBORN, N. R. 1991 In *Massive Stars in Starbursts* (eds. C. Leitherer, N. R. Walborn, T. M. Heckman, & C. A. Norman). STScI Symposium Ser. 5, p. 145. Cambridge University Press.

WALBORN, N. R. 2002 In *Hot Star Workshop III: The Earliest Phases of Massive Star Birth* (ed. P. A. Crowther). ASP Conf. Ser. 267, p. 111.

WALBORN, N. R., BARBÁ, R. H., BRANDNER, W., RUBIO, M., GREBEL, E. K., & PROBST, R. G. 1999 *AJ* **117**, 225.

WALBORN, N. R. & BLADES, J. C. 1997 *ApJS* **112**, 457.

WALBORN, N. R., HOWARTH, I. D., LENNON, D. J., MASSEY, P., OEY, M. S., MOFFAT, A. F. J., SKALKOWSKI, G., MORRELL, N. I., DRISSEN, L., & PARKER, J. W. 2002 *AJ* **123**, 2754.

WALBORN, N. R., MAÍZ-APELLÁNIZ, J., & BARBÁ, R. H. 2002 *AJ* **124**, 1601.

WALBORN, N. R. & PARKER, J. W. 1992 *ApJ* **399**, L87.

WANG, Q. D. 1999 *ApJ* **510**, L139.

WEAVER, R., MCCRAY, R., CASTOR, J., SHAPIRO, P., & MOORE, R. 1977 *ApJ* **218**, 377.

WHITMORE, B. C., ZHANG, Q., LEITHERER, C., FALL, S. M., SCHWEIZER, F., & MILLER, B. W. 1999 *AJ* **118**, 1551.

WILKINSON, M. I., HURLEY, J. R., MACKEY, A. D., GILMORE, G. F., & TOUT, C. A. 2003 *MNRAS* **343**, 1025.

WILSON, C. D. & SCOVILLE, N. 1992 *ApJ* **385**, 512.

WISEMAN, J. J. & HO, P. T. P. 1998 *ApJ* **502**, 676.

YANG, H., CHU, Y.-H., SKILLMAN, E. D., & TERLEVICH, R. 1996 *AJ* **112**, 146.

DE ZEEUW, P. T., HOOGERWERF, R., DE BRUIJNE, J. H. J., BROWNE, A. G. A., & BLAAUW, A. 1999 *AJ* **117**, 354.

Magellanic Cloud planetary nebulae as probes of stellar evolution and populations

By LETIZIA STANGHELLINI

Space Telescope Science Institute, 3700 San Martin Drive, Baltimore, MD, USA, and
European Space Agency

Planetary Nebulae (PNs) in the Magellanic Clouds offer the unique opportunity to study both the population and evolution of low- and intermediate-mass stars, in an environment that is free of the distance scale bias and the differential reddening that hinder the observations of the Galactic sample. The study of LMC and SMC PNs also offers the direct comparison of stellar populations with different metallicity. The relative proximity of the Magellanic Clouds allows detailed spectroscopic analysis of the PNs therein, while the *Hubble Space Telescope* (*HST*) is necessary to obtain their spatially-resolved images. In this paper we discuss the history and evolution of this relatively recent branch of stellar astrophysics by reviewing the pioneering studies, and the most recent ground- and space-based achievements. In particular, we present the results from our recent *HST* surveys, including the metallicity dependence of PN identification (and, ultimately, the metallicity dependence of PN counts in galaxies); the morphological analysis of Magellanic PNs, and the correlations between morphology and other nebular properties; the relations between morphology and progenitor mass and age; and the direct analysis of Magellanic central stars and their importance to stellar evolution. Our morphological results are broadly consistent with the predictions of stellar evolution if the progenitors of asymmetric PNs have on average larger masses than the progenitors of symmetric PNs, without any assumption or relation to binarity of the stellar progenitors.

1. Introduction

Planetary Nebulae (PNs) are the gaseous relics of the envelopes ejected by low- and intermediate-mass stars ($1 < M < 8\ M_\odot$) at the tip of the asymptotic giant branch (AGB), thus they are important probes of stellar evolution, stellar populations, and cosmic recycling. PNs have been observed in the Local Group (as well as in external galaxies), probing stellar evolution and populations in relation to their environment.

The details of the observations of Galactic PNs and their central stars (CSs) typically surpass the details of stellar and hydrodynamic models. Galactic PN studies are a necessary background toward the understanding the PN population in Local Group galaxies. Yet, the distance scale of Galactic PNs is uncertain to such a degree that the meaning of the comparison between observations and theory is hindered. By the same token, statistical studies of PN populations in the Galaxy suffer for the observational bias against the detection of Galactic disk PNs, and for the patchy interstellar extinction.

PNs in the Magellanic Clouds (LMC, SMC), hundreds of low-extinction planetaries at uniformly known distances, are a real bounty for the stellar evolution scientist. The composition gradient between the LMC, the SMC, and the Galaxy, afford the study of the effects of environment metallicity on PN evolution. The relative vicinity of the Clouds, and the spatial resolution that can be achieved with the *Hubble Space Telescope* (*HST*), allow the detection of PN morphology. Studying the PNs in the Magellanic Clouds is a perfect example of how the Local Group can be efficiently (and uniquely) used as an astrophysical laboratory. In this paper we review the history and the evolution of this field of study, with particular focus on the results from our recent *HST* Magellanic PN programs.

2. Pioneers in the field, and recent developments

PN studies in the Magellanic Clouds are relatively recent. The first detection and spectral identification of Magellanic PNs is due to Lindsay (1955). Studies of Magellanic PN samples became common, and their importance evident, in the early sixties (Aller 1961; Westerlund 1964). Ground based observations of Magellanic PNs suffer from the fact that the contributions of nebular and stellar radiation are superimposed. Attempts to measure the CS magnitudes have been hampered by the difficulties in separating stellar and nebular contributions (e.g., Webster 1969).

Observations with the *IUE*, combined with the optical spectra acquired from the ground, have allowed the abundance analysis of Magellanic PNs. Space observations in the UV range were used for the detection of the complete set of carbon lines at various ionization stages, and made the carbon abundance derivation much more reliable. The key results in abundance studies can be found, to name a few, in Peimbert (1984), Boroson & Liebert (1989), and Kaler & Jacoby (1991). Optical spectroscopy of large samples of Magellanic PNs have been carried out by Dopita and collaborators (Meatheringham & Dopita 1991ab; Vassiliadis, Dopita, Morgan, & Bell 1992). *IUE* observations were also used to measure the stellar luminosity beyond the Lyman limit for the CSs, giving an estimate of the total stellar luminosity, and approximate estimate to the mass (Aller et al. 1987). Several papers on Magellanic PN spectroscopy, abundances, and the connection of nebular and stellar evolution can be found in the proceedings of IAU Symposia on planetary nebulae (e.g., Westerlund 1968; Feast 1968; Webster 1978), while the most recent, complete review on Magellanic PNs is due to Barlow (1989).

Studies of the chemical content of Magellanic PNs have been active in the recent past as well. On the observational side, Leisy & Dennefeld (1996), and Costa, de Freitas Pacheco, & Idiart (2000), have estimated new chemical abundances for several Magellanic PNs from optical and UV observations, enriching the databases for studies on the dredge-up of post-AGB stars and on the ISM enrichment in galaxies. On the theoretical side, van den Hoek & Groenewegen (1997) calculated new chemical yields of the interstellar medium enrichment from synthetic evolution of intermediate-mass stars. With models from a wide range of initial masses and metallicities (including those appropriate for the Magellanic populations), van den Hoek & Groenewegen confirm that the yields of nitrogen and carbon change abruptly for M > 3 $M_\odot$, due to the hot-bottom burning effect.

To date, 277 LMC PNs (Leisy, Dennefeld, Alard, & Guibert 1997) and 55 SMC PNs (Meyssonnier & Azzoppardi 1993) are known. The total number of Magellanic PNs have more than doubled from the last count by Barlow (1989). In the last few years, several emission line surveys have been completed, or are near completion (e.g., UKST survey: Morgan 1998, Parker & Phillips 1998; UM/CTIO survey: Smith et al. 1996). Analysis of these surveys is essential for the future health of Magellanic PN research. We expect that the PN counts in the Clouds will increase significantly, improving the statistical significance of these studies. One important aspect of these surveys is the discovery of fainter PNs, that contributes to increasing the reliability of the faint end of the Magellanic PN luminosity function, and to enlarge the pool of known evolved PNs.

Related to the populations of Magellanic PNs is the 2MASS survey (Egan, Van Dyk, & Price 2001). The importance of this multi-wavelength infrared survey to LMC PNs is related to the spatial distribution of the different types of AGB stars. Egan et al. showed that low mass AGB stars occupy the whole of the LMC projected volume, while the higher mass, younger population, AGB stars populate preferentially the LMC bar. Chemical and morphological studies of large LMC PN samples should be compared to the

Program	Galaxy	PNs	Observing mode	Papers†
8271	LMC	29	G750M/G430M/50CCD	I, II
8663	SMC	29‡	G750M/G430M/50CCD	III, IV
9077	LMC	52	G750M/G430M/50CCD	V
9120	LMC	28	G140L/G230L	VI

† Papers V and VI are in preparation.
‡ Two PNs in this sample are misclassified H II regions.

TABLE 1. STIS observations of Magellanic PNs

AGB samples, to relate the PN populations in the LMC to their immediate evolutionary progenitors.

3. Early *Hubble Space Telescope* observations

Extended studies of Galactic PNs have shown that PN morphology is intimately related to the mass and evolution of their CSs, to their stellar progenitors, and to the nebular chemistry. In the case of the LMC and the SMC, morphological studies became possible with the use of the cameras on the *HST*. The *HST* has also the capability of spatially separate the image of the nebula and that of the CS, making direct stellar analysis possible.

The early narrow-band images of Magellanic PNs were obtained before the first *HST* servicing mission (i.e., before the installation of COSTAR on *HST*) with the Faint Object Camera (Blades et al. 1992). Other images by Blades et al. have later been published by Stanghellini et al. (1999), where the quality of the pre-COSTAR images was validated through their comparison with post-COSTAR images of the same objects. These papers have made available 15 Magellanic PN images usable for statistical and morphological studies, while another 15 LMC PNs have been observed with the Planetary Camera 1 (Dopita et al. 1996; Vassiliadis et al. 1998). Finally, an additional ten Wide Field and Planetary Camera 2 narrow-band images of LMC PNs are available in the Hubble Data Archive (program 6407).

4. Our Magellanic PN program

During the *HST* Cycle 8 we started a series of surveys aimed at obtain the size, morphology, and CS properties of all Magellanic PNs known to date. The *HST* was an obligatory choice, since the Magellanic PNs are typically half an arcsec across, thus they are generally not resolved with ground-based telescopes.

The medium-dispersion, slitless capability of STIS offers us a valuable opportunity to study the evolution and morphology of the Magellanic Cloud PNs and their CSs at once. We have applied this capability in several SNAPSHOT surveys, obtaining images in the light of up to seven of the most prominent low- and moderate-ionization optical, nebular emission lines. We also obtained direct continuum images to identify the correct CS (in spite of the crowded fields), and to measure the optical continuum emission.

In addition to the optical slitless spectra and broad band continuum images of the LMC and SMC PNs, we have acquired STIS UV spectra of 24 LMC PNs. In the cases where the CSs were hard to find in our STIS broad band images, we have also acquired WFPC2 Strömgren images (Program 8702). Throughout our investigations we have used the limited data available in the literature (see Stanghellini et al. 1999). In Table 1 we

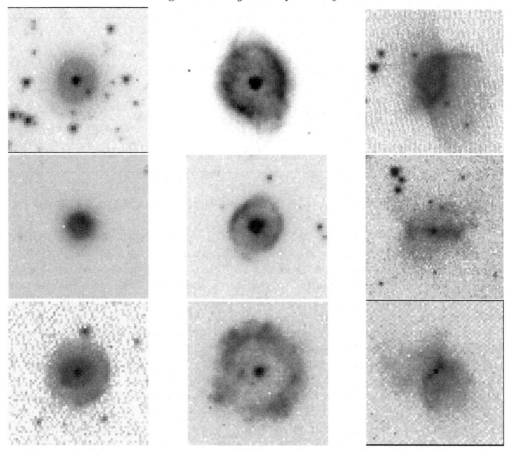

FIGURE 1. Magellanic PNs. *Left panel*: round PNs LMCJ 33, SMC−SP 34, and LMC−MG 40; *Center panel*: elliptical PNs LMC−SMP 101, SMC−MG 8, and SMC−MG 13; *Right panel*: bipolar PNs LMC−SMP 91, SMC−MA 1682, and LMC−MG 16. All thumbnails are 9 arcsec^2 sections of the STIS broad-band images.

list the major *HST* data sets of our Magellanic PN project. The actual data can be easily retrieved from the dedicated MAST Archival page *http://archive.stsci.edu/hst/mcpn*. In addition to the *HST* data, we have made extensive use of the spectra we acquired from the ground (papers in preparation), and available in the literature.

In Figure 1 we show a sampler of the most common morphological types of Magellanic PNs, from the broad-band images in our surveys. These PNs are more than 50 times farther away than the typical galactic PNs, yet the major morphological features are easily recognized, as are the location of their CSs, when visible.

In the following sections we describe some of the most important results we obtained from the analysis of our PN samples.

4.1. *Misclassified planetary nebulae*

Extra-Galactic PNs in general, and Magellanic PNs in particular, are typically discovered with Hα or [O III] λ5007 surveys, with the on- and off-band observations technique. The PN nature of the [O III] λ5007-bright object can be generally confirmed by means of spectroscopic follow-up. Marginal misclassification is possible even after spectroscopic

Morphological type	% LMC	% SMC
Round (R)	29	35
Elliptical (E)	17	29
R+E (symm.)	46	64
Bipolar (B)	34	6
Bipolar core (BC)	17	24
B+BC (asymm.)	51	30
Point-symmetric (P)	3	6

TABLE 2. Morphological distribution

analysis, and only the spatially resolved observations can ultimately confirm the PN nature.

Of the ∼230 known LMC PNs, we have observed about 80 with *HST*, and another 30 had been observed earlier by Dopita and Blades. In the SMC, we acquired *HST* observations of 29 targets of the 55 known SMC PNs. The spatially-resolved sample consists of the [O III] λ5007 bright half of the known Magellanic PNs, given that we had selected the brightest PNs in order to fit the observations within one *HST* orbit. Thus our sample is representative of the (bright) Magellanic PNs.

It is then extremely interesting to note that we confirmed 100% of the LMC targets to be PNs, while 10% of the SMC targets were, instead, ultra-compact H II regions around very young star clusters (Stanghellini, Villaver, Shaw, & Mutchler 2003b). This result bears on the validity of the PN luminosity function (PNLF) as a secondary distance scale indicator. In fact, the extra-Galactic PNLF is populated by the [O III] bright PNs of each studied population, and our result seems to indicate that such populations may be contaminated by H II regions in different ways depending on the galaxian metallicity.

4.2. *PN morphology*

The morphology of Galactic PNs has been studied rather thoroughly in the past decade, and it has been found that the morphological types correlate with the PN progenitor's evolutionary history and the stellar mass. There is strong evidence that asymmetric (e.g., bipolar) PNs are the progeny of the massive AGB progenitors (3–8 $M_\odot$). Bipolar PNs are nitrogen enriched and carbon poor (Stanghellini, Villaver, Manchado, & Guerrero 2002b). The analysis of the morphological types and their distribution in a PN population is then very useful to infer the age and history of a given stellar sample.

Galactic PNs have been classified as round, elliptical, bipolar (and quadrupolar), bipolar core (those bipolar PNs whose lobes are too faint to be detected, but whose equatorial ring is very evident), and point-symmetric. The majority of Galactic PNs are elliptical, but the actual number of bipolars could be underestimated, given that they typically lie in the Galactic plane (i.e., they may suffer high reddening). PNs in the Clouds, when spatially resolved, show the same admixture of morphological types than the Galactic PNs (see Fig. 1). While we do not attempt a statistical comparison of the MC and Galactic PN morphological types, given the selection effects that hamper Galactic PNs, we can meaningfully compare the LMC and SMC samples. Both samples suffer from low field extinction, and they have been preselected in more or less the same way.

The results of the morphological distribution in the Clouds are summarized in Table 2. Together with the percentage in each morphological class, we give the total of symmetric (round and elliptical) and asymmetric (bipolar and bipolar core) PNs. One striking difference between the two distributions is that the fraction of bipolar PNs in the LMC is

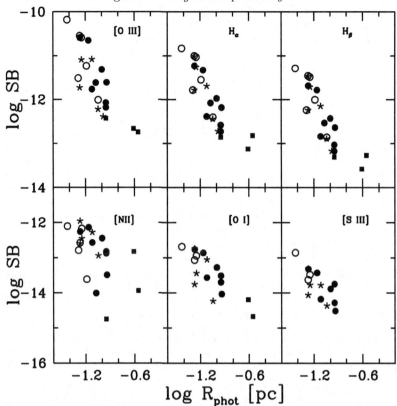

FIGURE 2. Surface brightness decline from the *HST* multiwavelength images of the LMC and SMC PNs (adapted from Stanghellini et al. 2002a). Emission lines in which the SB is derived are indicated in the panels (see also text). Symbols indicate morphological types: round (*open circles*), elliptical (*asterisks*), bipolar core (*filled circles*), and bipolar and quadrupolar (*filled squares*). The photometric radii are measured from the [O III] $\lambda5007$ images, where available.

almost six times that of the SMC. Bipolar PNs are easily recognized, thus this is a sound result. If we add to the asymmetric PN count the bipolar core PNs, we obtain that half of the LMC PNs are asymmetric, while only a third of the SMC PNs are asymmetric. Observational biases play in the same way for the two samples.

What insight can we get from the morphological results? First of all, it is clear that the set of processes involved in the formation of the different PN shapes are at work *in all galaxies where morphology has been studied*. Second, the SMC environment may not favor the onset of bipolarity in PNs. Otherwise, the different morphological statistics may indicate different populations of stellar progenitors in the two Clouds. While it seems reasonable to conclude that a low metallicity environment is unfavorable to bipolar evolution, the exact causes have not been studied yet. A detailed study of metallicity and mass loss may clarify this point. On the other hand, the different morphological statistics may simply be related to a lower average stellar mass of the PN progenitors in the SMC. If this was the case, we should also observe lower CS masses in the SMC PNs than in the LMC PNs. Our preliminary measurements seem to indicate that this is also the case (Villaver, Stanghellini, & Shaw 2003).

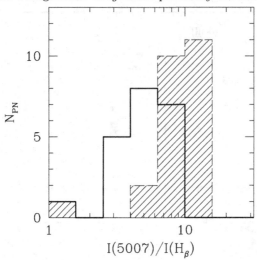

FIGURE 3. Distribution of the [O III] $\lambda5007$ over Hβ intensity ratios in the SMC (thick histogram) and LMC (shaded histogram) PNs, from Stanghellini et al. 2003a.

4.3. *Surface brightness evolution*

The surface brightness of LMC and SMC PNs correlates with the photometric radius, as illustrated in Figure 2. In the figure we plot the logarithmic surface brightness, derived from the observed total line fluxes and the apparent radii, versus the photometric radii of the nebulae, measured as the distance from the CS (or the geometrical center of the nebulae) where the enclosed flux is 85% of the total nebular flux (see Stanghellini et al. 1999). The six panels illustrate the surface brightness-radius relation for the major six emission lines: [O III] $\lambda5007$, Hα, Hβ, [N II] $\lambda6584$, [O I] $\lambda6300$, and [S III] $\lambda6312$.

The surface brightness-photometric radius relation is tight in all spectral lines, with the exception of the [N II] emission line, where a larger spread is present, particularly for bipolar PNs. A possible factor is the larger range of nitrogen abundances in bipolar and BC PNs. The surface brightness-photometric radius relations hold only in the cases in which the nebular density N_e is smaller than the critical density, N_{crit} (the density at which the collisional de-excitation rate balances the radiative transition rate).

A good eye-fit to the surface brightness-photometric radius relation is SB $\propto R_{phot}^{-3}$. This relation can be reproduced via hydrodynamic modeling of evolving PNs and their CSs (Villaver & Stanghellini, in preparation). The surface brightness-photometric radius relation in the light of Hα (or Hβ) is tight enough that it can be used to set the distance scale for Galactic PNs with intrinsic uncertainties of the order of 30% or less (Stanghellini et al. in preparation), while the current calibration of the Galactic PN distance scales carry errors of the order of 50% or more.

In Figure 2 we note that the symmetric (round and elliptical) PNs tend to cluster at high surface brightness and low radii, while the asymmetric PNs occupy the lower right parts of the diagrams. This separation can be interpreted with a slower evolutionary rate for the symmetric PNs, which agrees with the idea that symmetric PNs derive from lower mass progenitors (Stanghellini, Corradi, & Schwarz 1993; Stanghellini et al. 2002b).

4.4. *The [O III]/Hβ distribution*

In Figure 3 we plot a histogram of the ratio of (reddening-corrected) fluxes of the [O III] $\lambda5007$ and Hβ lines for the PNs of the SMC and the LMC. The median of the SMC distribution is a factor of two lower than for the corresponding LMC distribution

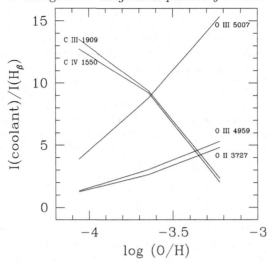

FIGURE 4. Intensity ratios of the major PN coolants over Hβ, versus oxygen abundances (from Stanghellini et al. 2003a).

($<$[O III]/H$\beta >_{\mathrm{SMC}}= 5.7 \pm 2.5$ and $<$[O III]/H$\beta >_{\mathrm{LMC}}= 9.4 \pm 3.1$). This result is free of object selection biases since both sets of targets were chosen in much the same way.

The [O III]/Hβ emissivity ratio is physically scaled linearly with the O/H abundance and the fractional ionization of O^{++}. Also it depends exponentially on the local electron excitation temperature, T$_e$(O^{++}) since electron collisions on the high-energy tail of the free energy distribution excite the transition. Of course, T$_e$(O^{++}) depends on O/H and O^{++}/O as well. So interpreting the differences between the [O III]/Hβ ratios of the SMC and the LMC is best done using ionization models.

Our Cloudy (Ferland 1996) models explore the major line emission in a set of Galactic, LMC, and SMC models with same gas density (1000 cm^{-3}) and different metallicities, adequately chosen to represent the *average* nebula in each studied galaxy, as explained in Stanghellini et al. (2003a). The stellar ionizing spectrum is assumed to be a blackbody with temperatures and luminosities from the H-burning evolutionary tracks for the appropriate galaxian population by Vassiliadis and Wood (1994).

In Figure 4 we show the line intensity relative to Hβ for the major coolants in the SMC, LMC, and Galactic PNs, versus the oxygen abundance, as derived from our simplified Cloudy models. While we have calculated the evolution of these intensity ratios following the evolution of the CS from the early post-AGB phase to the white dwarf stage, we only plot here the flux ratios corresponding to the models with the highest temperature, for each PN composition. In general, our target selection tends to favor targets with hottest CSs: T$_{\mathrm{eff}} \geqslant 50,000$ K both in the SMC and in the LMC, thus the set of high-temperature models is the most adequate to reproduce the observations for LMC and SMC PNs.

The cooling processes that determine T$_e$(O^{++}) in the SMC, LMC and Galactic PNs are noteworthy. In the Galaxy the primary coolants of PNs with hot CSs are the optical forbidden lines of [O III] λ5007 and other lines of O$^+$ and O^{++}. However, in environments in which O/H is as low as in the SMC, the primary coolants may become ultraviolet intercombination lines of C$^+$ and C^{++}. The simple models described here seem to reproduce very well the optical flux ratios of PNs in the Magellanic Clouds. It will be interesting to confirm these predictions with future UV observations.

4.5. *Chemical analysis of LMC PNs*

Studies of Galactic PNs have shown that the bipolar and, more in general, asymmetric PNs present lower carbon abundance (and higher nitrogen abundance) than their symmetric (round or elliptical) counterparts (Peimbert 1978; Stanghellini et al. 2002b). The reason for the carbon underabundance is to be found in late phases of AGB evolution, when the nuclear burning extends to the bottom of the stellar envelope (HBB, see Iben & Renzini 1983). The results seem to imply that asymmetric PNs are the progeny of the more massive stars (M > 3 $M_\odot$) in the PN progenitor range, since only these stars undergo the HBB process (for the yields of the LMC AGB stars see Van den Hoek & Groenevegen 1997). A statistically sound result is not possible in the Galactic environment, since the asymmetric PNs tend to be located near the Galactic plane, and their observation may be severely affected by the Galactic plane reddening. With our LMC observations we are in the position to test these findings by circumventing the selection effects. We found that carbon abundance is higher than 1×10^{-4} for all round and elliptical PNs, while it can be as low as $\approx 3 \times 10^{-6}$ for the asymmetric PNs (Stanghellini, Shaw, Balick, & Blades 2000, all abundances are by number, versus hydrogen), in agreement with the trends that were expected form the Galactic results.

We have also studied the abundance-versus-morphology relations for the elements whose abundance do not dramatically change during the evolution of the PN progenitors, such as sulfur, argon, oxygen, and neon (the so-called *alpha elements*). In Figure 5 we show the relations between neon and, from the top panel to the lower panel, oxygen, argon, and sulfur abundances. The LMC PNs plotted are coded for morphological type, as described in the label. The large circled symbols represent the *average* LMC abundance of the given elements in the H II regions. We see that there is morphological discrimination between the high and low alpha-element abundance PNs.

4.6. *Central stars of LMC PNs*

Our *HST* programs allowed, for the first time, the direct observation of CSs of extra-Galactic PNs. While the observations in slitless mode were specially crafted to detect the nebulae in the most prominent emission lines, the broad-band imagery showed the CSs in approximately 60% of the targets.

We have measured magnitudes via aperture photometry of the CSs in the LMC (Villaver, Stanghellini, & Shaw 2003) and SMC PNs. By means of the Zanstra analysis, and using the most recent, uniformly selected He II λ4686 fluxes, we have determined the (H I and He II) Zanstra stellar temperatures. Finally, by placing the stars on the $\log L$–$\log T_{\rm eff}$ plane, and comparing their loci with the theoretical evolutionary post-AGB tracks of Vassiliadis & Wood (1994), we have determined the stellar mass.

The median mass of LMC and SMC CSs, respectively 0.63 and 0.59 $M_\odot$, do not differ very much, indicating that the possible variance with metallicity of the factors involved (initial mass function, star formation rate and history, and mass-loss rate) probably average each other out (Villaver, Stanghellini, & Shaw 2003).

Approximately 25 CSs, whose masses are determined with confidence, have been analyzed so far. We will examine a larger sample of CSs in the Clouds to deepen the analysis of possible correlations between the core mass and the morphology of PNs in the future.

Knowing the position of the CSs on the $\log L$–$\log T_{\rm eff}$ diagram is essential, not only for the mass determination, but also to constrain the transition time (the time interval between the quenching of the envelope ejection at the AGB tip and the PN illumination; see Stanghellini & Renzini 2000). Stellar evolutionary models for post-AGB stars start with *ad hoc* models that imply strong (or semi-empirical) assumptions on the mass loss rate. From an observational viewpoint, the timing of post-AGB evolution is derived from

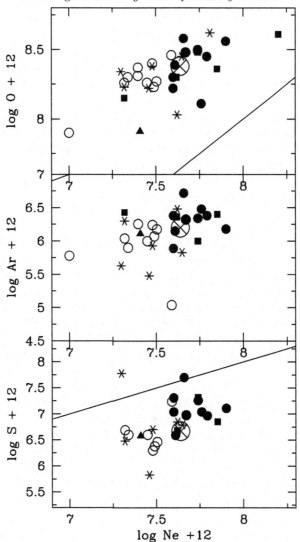

FIGURE 5. O, Ar, and S vs. Ne abundances of LMC PNs for morphological types round (*open circles*), elliptical (*asterisks*), quadrupolar (*filled triangles*), bipolar core (*filled circles*), and bipolar (*filled squares*). The large crossed circle represents the average for LMC H II regions (adapted from Stanghellini et al. 2000).

the dynamic time of the PN. For Galactic PNs, the comparison of the dynamic and evolutionary time-scales has the double offset of the (undetermined) transition time and the distance bias, while for LMC and SMC PNs the difference between the observed and theoretical evolutionary time allows the estimate of the transition time.

The transition time is relevant in post-AGB population synthesis, for example, in the studies of the PNLF as a secondary distance-scale indicator, and also in studies concerning the UV contribution from post-AGB stars in galaxies. Further constraints on the effective temperature of the CSs derived from the UV continuum fitting from our Cycle 10 UV spectra, will strengthen the results on stellar evolution.

5. Summary

Magellanic PNs are ideal probes to study stellar evolution and populations of low- and intermediate-mass stars. The use of the *HST* is fundamental for determining the PN shapes, the radii, and also to detect the CSs. Furthermore, only with the use of spatially resolved images can one identify the LMC and SMC PNs unambiguously, without the accidental inclusion of compact H II regions in the PN samples.

We have presented the results derived from our *HST* programs against the background of the important work that has been done in LMC and SMC PNs previously.

We found that PNs have the same morphological types in the Galaxy, the LMC, and the SMC. We also found that the distribution of the morphological types is noticeably different in the SMC and the LMC, and that the LMC seems to be populated by PNs whose progenitors are, on average, more massive.

We analyzed morphology and chemical composition of a sample of LMC PNs, and found that asymmetric (bipolar and bipolar core) PNs have higher Ne, Ar, and S abundance than symmetric (round and elliptical) PNs in the LMC, a confirmation that they trace more recent stellar populations.

An empirical relation between the nebular radii and the surface brightness is found to hold in both SMC and LMC PNs, independent of morphological type. The relation, once calibrated, will be used to determine the distance scale for Galactic extended PNs.

The PN cooling is affected by metallicity, and it seems that the [O III] λ5007 emission is not always the ideal line to detect bright PNs in all Galaxies, since the strongest cooling lines in very low metallicity PNs seem to be the UV [C III] (and [C IV]) semi-forbidden emission.

The observed Magellanic CSs that we discussed here constitute the first sizable CS sample beyond the Milky Way that has been directly observed. While we found only marginal differences between median masses of the LMC and the SMC CSs, we need to enlarge the sample of CSs whose masses can be reliably measured, given the importance of knowing initial- to final-mass relation in different metallicity environments.

The work presented in §4 has been developed in collaboration with Dick Shaw, Eva Villaver, Bruce Balick, and Chris Blades.

REFERENCES

ALLER, L. H. 1961 *AJ* **66**, 37.

ALLER, L. H., KEYES, C. D., MARAN, S. P., GULL, T. R., MICHALITSIANOS, A. G., & STECHER, T. P. 1987 *ApJ* **320**, 159.

BARLOW, M. J. 1989, in *Planetary Nebulae* (Ed. S. Torres-Peimbert), IAU Symp. 131, p. 319.

BLADES, J. C., ET AL. 1992 *ApJ* **398**, L41.

BOROSON, T. A. & LIEBERT, J. 1989 *ApJ* **339**, 844.

COSTA, R. D. D., DE FREITAS PACHECO, J. A., & IDIART, T. P. 2000 *A&AS* **145**, 467.

DOPITA, M. A., ET AL. 1996 *ApJ* **460**, 320.

EGAN, M. P., VAN DYK, S. D., & PRICE, S. D. 2001 *AJ* **122**, 1844.

FEAST, M. W. 1968, in *Planetary Nebulae* (Eds. D. E. Osterbrock & C. R. O'Dell), IAU Symp. 34, p. 34. Reidel.

IBEN, I., JR. & RENZINI, A. 1983 *ARA&A* **21**, 271.

KALER, J. B. & JACOBY, G. H. 1991 *ApJ* **382**, 134.

LEISY, P. & DENNEFELD, M. 1996 *A&AS* **116**, 95.

LEISY, P., DENNEFELD, M., ALARD, C., & GUIBERT, J. 1997 *A&AS* **121**, 407.

LINDSAY, E. M. 1955 *MNRAS* **115**, 241.

MEATHERINGHAM, S. J. & DOPITA, M. A. 1991a *ApJS* **75**, 407.

MEATHERINGHAM, S. J. & DOPITA, M. A. 1991b *ApJS* **76**, 1085.

FERLAND, G. J. 1996 *Hazy, a Brief Introduction to Cloudy*, University of Kentucky Dept. of Physics and Astronomy Internal Report.

MEYSSONNIER, N. & AZZOPARDI, M. 1993 *A&AS* **102**, 451.

MORGAN, D. H. 1998 *Publications of the Astronomical Society of Australia* **15**, 123.

MURPHY, M. T. & BESSELL, M. S. 2000 *MNRAS* **311**, 741.

PARKER, Q. A. & PHILLIPPS, S. 1998 *Publications of the Astronomical Society of Australia* **15**, 28.

PEIMBERT, M. 1978, in *Planetary Nebulae* (Ed. Y. Terzian). IAU Symp. 76, p. 215. Reidel.

PEIMBERT, M. 1984, in *Structure and Evolution of the Magellanic Clouds*. IAU Symp. 108, p. 363. Reidel.

PEIMBERT, M. 1997, in *Planetary Nebulae* (Eds. H. Habing & H. Lamers). IAU Symp. 180, p. 175. Kluwer.

SHAW, R. A., STANGHELLINI, L., MUTCHLER, M., BALICK, B., & BLADES, J. C. 2001 *ApJ* **548**, 727 (Paper I).

SMITH, R. C., ET AL. 1996 *BAAS* **28**, 900.

STANGHELLINI, L., BLADES, J. C., OSMER, S. J., BARLOW, M. J., & LIU, X.-W. 1999 *ApJ* **510**, 687.

STANGHELLINI, L., CORRADI, R. L. M., & SCHWARZ, H. E. 1993 *A&A* **279**, 521.

STANGHELLINI, L., SHAW, R. A., BALICK, B., & BLADES, J. C. 2000 *ApJ* **534**, L167.

STANGHELLINI, L., SHAW, R. A., BALICK, B., MUTCHLER, M., BLADES, J. C., & VILLAVER, E. 2003a *ApJ* **596**, 997 (Paper III).

STANGHELLINI, L., SHAW, R. A., MUTCHLER, M., PALEN, S., BALICK, B., & BLADES, J. C. 2002a *ApJ* **575**, 178 (Paper II).

STANGHELLINI, L., VILLAVER, E., MANCHADO, A., & GUERRERO, M. A. 2002b *ApJ* **576**, 285.

STANGHELLINI, L., VILLAVER, E., SHAW, R. A., & MUTCHLER, M. 2003b *ApJ* **598**, 1000 (Paper IV).

VAN DEN HOEK, L. B. & GROENEWEGEN, M. A. T. 1997 *A&AS* **123**, 305.

VASSILIADIS, E., DOPITA, M. A., MORGAN, D. H., & BELL, J. F. 1992 *ApJS* **83**, 87.

VASSILIADIS, E. & WOOD, P. R. 1994 *ApJS* **92**, 125.

VASSILIADIS, E., ET AL. 1998 *ApJ* **503**, 253.

VILLAVER, E., STANGHELLINI, L., & SHAW, R. A. 2003 *ApJ* **597**, 298.

WEBSTER, B. L. 1969 *MNRAS* **143**, 113.

WEBSTER, B. L. 1978, in *Planetary Nebulae*. IAU Symp. 76, p. 11. Reidel.

WESTERLUND, B. E. 1964, in *The Galaxy and the Magellanic Clouds* (Ed. F. J. Kerr). IAU Symp. 20, p. 316. Australian Academy of Science.

WESTERLUND, B. E. 1968, in *Planetary Nebulae* (Eds. D. E. Osterbrock & C. R. O'Dell). IAU Symp. 34, p. 23. Reidel.

The old globular clusters: Or, life among the ruins

By WILLIAM E. HARRIS

Department of Physics and Astronomy, McMaster University, Hamilton ON L8S 4M1 Canada

Recent progress on the astrophysics of globular clusters is discussed. Highlights are (a) developments in color-magnitude survey work, (b) globular cluster structures and the "fundamental plane," and (c) the relation between globular clusters and the halo field stars in the same host galaxy.

---

1. Color-Magnitude studies: The beginning and end of an era

Above almost all other types of astronomical systems, globular clusters offer the chance to take a broad historical perspective. Much of the history of astrophysics in the twentieth century—stellar structure, stellar evolution, the distance scale, galactic structure and evolution, stellar populations, variable stars, high-energy sources—was driven by the need to understand the complex array of phenomena taking place inside these dense stellar systems.

No review of this kind should fail to mention the continuing efforts to understand the stellar content of these elegant systems through color-magnitude studies (CMDs), which are penetrating to ever-greater detail and depth. The very first such studies (see, for example, Shapley & Davis 1920) barely showed the red-giant stars and brightest horizontal-branch stars for the nearest clusters. This past year, a watershed in this kind of basic color-magnitude survey work was reached with the publication of the monumental survey project of Piotto et al. (2002), who used the WPFC2 camera on board *HST* to obtain (B, V) CMDs for 74 globular clusters, very nearly half of the entire Milky Way globular cluster population. The objects range from nearby, high-latitude clusters with beautifully precise, classic CMD sequences, down to sparse objects deeply embedded in the heavy field contamination and differential reddening of the Galactic bulge. At this point, all but two or three of the 150 known Milky Way clusters (Harris 1996) have published CMDs, and it can reasonably be claimed that the survey era is finally over, almost a century after it began with a long-vanished and entirely different generation of telescopes and detectors.

The photometric work on globular clusters has, in the meantime, shifted its ground to studies of other kinds such as the delineation of the faintest parts of the CMD—the lower ends of the main sequence and white dwarf sequence (e.g., Richer et al. 2002a,b; Bedin et al. 2001). Here again, the depth and high resolution of *HST* have enabled major breakthroughs in long-standing problems, with new confidence that the ultimate endpoints of these faintest of stellar sequences have (just barely) been reached. There is now remarkable agreement in age calibrations between the white-dwarf sequence cooling times and the main-sequence turnoff ages, and with the WMAP sharpening of the age of the universe, for the first time we can work forward with some confidence that the first round of globular cluster formation took place well within a Gigayear after the Big Bang. Work on the fine structure of the horizontal branch continues, though with fully plausible interpretations for the various gaps and inhomogeneous distributions of stars still not within our grasp (Moehler 2001; Soker et al. 2001; Baev et al. 2001; Brown et al. 2001; Cassisi et al. 2003). And, the true complexity of the processes within the cores of clusters is still being elucidated. In these dense and highly evolved regions we are

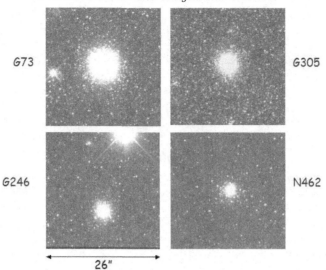

FIGURE 1. Images of a selection of globular clusters in M31, from Harris et al. (2003). Shown are 400-second exposures with the STIS/50CCD camera (approximately like V-band); each field is 26″ across.

seeing the results of interplay among stellar dynamics, stellar evolution, and collisions and hydrodynamics, that are unique among stellar systems (e.g., Sills et al. 2002).

2. Globular cluster structures: Steps toward universality?

One of the most striking features of globular clusters—the essential thing, in fact, that makes them look so classically simple—is that their large-scale structural properties inhabit only a narrow range of parameter space called the fundamental plane (Djorgovski 1995; McLaughlin 2000). Expressing this narrow range of properties in a physically simple way takes various forms: in McLaughlin's version, the binding energy E_b of the entire cluster is

$$E_b = G \left(\frac{4\pi}{9} \right)^2 \left(\frac{M}{L} \right)^2 \frac{L^2}{r_c} \frac{\mathcal{E}(c)}{\mathcal{L}(c)^2} \quad , \tag{2.1}$$

where $\mathcal{E}, \mathcal{L}$ are dimensionless functions of the King-model concentration parameter $c \equiv \log(r_t/r_c)$. For the old-halo clusters with Galactocentric distances $r_{gc} > 8$ kpc, which are the ones whose structures have been least modified by dynamical evolution from the Galactic tidal field, E_b is almost exactly proportional to cluster luminosity L^2, which indicates that the *combination* of all the other factors in the equation above is nearly independent of cluster mass. Even for clusters closer in to the Galactic center, the same correlation is clearly present, though with noticeably increased scatter (see McLaughlin's discussion) which is probably due to advanced dynamical evolution (the core-collapsed clusters in particular fall away from the standard FP line).

There is much evidence on other grounds that the old globular cluster populations, in broad terms, have very similar properties in most (or perhaps even all) other large galaxies (Harris 2001). A natural question is whether or not the structural fundamental plane is the same elsewhere, or indeed whether or not it is driven mostly by formation or by later evolution. Here is where the other Local Group galaxies can play a new role: the high spatial resolution and imaging tools of *HST* allow us to obtain high-quality profile

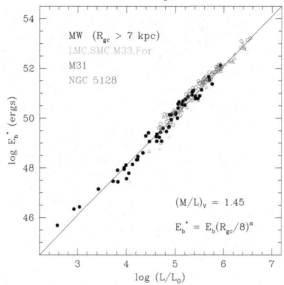

FIGURE 2. Structural fundamental plane (Binding energy vs. luminosity) for old globular clusters in seven galaxies. The solid line represents $E_b \sim L^2$. *Solid dots* are Milky Way clusters: *open circles* are M31; *crosses* are the dwarf galaxies LMC, SMC, M33, and Fornax; and *triangles* are NGC 5128.

measurements of the globular clusters in a wide variety of galaxies, and thus to directly address the hypothesis that the fundamental plane is universal.

New *HST*-based imaging results for these systems have now been obtained and allow us to extend the work within the Milky Way in a major way. New and very important observational material includes the clusters in the Magellanic Clouds, Fornax, and Sagittarius (Mackey & Gilmore 2003a,b,c); M33 (Larsen et al. 2002); M31 (Meylan et al. 2001; Barmby et al. 2002; Harris et al. 2003), and NGC 5128 (G. Harris et al. 1998; Holland et al. 1999; W. Harris et al. 2002).† These studies show that the classic King-Michie cluster models (lowered Maxwellian stellar velocity distributions) fit the old clusters in these other galaxies just as well as in the Milky Way. From the King-model fits, the binding energy E_b can be calculated readily (see the prescriptions in McLaughlin 2000 and Harris et al. 2002) and their fundamental-plane distribution (E_b vs. L) plotted.

Some preliminary results are shown in Figure 2. We find that across *four orders of magnitude in cluster mass*, the FP relation is accurately described by $E_b \sim L^2$, with all seven galaxies defining very much the same line. The equation matching the entire set of data is

$$\log E_b^\star = 40.10 + 2.00 \log(L/L_\odot) \quad , \tag{2.2}$$

where $E_b^\star$ is the binding energy normalized to a fiducial 8-kpc distance from the galactic center (to take account of the predictable increase in cluster scale size with increasing R_{gc}). The mean scatter around this relation is less than 0.3 dex, or a factor of two. It should be noted that to calculate E_b, we have to assume a mass-to-light ratio. To do this,

---

† Why is NGC 5128 included in a discussion on the Local Group? The reason is simply that NGC 5128, the central giant galaxy in the Centaurus group 4 Megaparsecs distant, is by far the closest accessible giant elliptical, and this is the one type of host galaxy not contained within the Local Group. In other words, to understand the cluster characteristics in the full range of galaxy environments we simply have to stretch our boundaries a bit. For the present purposes then, we can consider NGC 5128 as an "honorary member" of the Local Group.

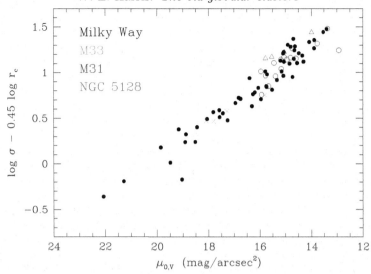

FIGURE 3. Globular cluster fundamental plane in the version used by Djorgovski (1995): the log of $\sigma_v r^{-0.45}$ is plotted against central surface brightness μ_0, V. Symbol types are as in the previous figure.

we have simply used $(M/L)_V = 1.45$, the McLaughlin (2000) mean value for the Milky Way clusters. This assumption could also be removed if we could use measured internal velocity dispersions σ_v to calculate (M/L) explicitly, but the required velocity data are much harder to obtain and are not yet available for more than a relatively small sample of objects. The Djorgovski (1995) formulation of the FP (in one of its versions, a plot of $\sigma_v \cdot r_c^{-0.45}$ versus central surface brightness μ_0) is shown in Figure 3 for those objects with known σs. Again, the data from different galaxies fall within the general trend defined by the Milky Way. In short, this material provides strong new evidence that we are looking at very much the same type of star cluster in all these host galaxies, and that the old globular clusters do indeed represent a common thread of early star formation.

McLaughlin has shown that the various FP formulations in his paper and in Djorgovski's are equivalent for a certain (M/L). A still simpler way to look at the underlying nature of the FP can, however, be gleaned if we rewrite the binding energy equation in terms of the *half-mass radius* r_h, which is nearly immune to dynamical evolution and thus is an excellent representation of the cluster 'scale size.' This turns out to be

$$E_b^\star = G \left(\frac{4\pi}{9} \right)^2 \left(\frac{M}{L} \right)^2 \frac{L^2}{r_h^\star} \left(\frac{\mathcal{R}(c)\mathcal{E}(c)}{\mathcal{L}(c)^2} \right) \quad , \tag{2.3}$$

where $r_h^\star$ is the half-mass radius normalized to a fiducial Galactocentric distance as described above. The last quantity in brackets, $(\mathcal{R}\mathcal{E}/\mathcal{L}^2)$, turns out to be constant to within about 10% over the range of c-values for real clusters. Then, if we take the numerical equation for Figure 2 as given above, we find that this reduces to a single scaling relation for the half-mass radius,

$$r_h^\star \simeq 2.1 \text{ pc} \times (M/L)_V^2 \quad . \tag{2.4}$$

In other words, the half-mass radius (normalized to a given Galactocentric location) plays the role of a *universal constant* for the structures of all globular clusters, insofar as the mass-to-light ratio is also similar from cluster to cluster. If $(M/L) \simeq 1.5$ for real clusters,

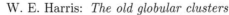

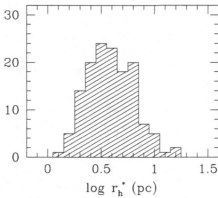

FIGURE 4. Histogram of half-mass radii for Milky Way globular clusters, normalized to $R_{gc} = 8$ kpc. The standard deviation of the distribution is 0.23 dex and the mean is 5 pc.

then we have $r_h \simeq 5$ pc, which closely matches the true scale sizes of the known clusters, with an intrinsic scatter of less than a factor of two (Figure 4).

An elegant way to recast what the FP is telling us is that the old globular clusters have a near-universal characteristic scale size and (M/L) which is independent of cluster mass, metallicity, and (apparently) host galaxy type. These two properties together appear to have been built into their structures at a very early stage, and are the underlying source of the other FP representations.

3. Metallicity distributions

The metallicity distribution (MDF) of the globular clusters that we see today traces their enrichment history during early galaxy evolution. Many recent studies show repeatedly that the globular cluster MDF has a distinctly bimodal form with very roughly equal numbers of metal-poor ([Fe/H] < -1) and metal-rich ([Fe/H] > -1) clusters. These MDFs have usually been taken to represent some kind of two-stage history of hierarchical galaxy formation or late major mergers, or perhaps the stochastic outcome of dissipationless merging of large and small progenitor galaxies (Larsen et al. 2001; Ashman & Zepf 1992; Forbes et al. 1997; Côté et al. 2002; Beasley et al. 2002, among others). Each of these scenarios has its interpretive problems, however.

A new observational piece of the puzzle is now our ability to compare the MDF of the globular cluster system with the MDF of the halo stars *in the same parent galaxy*. The galaxies in the Local Group and other very nearby groups—such as NGC 5128—have played the crucial initial roles in such comparisons, and include work in M31 (Rich et al. 1996; Durrell et al. 2001; Reitzel & Guhathakurta 2002), the LMC (Cole et al. 2000; Larsen et al. 2000; Harris & Harris 2001), M33 (Sarajedini et al. 2003; Brooks et al. 2003); and NGC 5128 (Harris & Harris 2002).

The key result from these comparisons is that the MDFs of the halo field-star populations are in most cases very different from the globular cluster MDFs: they are weighted much more strongly to the metal-rich end, and bimodality does not obviously show up. That is, the oldest field stars in many types of galaxies are routinely found to be moderately metal-rich, with very small relative numbers of metal-poor stars. *There are are large numbers of metal-poor stars where there are very small numbers of field stars.* It is possible to interpret the halo and bulge MDFs within the context of relatively uncomplicated chemical evolution models, either simple closed-box ones (M31 or M33; see

Durrell et al. 2001 and Brooks et al. 2003) or accreting-box models where star formation goes on simultaneously with a modest infall rate of pristine, primordial gas (NGC 5128 and the LMC; see Harris & Harris 2002). A hierarchical-merging scheme where a large population of primordial gas clouds merges successively into a final large galaxy can also produce MDFs matching the observations (Beasley et al. 2003).

These new studies force us to confront the possibility that the formation story line for globular clusters—which in the present-day universe make up less than one percent of the old stellar population of a big galaxy—is *not* representative of the entire galaxy's early history. That is, the bimodality of the globular cluster MDF and the very large number of metal-poor objects (compared with the field stars which, by definition, make up the bulk of the galaxy) point strongly towards two major and distinct cluster formation epochs rather than a single continuous formation sequence.

Within the context of a hierarchical-merging scheme, it has been suggested that the metal-poor clusters formed quite early in the sequence but that their formation was then truncated—perhaps by the rapid gas loss driven by the first rounds of star formation from the small potential wells of the first protogalactic clouds (e.g., Beasley et al. 2002; Harris & Harris 2002). In the formation of a giant galaxy, there would then have been a later round of major star formation which built the bulk of the galaxy. In this scenario, the metal-*rich* clusters and most of the field stars would have formed together. Important and ill-understood elements of such a picture are the relative formation efficiencies of the metal-poor and metal-rich clusters (which probably need to be quite different) and the exact redshift of truncation.

Within the context of a major-merger scheme for giant ellipticals (Ashman & Zepf 1992), the metal-poor clusters formed within the progenitor spirals, whereas most of the metal-rich clusters in the present-day elliptical formed from gas during the merger. This process appears to work satisfactorily for many "field" ellipticals with spiral-like low specific frequencies (e.g., Whitmore et al. 2002; Schweizer et al. 1996), but has traditional difficulties in explaining high-specific-frequency systems like those in Virgo, Fornax, and other rich galaxy environments (see Bekki et al. 2003; Harris et al. 2004).

4. Summary

Much recent progress has been made through the Local Group members and other nearby galaxies in learning about the systemic properties of globular clusters, but major pieces of their formation history remain tantalizingly absent.

• The fundamental plane of structural properties for old globular clusters appears to be a nearly universal feature in all types of galaxies we have been able to investigate. What happened in the first few Myr of protocluster history to create this standard structural form?

• Are the MDFs of globular cluster systems telling us more about the sequence of cluster formation events in particular, than about galaxy formation in general? In particular, what happened to make the metal-poor cluster subpopulation stand out relative to the comparably old field-star populations in many galaxies?

• We are just beginning to exploit the diversity in MDFs for the *halo star* populations in nearby galaxies, and to connect them to the globular cluster systems in the same host galaxies.

214   W. E. Harris: *The old globular clusters*

REFERENCES

ASHMAN, K. M. & ZEPF, S. E. 1992 *ApJ* **384**, 50.

BAEV, P. V., MARKOV, H., & SPASSOVA, N. 2001 *MNRAS* **328**, 944.

BARMBY, P., HOLLAND, S., & HUCHRA, J. P. 2002 *AJ* **123**, 1937.

BEASLEY, M., BAUGH, C. M., FORBES, D. A., SHARPLES, R. M., & FRENK, C. S. 2002 *MNRAS* **333**, 383.

BEASLEY, M., HARRIS, W. E., HARRIS, G. L. H., & FORBES, D. 2003 *MNRAS* **340**, 341.

BEDIN, L. R., ANDERSON, J., KING, I. R., & PIOTTO, G. 2001 *ApJ* **560**, L75.

BEKKI, K., HARRIS, W. E., & HARRIS G. L. H. 2003 *MNRAS* **338**, 587.

BROOKS, S., WILSON, C., & HARRIS, W. E. 2003, in preparation.

BROWN, T. M., SWEIGART, A. V., LANZ, T., LANDSMAN, W. B., & HUBENY, I. 2001 *ApJ* **562**, 368.

CASSISI, S., SCHLATTL, H., SALARIS, M., & WEISS, A. 2003 *ApJ* **582**, L43.

COLE, A. A., SMECKER-HANE, T. A., & GALLAGHER, J. S. 2000 *AJ* **120**, 1808.

CÔTÉ, P., WEST, M. J., & MARZKE, R. O. 2002 *ApJ* **567**, 853.

DJORGOVSKI, S. 1995 *ApJ* **438**, L29.

DURRELL, P. R., HARRIS, W. E., & PRITCHET, C. J. 2001 *AJ* **121**, 2557.

FORBES, D. A., BRODIE, J. P., & GRILLMAIR, C. J. 1997 *AJ* **113**, 1625.

LARSEN, S. S., BRODIE, J. P., HUCHRA, J. P., FORBES, D. A., & GRILLMAIR, C. J. 2001 *AJ* **121**, 2947.

HARRIS, W. E. 1996 *AJ* **112**, 1487.

HARRIS, W. E. 2001. In *Star Clusters, Saas-Fee Advanced Course 28* (eds. L. Labhardt & B. Binggeli). p. 223. Springer.

HARRIS, W. E., BARMBY, P., HARRIS, G. L. H., & MCLAUGHLIN, D. E. 2003, in preparation.

HARRIS, W. E. & HARRIS, G. L. H. 2001 *AJ* **122**, 3065.

HARRIS, W. E. & HARRIS, G. L. H. 2002 *AJ* **123**, 3108.

HARRIS, W. E., HARRIS, G. L. H., & GEISLER, D. 2004 *AJ*, submitted.

HARRIS, W. E., HARRIS, G. L. H., HOLLAND, S. T., & MCLAUGHLIN, D. E. 2002 *AJ* **124**, 1435.

HOLLAND, S., CÔTÉ, P., & HESSER, J. E. 1999 *AAp* **348**, 418.

LARSEN, S. S., BRODIE, J. P., SARAJEDINI, A., & HUCHRA, J. P. 2002 *AJ* **124**, 2615.

LARSEN, S. S., CLAUSEN, J. V., & STORM, J. 2000 *AAp* **364**, 455.

MACKEY, A. D. & GILMORE, G. F. 2003a *MNRAS* **338**, 85.

MACKEY, A. D. & GILMORE, G. F. 2003b *MNRAS* **338**, 120.

MACKEY, A. D. & GILMORE, G. F. 2003c *MNRAS* **340**, 175.

MCLAUGHLIN, D. E. 2000 *ApJ* **539**, 618.

MEYLAN, G., SARAJEDINI, A., JABLONKA, P., DJORGOVSKI, S. G., BRIDGES, T. J., & RICH, R. M. 2001 *AJ* **112**, 830.

MOEHLER, S. 2001 *PASP* **113**, 1162.

PIOTTO, G., ET AL. 2002 *AAp* **391**, 945.

REITZEL, D. B. & GUHATHAKURTA, P. 2002 *AJ* **124**, 234.

RICH, R. M., MIGHELL, K. J., FREEDMAN, W. L., & NEILL, J. D. 1996 *AJ* **111**, 768.

RICHER, H. B., ET AL. 2002a *ApJ* **574**, L151.

RICHER, H. B., ET AL. 2002b *ApJ* **574**, L155.

SARAJEDINI, A., ET AL. 2003, poster at this conference.

SCHWEIZER, F., MILLER, B. W., WHITMORE, B. C., & FALL, S. M. 1996 *AJ* 112, 1839.

SHAPLEY, H. & DAVIS, H. N. 1920 *Proc. National Academy of Sciences* **6**, 486.

SILLS, A., ADAMS, T., DAVIES, M. B., & BATE, M. R. 2002 *MNRAS* *332*, 49.

SOKER, N., CATELAN, M., ROOD, R. T., & HARPAZ, A. 2001 *ApJ* **563**, L69.

WHITMORE, B. C., SCHWEIZER, F., KUNDU, A., & MILLER, B. W. 2002 *AJ* **124**, 147.

Chemical evolution models of Local Group galaxies

By MONICA TOSI

INAF - Osservatorio Astronomico di Bologna, Via Ranzani 1, I-40127 Bologna, Italy

Status quo and perspectives of standard chemical evolution models of Local Group galaxies are summarized, and what we have learned from them is discussed, as well as what we have not learned yet, and what I think will be learned in the near future. Galactic chemical evolution models have shown that: i) stringent constraints on primordial nucleosynthesis can be derived from the observed Galactic abundances of the light elements; ii) the Milky Way has been accreting external gas from early epochs to the present time; and iii) the vast majority of Galactic halo stars have formed quite rapidly at early epochs. Chemical evolution models for the closest dwarf galaxies, although still uncertain, are expected to become extremely reliable in the immediate future, thanks to the quality of new generation photometric and spectroscopic data which are currently being acquired.

1. Introduction

The proximity of Local Group galaxies makes them the ideal benchmarks to study galaxy formation and evolution, because they are the only systems where the accuracy and the wealth of observational data allows us to understand them in a sufficiently reliable way. In fact, to understand the evolution of galaxies, astronomers must follow two distinct and complementary approaches: on one hand they must develop theoretical models of galaxy formation, of chemical and dynamical evolution, and on the other hand, they must collect accurate observational data to constrain the models. It is of particular importance to acquire reliable data on chemical abundances, masses and kinematics of galactic components (gas, stars, dark matter), star formation (SF) regimes, and stellar initial mass function (IMF)—quantities that are best derived in nearby systems.

Following Socrates' indication, $\gamma\nu\hat{\omega}\theta\iota\ \sigma\alpha\upsilon\tau\acute{o}\nu$ (*know thyself*), that the knowledge of truth must be derived not from metaphysics but from critical analysis of the reality, here I try to critically review *status quo* and perspectives of standard chemical evolution models—with the warning that these kinds of models, although quite successful, refer only to large-scale, long-term phenomena, and cannot account for the small-scale, short-term variations often observed in the chemical and dynamical properties of galaxies.

2. Parameters

The major parameters involved in standard chemical evolution models are:
- SF law and rate (often simplistically approximated either as exponentially decreasing functions of time, SFR $\propto e^{-t/\tau}$, or as power laws of the gas density, e.g., SFR $\propto \Sigma_{\text{gas}}^n$);
- Gas flows in and out of the considered region (the infalling gas rate usually being approximated with an exponentially decreasing function of time, $f_i \propto e^{-t/\theta}$, and the galactic outflows (or winds) assumed to be proportional to the energy released by supernova explosions, $f_w \propto E_{SN}$);
- IMF (usually represented as a power law, $\phi \propto m^{-\alpha}$ with one or more exponents, α, for different mass ranges, see Gallagher & Grebel, this volume); and
- Stellar lifetimes and nucleosynthesis yields.

Some of them, however, are implicitly linked to other parameters, such as the amount of mixing occurring in stellar interiors or the stellar mass loss rates.

Since there are many parameters for chemical evolution modeling, the crucial prescription to avoid misleading results is to **always compare the model predictions with all the available constraints**, not just those relative to the examined quantities. This prescription implies that, until now, Galactic models are much better constrained than those of external, less-studied systems. In the following section, they are discussed separately.

3. The Galaxy

In the case of the Galaxy, the observational constraints formally outnumber the model parameters. Indeed, in the last two decades, an increasing number of accurate and reliable data have been accumulated that allow us to put stringent limits on the evolution of the Milky Way. The *minimal* list of data that should always be compared with the model predictions (see also Boissier & Prantzos 1999) includes:

• the current distribution with Galactocentric distance of the SFR (e.g., as compiled by Lacey & Fall, 1985);

• the current distribution with Galactocentric distance of the gas and star densities (see e.g., Tosi 1996; Boissier & Prantzos 1999 and references therein);

• the current distribution with Galactocentric distance of element abundances as derived from H II regions and from B-stars (e.g., Shaver et al. 1983; Smartt & Rollerston 1997);

• the distribution with Galactocentric distance of element abundances at slightly older epochs, as derived from PNe II (e.g., Pasquali & Perinotto 1993; Maciel & Chiappini 1994; Maciel & Köppen 1994; Maciel et al. 2003);

• the age-metallicity relation (AMR), not only in the solar neighborhood, but also at other distances from the center (e.g., Edvardsson et al. 1993);

• the metallicity distribution of G-dwarfs in the solar neighborhood (e.g., Rocha-Pinto & Maciel 1996);

• the local Present-Day-Mass-Function (PDMF, e.g., Scalo 1986; Kroupa et al. 1993); and

• the relative abundance ratios (e.g., [O/Fe] vs. [Fe/H]) in disk and halo stars (e.g., Barbuy 1988; Edvardsson et al. 1993).

There are now several models able to reproduce all these observed properties. Even if this circumstance does not yet provide a unique detailed scenario for the formation and evolution of the Milky Way, it allows us to make robust predictions on several important issues.

An example of robust prediction resulting from the requirement of reproducing all the above-listed data is the Galactic evolution of deuterium, one of the elements produced during the Big Bang and one of the best baryometers (see Steigman, this volume). When people began to study the evolution of D (e.g., Steigman & Tosi 1992; Galli et al. 1995, Prantzos 1996), observational data on the D abundance were available only for the local interstellar medium (ISM) and for the Protosolar Cloud (Linsky 1998; Geiss & Gloeckler 1998 and references therein). All of the models able to reproduce the above list of constraints (see Tosi 1996 and references therein) predicted only a moderate depletion of D from its primordial value to the present one: a factor of three at most (see Fig.1). This implies a primordial number ratio to hydrogen $((D/H)_p \leq (4-5) \times 10^{-5})$, which is impossible to reconcile within the framework of standard Big Bang nucleosynthesis (SBBN) due to the low primordial abundance by mass of ^{4}He, $Y_p \simeq 0.23$. This was inferred earlier

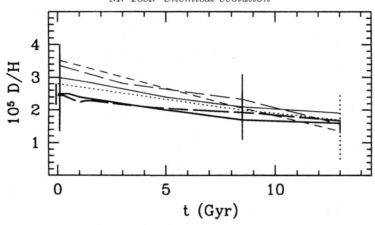

FIGURE 1. Deuterium evolution in the solar ring as predicted by various models: solid line, Steigman & Tosi 1992; short-dashed line, Galli et al. (1995); dotted line, Prantzos (1996); long-dashed line, Chiappini et al. (1997). The thick solid line and the thick long-dashed line are both from Romano et al. 2003. All the vertical lines correspond to data derived from observations; from left to right they are: the primordial D/H derived from *WMAP*, the range of abundances inferred from high-redshift QSO absorbers, the pre-solar value, and the local ISM value (see text for references). The dotted line at 13 Gyr shows the range of D/H ratios derived along different lines of sight (e.g., Vidal-Madjar et al. 1998; Sonneborn et al. 2000).

on by several groups from low metallicity H II regions and globular clusters. These Y_p determinations have subsequently become the subject of hot debates (see Steigman, this volume, and references therein, for a critical discussion of this result), but, at the time, this inconsistency led some people to think that the Galactic models were wrong and that we should find a way to deplete much more D during Galaxy evolution to allow for a higher primordial abundance. Then high-redshift, low-metallicity QSO absorbers started to be observed and provided D/H always lower than 4×10^{-5} (e.g., Burles & Tytler 1998 and references therein), perfectly consistent with the predictions of the Galactic models with low D depletion (see Fig.1; Tosi et al. 1998; Chiappini et al. 2003). However, concerns remained that, despite their low metallicity, high-redshift absorbers might have D contents lower than primordial, where some stellar activity could have already taken place there and burnt some of the original D. A few months ago, the microwave satellite *WMAP* provided a direct estimate of the baryon-to-photon ratio, which corresponds, within the SBBN framework, to a primordial $(D/H)_p = (2.62 \pm 0.30) \times 10^{-5}$ (Spergel et al. 2003). This value is again in excellent agreement with the predictions of Galactic chemical evolution models aimed at reproducing the complete set of observational constraints (cf. thick lines in Fig.1, Romano et al. 2003), and shows how robust model predictions could be when sufficiently constrained. On the other hand, the significant variations of the D abundances measured along different lines of sight in the local ISM (dotted vertical line in Fig.1) indicate that more sophisticated models would also be needed to reproduce local fluctuations (see also Pilyugin & Edmunds 1996).

What have we learned of Galaxy formation and evolution from chemical evolution models? One of the main results, in my opinion, is that the Milky Way has not formed from a very rapid monolithic collapse of a single proto-galaxy. We have recent observational evidence that the Galaxy is accreting the Sagittarius dwarf (e.g., Ibata et al. 1995); it is likely that it will accrete the Magellanic Clouds in the future, and some (e.g., Dinescu et al. 1999; Hilker & Richtler 2000; Ferraro et al. 2002) think that ωCentauri is also an accreted small dwarf rather than a real globular cluster. In addition, earlier

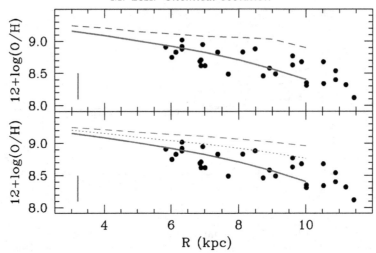

FIGURE 2. Radial distribution of the Galactic oxygen abundance at the present epoch, as derived from observations of H II regions (dots) and as predicted by chemical evolution models (Tosi 1988a,b). *Top panel*: models with primordial infall (solid line) and without any infall after disk formation (dashed line). *Bottom panel*: models with infall of different metallicity (solid line for $Z_i = 0$, dotted for $Z_i = 1/2\ Z_\odot$, short-dashed for $Z_i = Z_\odot$). See text for details.

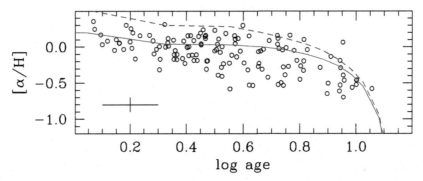

FIGURE 3. Age-metallicity relation in the solar neighborhood as derived from field star observations (dots) and as predicted by the models (lines) of the top panel of Fig. 2. See text for details.

chemical evolution models (e.g., Tinsley 1980; Tosi 1988a,b; Matteucci & François 1989; Chiappini et al. 1997; Boissier & Prantzos 1999) showed that the Milky Way must have kept accreting metal-poor gas at a relatively steady rate.

Historically, one of the first reasons to invoke a continuous infall of metal-poor gas on the Galactic disk was the so-called G-dwarf problem (see Wyse, this volume), i.e., the fact that, without infall, chemical evolution models overpredict the number of low-metallicity long-lived stars. There are, however, other observed properties that need infall to be reproduced. Figs. 2 and 3 show two examples of this need. In Fig. 2 the radial distribution of the current oxygen abundance in the disk is plotted as a function of Galactocentric distance. In both panels the dots correspond to the H II region's values and the curves to the predictions of Tosi (1988a and b) models. When a constant equidense infall of primordial gas is assumed (solid line in the top panel), the models reproduce the observed distribution very well, while without any infall after the disk formation (dashed line in the top panel) they predict an abundance gradient flatter than observed

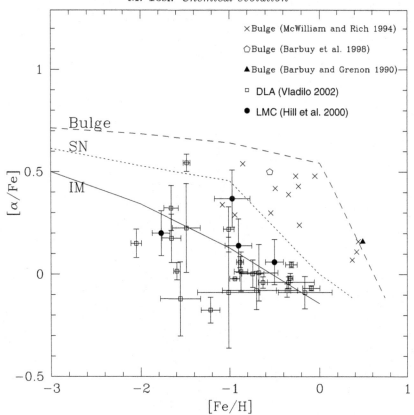

FIGURE 4. The lines show Matteucci's model predictions for the α/Fe ratios for different SF regimes. See Matteucci (2003) for references and details.

and overproduce oxygen at the present epoch at all galactocentric distances. The bottom panel of Fig. 2 illustrates why the accreted gas should be metal poor: the solid line, as in the top panel, corresponds to a metal-free infall and is in excellent agreement with the data. The dashed line corresponds to the same model, but assumes a solar-infall metallicity, and clearly overpredicts oxygen. The dotted line assumes a half-solar metallicity and also overpredicts oxygen.

Fig. 3 shows the age-metallicity relation in the solar neighborhood as derived by Edvardsson et al. (1993) from observations of F and G dwarfs and as predicted by the same models in the top panel of Fig. 2. Again, the model assuming a constant infall of primordial gas (solid line) fits the data well, while the model with no infall after the disk formation (dashed line) overproduces the metallicity since quite early epochs. To reproduce both the observed radial gradients and abundances, nowadays all chemical evolution models assume a fairly conspicuous amount of steady gas accretion (but see also Lacey & Fall 1985). Most authors assume this gas to be primordial, but it was shown (Tosi 1988b) that it could reach a metallicity up to 0.2 $Z_\odot$ without losing its diluting effects on the predicted abundances. This metallicity happens to be that derived for the few high-velocity clouds falling on the Galactic disk where abundance measurements are possible (e.g., DeBoer & Savage 1983).

Another important result is that both the observations and then chemical evolution models suggest that the halo of the Galaxy must have created most of its present stars fairly rapidly. One of the main reasons to reach this conclusion comes from the abundance

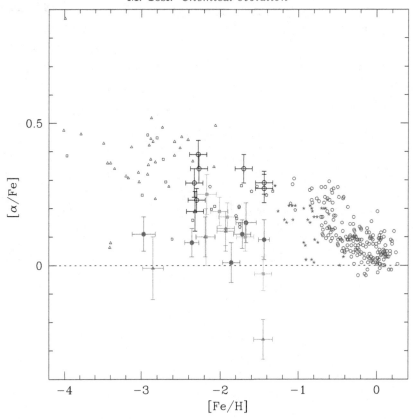

FIGURE 5. α/Fe observed in different environments. Open circles refer to local disk stars, open triangles to local halo stars, open symbols with error bars to Galactic globular clusters, and filled symbols with error bars to dSphs (Shetrone et al. 2001 and references therein).

ratios between alpha elements (those produced directly from ^{4}He burning, like oxygen and magnesium) and iron. Iron is mostly produced by supernovae of type Ia (intermediate mass stars in binary systems), while alpha elements are essentially synthesized in massive stars. The different lifetimes of their main producers imply that enrichment of the alpha elements is very rapid (within a few Myr), while that of iron starts to occur after $\sim$100 Myr and has its bulk about 1 Gyr after the onset of the SF activity. As shown by Matteucci (1992), this circumstance makes the stellar $[\alpha/\text{Fe}]$ measured in a region an excellent indicator of its SF regime. The curves in Fig. 4 display the different behaviors predicted by Matteucci's models for different SF regimes. In regions where the SF is supposed to be continuous over several Gyrs, stars can form when iron has had plenty of time to enrich the medium and contribute to their initial metallicity. Hence, the stellar $[\alpha/\text{Fe}]$ is predicted to steadily decrease when iron increases (solid line in Fig. 4). Alternatively, in regions like bulges, where the SF is very rapid and intense, all the stars form prior to the release of iron by SNe Ia, and contain much α but little iron, thus showing a plateau (dashed line) with high $[\alpha/\text{Fe}]$. In the solar neighborhood, where some stars belong to the halo and others to the disk, Matteucci's models predict (dotted line) a fairly flat and high $[\alpha/\text{Fe}]$ vs. $[\text{Fe/H}]$ for the halo stars with low $[\text{Fe/H}]$, and a decreasing $[\alpha/\text{Fe}]$ vs. $[\text{Fe/H}]$ for disk stars with higher $[\text{Fe/H}]$. Since this is indeed the observed behavior (see Fig. 5), this indicates that the halo must have had a strong and rapid initial SF activity and the disk a continuous one.

Fig. 5 shows [α/Fe] vs. [Fe/H] as derived from spectroscopy of stars in various environments (Shetrone et al. 2001, and references therein). The small open circles correspond to solar neighborhood disk stars, the triangles to halo field stars, the larger open circles with error bars to halo clusters, and they show an overall distribution quite similar to that predicted by Matteucci's models. The filled symbols refer instead to stars observed in nearby dwarf spheroidals (dSphs) and they all show [α/Fe] systematically lower than those measured at the same [Fe/H] in Galactic stars. Further high-resolution spectroscopy of stars in other nearby dSphs (Tolstoy et al. 2003), including Sagittarius (Vladilo et al., this conference and Bonifacio et al. 2003), has confirmed this systematic difference. This evidence makes it extremely unlikely that our halo formed mostly from the merging of dwarf galaxies like these, because there is no conceivable mechanism able to make iron-poor, alpha-rich stars assemble alpha-poor, iron-rich ones. From this and other arguments (see Tosi 2003, and Wyse, this volume), it seems more likely that our Galaxy was formed mostly from gaseous building blocks and within a relatively short timescale (a couple of Gyr, at most).

Despite the numerous important achievements of chemical evolution modeling, there are important aspects of the Galaxy formation and evolution not well understood yet. Has the thin disk formed before or after the thick disk? Where is the infalling gas coming from? Is it actually as metal poor and as steady as required?

A clear example of our lack of detailed knowledge of the processes leading to the observed Galactic properties is the evolution of the abundance gradients in the disk. There are several chemical evolution models able to reproduce the present metallicity gradients derived from H II regions and young stars (e.g., Fig. 2), along with the whole set of constraints listed above. These models, however, differ from each other in several assumptions, and one of the major effects of such differences is that they predict quite different evolutions of the gradient (see e.g., Tosi 1996 and references therein): some predict the gradient to steepen with time, while others predict it to flatten. It is still difficult to understand what the actual evolution is, because the available data mostly refer to relatively young objects. Older single stars are, in fact, fainter and hence more difficult to measure, specially at the large distances required to derive the gradient with sufficient radial baseline. PNe in principle are good tools for this purpose, but the interpretation of their data in terms of progenitor age is still rather uncertain, and the progenitors of the safest ones are stars a few Gyr old. To get a reliable gradient back to the earliest epochs, the best targets are open clusters, since they are much less affected than individual objects by uncertainties of age, distance and metallicity. Several people have derived the abundance gradient from open cluster data (see e.g., Friel 1995 and references therein), but what is needed for a robust result on the gradient evolution is a cluster sample large enough to provide significant results in each age bin, with ages, metallicities and distances derived in a homogeneous way to avoid spurious effects (e.g., Bragaglia et al. 2002). Building up such a sample clearly takes time, but we are confident that the results will be rewarding.

4. Other Local Group galaxies

The chemical evolution of M31 and M33 (as well as that of other spirals outside the Group) has been modeled by a few groups with approaches similar to those applied to the Milky Way (e.g., Diaz & Tosi 1984; Mollà et al. 1996). These models predict that the disks of these systems have a roughly continuous SF, with shorter timescales for earlier-type spirals and longer ones for later-type spirals. Infall of metal-poor gas is required also for M31 (and other massive spirals), while it is not necessary in low-mass spirals like

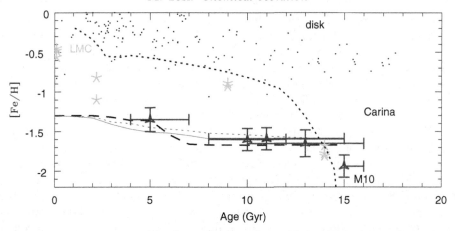

FIGURE 6. AMR derived from VLT/UVES spectroscopy of the dSph Carina (triangles with error bars) by Tolstoy et al. (2003). Shown for comparison are the data relative to LMC clusters (asterisks), Pagel & Tautvaisiene (1998), AMR from LMC clusters (dotted line) and the data by Edvardsson et al. (1993) for solar neighborhood stars (dots).

M33. These models, however, are not as well constrained as those for the Galaxy, since, at least until recently, the only data useful for chemical evolution modeling were the H II region abundances and the gas and star density distributions in the disks.

The data available for dwarf galaxies were equally scarce, but this kind of system has intrigued many more scientists. In the last 25 years there has been a wealth of papers dealing with the chemical evolution of dwarfs (starting e.g., with Lequeux et al. 1979 and Matteucci & Chiosi 1983): a rather frustrating challenge, if one considers how inconsistent with each other the results of these papers have been. While most authors agreed that the IMF in these galaxies is fairly similar to Salpeter's both on the SF regimes and on the existence of galactic winds, different groups have reached very different conclusions. For instance, many groups suggested that the SF in all dwarfs is episodic (e.g., Matteucci & Chiosi 1983; Pilyugin 1993; Larsen et al. 2001), but others argue that in late-type dwarfs the SF is continuous (e.g., Carigi et al. 1995; Legrand 2001). For the gas flows, some authors (e.g., Gilmore & Wyse 1991) concluded that winds triggered by SN explosions are not needed to explain the observed chemical abundances. However, other authors reached the opposite conclusion—that the winds are the only viable mechanism to predict abundances as low as observed—some (e.g., Matteucci & Tosi 1985; Pilyugin 1993; Recchi et al. 2002) arguing that the outflowing gas must be enriched in the elements produced by SNe, and others (e.g., Pagel & Tautvaisiene 1998; Larsen et al. 2001) arguing instead that the outflowing gas has the same composition as the galaxy medium. In other words, all kinds of possible scenarios have been attributed to the evolution of dwarf galaxies.

These inconsistencies are due to the lack of adequate observational data. Many groups, for instance, have adopted the age-metallicity relation presented by Pagel & Tautvaisiene (1998) for the LMC (dotted line in Fig. 6) to model dwarf galaxy evolution. However, the LMC is not a prototype for any kind of dwarf, and the AMR was derived from clusters data, which do not necessarily have the same evolution of field stars (see, in fact, Fig. 7). I believe, however, that the quantity and quality of the observational data on nearby dwarfs is so dramatically improving today that our understanding of the dwarfs' evolution will make an impressive step forward in the near future. Indeed:

1) High resolution spectrographs at 10 m class telescopes are providing a wealth of new, accurate abundances for field stars in nearby dwarfs. The ages of these stars can be

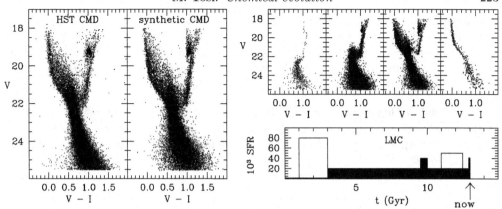

FIGURE 7. From the Coimbra experiment (Skillman & Gallart 2002). *Left-hand panel*: CMD of a field of the LMC bar derived by Smecker-Hane et al. (2002) from *HST*/WFPC2 photometry; *second panel from left*: the corresponding best synthetic CMD by Tosi et al. (2002). *Top right panels*: the synthetic CMD of the previous panel split in its four episode components, the oldest to the youngest one (from left to right). *Bottom-right panel*: the resulting SF history, i.e., SF rate per unit area (in units of 10^3 $M_\odot$ yr^{-1} kpc^{-2}) vs. time is shown as a filled histogram. The empty histogram refers to the SF history derived from LMC star clusters.

derived from the color-magnitude diagrams (CMDs) resulting from deep, high-resolution photometry, both from ground and space. Fig. 6 draws the first results of Tolstoy et al. (2003) from VLT/UVES observations and shows how different the AMR of LMC clusters and of Milky Way field stars is from the AMR of the stars in the Carina dwarf galaxy (as well as in the other dwarfs of their program).

2) Deep and tight CMDs provide reliable information on the IMF of the observed regions (see e.g., Gallagher & Grebel, this volume), and on the SF history of the observed regions.

The SF history is quite reliably derived by interpreting the observational CMDs with the synthetic CMD method (e.g., Tosi et al. 1991), which has proven a powerful and robust tool. It was recently tested on an LMC field, comparing the scenarios obtained by different groups (the so-called *Coimbra experiment*, see Skillman & Gallart 2002 and references therein). Fig. 7 shows (in the bottom-right panel) the SF history of the field on the bar of the LMC as derived by our group (Tosi et al. 2002) applying the synthetic CMD method to the CMD obtained by Smecker-Hane et al. (2002) from *HST*/WFPC2 photometry, and kindly provided for the *Coimbra experiment*. It is apparent that such beautiful data let them measure with sufficient accuracy even the oldest/faintest stars, thus allowing for the derivation of the SF history back to the earliest epochs. The SF history in this field of the LMC (filled histogram in Fig. 7) is fairly continuous, although with significant variations in the rate, and quite different from that inferred by Pagel & Tautvaisiene (1998) from cluster data (empty histogram).

The combination of these high-quality photometric and spectroscopic data with appropriate interpretation tools will soon allow us to know the AMR, the SF history and the IMF of nearby galaxies. Being the closest galaxies, the Magellanic Clouds are in the best position to allow for accurate and extensive data sets on these quantities. Hence, in a few years, *HST* photometry and 10 m class telescope spectroscopy may enable us to model their chemical evolution even more safely than that of the solar neighborhood. This opens promising new horizons to chemical evolution studies of dwarf galaxies in general, and of late-type dwarfs in particular. Taking into account that the SMC can be considered a prototype for this kind of galaxy because it has their typical mass, gas fraction, and metallicity, this would imply an unprecedented step toward understanding the evolution of galaxies that are not only the most numerous, but also can provide better clues to galaxy formation processes. This will confirm, once again, that the Local Group is the best astrophysical laboratory to understand galaxy evolution.

I'm grateful to Johannes Geiss and all the colleagues of the ISSI LOLA–GE team for the enlightening discussions. Mat Shetrone kindly provided his figure. This work has been partially supported by the Italian ASI and MIUR through grants IR11301ZAM and Cofin-2002028935.

REFERENCES

BARBUY, B. 1988 *A&A* **191**, 121.

BOISSIER, S. & PRANTZOS, N. 1999 *MNRAS* **307**, 857.

BONIFACIO, P., SBORDONE, L., MARCONI, G., PASQUINI, L., & HILL, V. 2003 **414**, 503.

BRAGAGLIA, A., TOSI, M., MARCONI, G., & DI FABRIZIO, L. 2002. In *Observed HR Diagrams and Stellar Evolution*, (eds. T. Lejeune & J. Fernandes). ASP Conf. Ser. No. 274, p. 385. ASP.

BURLES, S. & TYTLER, D. 1998 *Sp.Sc.Rev.* **84**, 65.

CARIGI, L., COLIN, P., PEIMBERT, M., & SARMIENTO, A. 1995 *ApJ* **445**, 98.

CHIAPPINI, C., RENDA, A., & MATTEUCCI, F. 2002 *A&A* **395** 789.

CHIAPPINI, C., MATTEUCCI, F., & GRATTON, R. 1997 *ApJ* **477**, 765.

DEBOER, K. S. & SAVAGE, B. D. 1983 *ApJ* **265**, 210.

DIAZ, A. I. & TOSI, M. 1984 *MNRAS* **208**, 365.

DINESCU, D. I., VAN ALTENA, W. F., GIRARD, T. M., & LÓPEZ, C. E. 1999 *ApJ* **117**, 277.

EDVARDSSON, B., ANDERSEN, J., GUSTAFSSON, B., LAMBERT, D. L., NISSEN, P. E., & TOMKIN, J. 1993 *A&A* **275**, 101.

FERRARO, F. R., BELLAZZINI, M., & PANCINO, E. 2002 *ApJ* **573**, L95.

GALLAGHER, J. S. III & GREBEL, E. 2003, this volume.

GALLI, D., PALLA, F., FERRINI, F., & PENCO, U. 1995 *ApJ* **443**, 536.

GEISS, J. & GLOECKLER, G. 1998 *Sp.Sc.Rev.* **84**, 239.

GILMORE, G. & WYSE, R. F. G. 1991 *ApJ* **367**, L55.

HILKER, M. & RICHTLER, T. 2000 *A&A* **362**, 895.

IBATA, R. A., GILMORE, G., & IRWIN, M. J. 1995 *MNRAS* **277**, 781.

KROUPA, P., TOUT, C. A., & GILMORE, G. 1993 *MNRAS* **262**, 545.

LACEY, C. G. & FALL, S. M. 1985 *ApJ* **290**, 154.

LARSEN, T. I., SOMMER-LARSEN, J., & PAGEL, B. E. J. 2001 *MNRAS* **323**, 555.

LEGRAND, F. 2001 *ApSSS* **277**, 287.

LEQUEUX, J., RAYO, J., SERRANO, A., PEIMBERT, M., & TORRES-PEIMBERT, S. 1979 *A&A* **80**, 155.

LINSKY, J. L. 1998 *Sp.Sc.Rev.* **84**, 285.

MACIEL, W. J. & CHIAPPINI, C. 1994 *Astrophys.& Sp.Sc.* **219**, 231.

MACIEL, W. J., COSTA, R. D. D., & UCHIDA, M. M. M. 2003 *A&A* **397**, 667.

MACIEL, W. J. & KÖPPEN, J. 1993 *A&A* **282**, 436.

MATTEUCCI, F. 1992 *Mem.S.A.It.* **63**, 301.

MATTEUCCI, F. 2003 *Ap&SS* **284**, 539.

MATTEUCCI, F. & CHIOSI, C. 1983 *A&A* **123**, 121.

MATTEUCCI, F. & FRANÇOIS, P. 1989 *MNRAS* **239**, 885.

MATTEUCCI, F. & TOSI, M. 1985 *MNRAS* **217**, 391.

MOLLA, M., FERRINI, F., & DIAZ, A.I. 1996 *ApJ* **466**, 668.

PAGEL, B. E. J. & TAUTVAISIENE, G. 1998 *MNRAS* **299**, 535.

PASQUALI, A. & PERINOTTO, M. 1993 *A&A* **280**, 581.

PILYUGIN, L. S. 1993 *A&A* **277**, 42.

PILYUGIN, L. S. & EDMUNDS, M. G. 1996 *A&A* **313**, 792.

PRANTZOS, N. 1996 *A&A* **310**, 106.

RECCHI, S., MATTEUCCI, F., D'ERCOLE, A., & TOSI, M. 2002, *A&A* **384**, 799.

ROCHA-PINTO, H. J. & MACIEL, W. J. 1996 *MNRAS* **279**, 447.

ROMANO, D., TOSI, M., MATTEUCCI, F., & CHIAPPINI, C. 2003 *MNRAS* **346**, 295.

SCALO, J. M. 1986 *Fund. Cosmic Phys.* **11**, 1.

SHAVER, P. A., MCGEE, R. X., NEWTON, L. M., DANKS, A. C., & POTTASCH, S. R. 1983 *MNRAS* **204**, 53.

SHETRONE, M. D., CÔTÉ, P., & SARGENT, W. L. W. 2001 *ApJ* **548**, 592.

SKILLMAN, E. D. & GALLART, C. 2002 *ASP Conf.Ser.* **274**, 535.

SMARTT, S. & ROLLERSTONE, W. 1997 *ApJ* **481**, L47.

SMECKER-HANE, T. A., COLE, A. A., GALLAGHER, J. S. III, & STETSON, P. B. 2002 *ApJ* **566**, 239.

SONNEBORN, G., TRIPP, T. M., FERLET, R., JENKINS, E. B., SOFIA, U. J., VIDAL-MADJAR, A., & WOZNIAK, P. R. 2000 *ApJ* **545**, 277.

SPERGEL, D. N., ET AL. 2003 *ApJS* **148**, 175.

STEIGMAN, G. 2004, this volume

STEIGMAN, G. & TOSI, M. 1992 *ApJ* **401**, 150.

TINSLEY, B. M. 1980 *Fund. Cosmic Phys.* **5**, 287.

TOLSTOY, E., VENN, K. A., SHETRONE, M., PRIMAS, F., HILL, V., KAUFER, A., & SZEIFERT, T. 2003 *AJ* **125**, 707.

TOSI, M. 1988a *A&A* **197**, 33.

TOSI, M. 1988b *A&A* **197**, 47.

TOSI, M. 1996 in *From Stars to Galaxies: The Impact of Stellar Physics on Galaxy Evolution* (eds. C. Leitherer, U. Fritze-von-Alvensleben, & J. Huchra). ASP Conf. Ser. 98, p. 229. ASP.

TOSI, M.2003 *Ap&SS* **284**, 651.

TOSI, M., GREGGIO, L., & ANNIBALI, F. 2002 *ASP Conf.Ser.* **274**, 529.

TOSI, M., GREGGIO, L., MARCONI, G., & FOCARDI, P. 1991 *AJ* **102**, 951.

TOSI, M., STEIGMAN, G., MATTEUCCI, F., & CHIAPPINI, C. 1998 *ApJ* **498**, 226.

VIDAL-MADJAR, A., LEMOINE, M., FERLET, R., HEBRARD, G., KOESTER, D., AUDOUZE, J., CASSE, M., VANGIONI-FLAM, E., WEBB, J. 1998 *A&A* **338**, 694.

WYSE, R. F. G. 2003, this volume.